China's Geography

China's Geography

Globalization and the Dynamics of Political, Economic, and Social Change

Second Edition

GREGORY VEECK, CLIFTON W. PANNELL,
CHRISTOPHER J. SMITH, AND YOUQIN HUANG

ROWMAN & LITTLEFIELD PUBLISHERS, INC.
Lanham • Boulder • New York • Toronto • Plymouth, UK

Cartography by

Dr. Thomas R. Hodler, Professor Emeritus
Department of Geography
The University of Georgia

Published by Rowman & Littlefield Publishers, Inc.
A wholly owned subsidiary of The Rowman & Littlefield Publishing Group, Inc.
4501 Forbes Boulevard, Suite 200, Lanham, Maryland 20706
http://www.rowmanlittlefield.com

Estover Road, Plymouth PL6 7PY, United Kingdom

British Library Cataloguing in Publication Information Available

Library of Congress Cataloging-in-Publication Data

China's geography : globalization and the dynamics of political, economic, and social change /
    Gregory Veeck . . . [et al.]. — 2nd ed.
        p. cm.
    Includes bibliographical references and index.
    ISBN 978-0-7425-6782-5 (cloth : alk. paper) — ISBN 978-0-7425-6783-2 (pbk. : alk. paper) —
ISBN 978-0-7425-6784-9 (electronic)
    1. China. I. Veeck, Gregory, 1956– II. Title.

DS706.C51138 2011
915.1—dc22

2011002802

♾™ The paper used in this publication meets the minimum requirements of American
National Standard for Information Sciences—Permanence of Paper for Printed Library
Materials, ANSI/NISO Z39.48-1992.

Printed in the United States of America

Contents

Images and Tables

Figures

Photographs

Tables

Acknowledgments

Acknowledgments tend to be either very short or very long. Those who know me should hardly be surprised that I have opted for the latter. There are many people who have my respect and gratitude for help with this book and the career on which it is partly based, and it is a pleasure to note these kindnesses. The book is the product of true collaboration among four scholars with markedly different areas of expertise and perspective. But we are joined through our enduring respect for China's great history, culture, and peoples as well as through our shared belief in the importance of presenting the complexities of China as clearly as we possibly can at this particular time. Working collectively on a manuscript of this length is never easy, and I will always be grateful for my collaborators' willingness to meet deadlines and share their opinions and data. It has been a privilege to work with all my coauthors, but especially my friend and mentor Clifton W. Pannell.

Many people have reviewed the manuscript during its longish journey. From an early discussion at the Billy Goat in Chicago, Susan McEachern of Rowman & Littlefield has displayed patience and enthusiastic support for the initial project and, surprisingly, this second edition. Many others at Rowman & Littlefield have helped and guided us through both editions, including Carrie Broadwell-Tkach, Sarah Wood, Jessica Gribble, Alden Perkins, Bruce Owens, John Shanabrook, and David Luljak. I know they go home at night and wonder how we get fed and dressed each day. We gratefully recognize the important contributions of several anonymous reviewers and our series editors Alexander Murphy and David Keeling. Thomas Hodler (University of Georgia Geography Emeritus) took over the cartographic duties in the second edition, and we are all grateful for his careful and creative work. Mary Lee Eggart's cartographic and artistic work is still very much alive in the second edition as well. The librarians at Western Michigan University were always able to find what I needed in a timely manner, no matter how obscure, and this holds true for the librarians at the University of Georgia and the State University of New York, Albany, as well. I also recognize the many hundreds of undergraduate students who have participated in our class discussions and who often raise the most challenging questions while exhibiting the least tolerance for artifice. The great group of scholars, who participate in the China Specialty Group of the Association of American Geographers, now more than

240 strong, has annually provided stimulating lectures and articles related to where China has been, where the nation is going, and what has happened along the way. Many were kind enough to make suggestions for the second edition, and we hope we have satisfied at least most of these concerns. Thanks especially to John Shaw and Jerry Volatile.

One measure of globalization is the ever-increasing ease of communication and collaboration between Western scholars and those living and working in China. Simply put, this book could not have been written without the help of our Chinese friends and colleagues in universities, research institutes, nongovernmental institutions, and government agencies. In my own case, researchers and students at the Rural Development Institute of the Chinese Academy of Social Sciences, Nanjing Agricultural University, Nanjing Institute of Geography, Nanjing Forestry University, Lanzhou University, Northeast Agricultural University, Qingdao Academy of Social Sciences, and Jilin University have all been important in my education. My coauthors would add another dozen or so institutions that have been equally important to them. Further, Chinese friends far from the academy whom I so admire and respect have also played a role—sharing their opinions and concerns about contemporary issues and problems. I am not the most traveled of people, but never have I felt as at home as when I am in China, largely because of the gracious way my friends in China give of their time and talents and their enthusiasm for an adventure no matter how large or small.

None of the research projects that provide the foundation for this book would have been possible without funding from agencies such as the National Science Foundation, National Geographic, Fulbright-Hays, the Fulbright Program, and the American Council of Learned Societies, including the Committee on Scholarly Communication with the People's Republic of China. The Lucia Harrison Fund and the Milton E. and Ruth M. Scherer Fund of the Department of Geography of Western Michigan University have also proven several times to provide critical support for my own work. We are grateful for all of our funding sources, and I wish to underscore the importance of such funding at the present time. Never was there more a time to understand China than the present.

I wish especially to thank my wife, Annie, and my children, Sarah and Robin. Daily, I count my blessings for my family, including the pleasure and memories of our many trips to China for more than a quarter century, as well as their steady love and friendship. Fred and Kay Krehbiel's infectious enthusiasm for travel started us down this road lo these many years ago, and I will always remember this early support when support was much more difficult to locate. And, finally, everyone should know of my gratitude and admiration for my parents, Mary-Frances and Bill Veeck, and all their fine children. This second edition is dedicated to the memory of my wonderful sister, Juliana, who taught me pretty much all I need to know about living. Julie, I miss you.

—Gregory Veeck

This book had its origins and evolution over a long period, and I wish to thank Greg Veeck for his enduring commitment and patience and for pushing to advance this work to its conclusion. Colleagues and students in Hong Kong special administrative region, mainland China, the United States, Canada, and Europe have offered much wisdom and knowledge, which have contributed to my work and for which I am most grateful. Any errors of fact or interpretation are mine. Thanks to my wife, Sylvia, for sharing in this writing, work, and travel; she has been a great and most supportive partner in our China adventure. My sons—Alex, Rich, Charles, and Tom—and their families have participated in my passion for things Chinese and Asian, and I greatly appreciate their enthusiasm and interest in visiting China and Asia.

—Clifton Pannell

CHAPTER 1

China's Path and Progress

The current pace of economic, sociocultural, and political change in China is stunning. No matter where you go, every city, town, and village seems to be under construction. The highway network has expanded more than fourfold from 1978 to 2008, expressways have gone from almost none to 60,302 km, rail freight has more than tripled in the same time period, and air passengers have increased from 2.3 million to 192.5 million (National Bureau of Statistics 2009, 610–14). On the back of increased industrial output (especially in manufacturing), the economy has grown at 8 to 10 percent a year for more than three decades, outpacing all other nations during this time (Jefferson and Rawski 1999; National Bureau of Statistics 2009; World Bank 2009). Foreign trade increased a staggering 500-fold from 1978 to 2008 (National Bureau of Statistics 2009, 724). In a word, China is booming—a speedy and potent work in progress! Its 1.34 billion citizens must continue to adjust to fast-changing economic, political, and social conditions while seeking to maintain connections to the threatened traditional values that many feel lie at the core of being Chinese.

How does this happen, and what is its significance? In this book, we seek to explore and explain the truly extraordinary changes under way in China and their impact on China's people as well as their implications and consequences for those beyond the border. Geographic approaches that focus on people and places by employing various scales of analysis are used to examine and determine the many features of this change. In this way, we present a fresh and different view and perspective of rapidly changing China.

Global Forces and China's Development

Part of what is driving this rapid change is the force of globalization (Dicken 2007). Powerful new economic, political, and technological pulses are at work, and China has been quick to adapt and adjust to them. This adaptation in turn is linked to a much more open economic system, what the Chinese now refer to as "socialism with Chinese characteristics." The result is not only astounding growth in trade with and exports to major economic powers such as Japan, the European Union, and the United States but at once a more influential and powerful nation on the world stage as well. Based on low labor costs and related land, environmental, and distribution costs, China's ability to attract investment in manufacturing is supported by a carefully crafted state policy that emphasizes innovation and high technology and has led to rapid development of a broad manufacturing base. Beginning with low-end consumer products, China has rapidly expanded this base to include a vast array of consumer and producer goods as well as electronics, transport equipment, and weapons and military equipment—all products sought by consumers throughout the world.

China's growing manufacturing capacity, then, is paralleled by rapidly increasing trade and investment linkages with other Asian economic tigers, such as South Korea and Taiwan. These new trade and investment trends have led in turn to new spatial arrangements as economic regions that reflect new global economies and technologies emerge in China. The arrangements include not only key trading patterns but also equally important new technologies in transportation and logistics that underpin China's ability to move and distribute goods now that it has become so skilled at manufacturing. A whole array of new transport systems is under construction as China strives to integrate its disparate regions and provinces into a greater whole that will fulfill its destiny as a functioning modern economic and political system. The cost of building this new infrastructure of highways, railways, pipelines, ports and harbors, and electri-

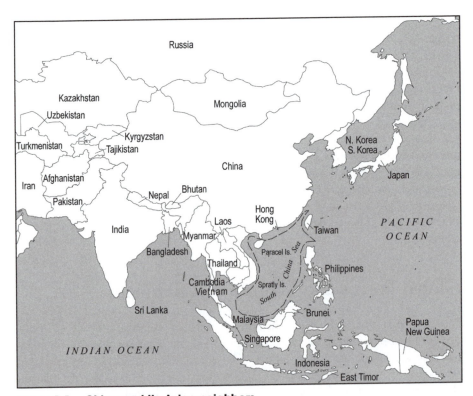

Figure 1.1. China and its Asian neighbors

cal and communications grids is enormous, and the country must continue to expand this basic structure in support of national development over the coming decades.

The spatial redeployment associated with the rapid economic growth of the past three decades has been felt most intensely in China's coastal regions, the areas most closely tied to the global trading system, but, as will unfold in the book, all of China has been impacted in myriad ways by the pace and magnitude of the nation's remarkable transformation (see figures 1.1 and 1.2). Consider the four key emerging regions noted in figure 1.2. The foremost example among these is the booming Pearl River Delta region on China's southeast coast, focused within the triangle of Hong Kong/Shenzhen, Guangzhou/Dongguan, and Macao/Zhuhai, with its long-standing shipping, commercial, and capital links to Southeast Asia and the world economy. Hong Kong remains one of the world's largest container ports and a major banking center, and it serves as China's linchpin in a long-term tradition of doing business based on a legal and commercial system that has the confidence of the global business community. A second emerging region is the Lower Chang Jiang or Shanghai/Sunan region with its powerful manufacturing and trade economy and strong links to Taiwan and the global trading system. Shanghai has attracted substantial Taiwan and foreign investment and has emerged as a great manufacturing center, but it has not yet recaptured the banking and commercial functions that made it so famous as a Chinese business center early in the twentieth century.

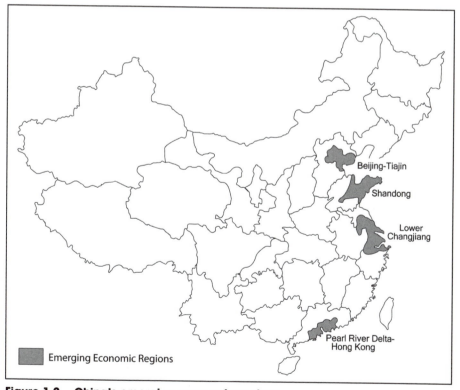

Figure 1.2. China's emerging economic regions

The Beijing-Tianjin region is the political power center of China as well as being a major commercial and manufacturing region with strong banking and high-tech industrial and research functions. Beijing is a key educational center with a powerful research-and-development thrust that has emerged in parallel with its prestigious universities and state-funded scientific and engineering centers, such as the various branches of the Chinese Academy of Sciences.

The Shandong economic region, though smaller and less active than the above three, nevertheless offers a good example of China's vibrant economy and shifting foreign policy in its strong and growing trade and commercial ties to South Korea. While the primary motivation behind these ties appears to be economic, there are clearly significant political and security ramifications in such new relationships and evolving spatial patterns. Consider that in the Shandong province city of Qingdao alone, there are over 70,000 South Korean residents, often operating small to medium-size electronics firms that retain competitiveness because of the availability of an educated, low-cost Chinese workforce.

There are numerous other large and small regions that reflect a dynamic pace of spatial rearrangement on various scales and the remarkable processes of economic growth at local, provincial, and subnational levels that are under way in China today. Many of these will be explored as we examine China in greater detail. While it is important to recognize the powerful impulse of the shifting global system and its opportunities for and effects on China, we must remember that China's socialist revolution was in part about reasserting the primacy of China's integrity as a sovereign and independent state. Consequently, in assessing the significance and role of global forces, we must call to mind China's remarkable historical record as a central state and empire and its growing confidence as it assumes a role as a responsible and rising player in the global economic, political, and security systems of the twenty-first century. Indeed, some observers have asserted that China is destined in the twenty-first century to become the most powerful and influential state in the world, possibly replacing the United States as the center of global influence and economic power while also becoming a dominant military force (Jacques 2009).

While the central state and China's Communist Party will receive most of the attention from foreigners and external observers, it is the interplay among China's central government, the provincial governments, local officials, local enterprises, and multinational firms and players that comprises the framework and context in which the new economic and political geography of today's rapidly rising China appears. We must search here for a more profound explanation and understanding of the extraordinary and far-reaching changes under way in China today, and to do this, we must leave the stereotype of a monolithic homogeneous China behind.

The now famous 1978 reforms championed by then paramount leader Deng Xiaoping that opened up the nation and radically altered its entire socioeconomic system are now more than thirty years old, yet there is little indication of any slowdown. It seems that all of China's people can hardly catch their breath and can hardly keep up with the myriad changes to their everyday lives. Maintaining values and perspective is not easy in these unpredictable if sometimes gilded times when fortunes are so often influenced by social connections, access to college, socioeconomic status, ethnicity, gender, and certainly where one lives. As we would expect, in any place, at any time, and because social and economic systems shift rapidly, new conditions provide opportunities for some to succeed beyond their wildest dreams but cause others to slip from their already precarious places in society. Many people feel lost. Imagine your thoughts if you were over forty, had not gone to college, and had worked loyally for a state-owned factory that then closed fifteen years ago and reported a bankrupt pension fund three years after the doors were locked for the last time. Remember that many older people in China today grew up in the propagandized "Worker's Paradise," where everyone supped from Mao's "iron rice bowl"—where all were provided for "according to their needs" and where everyone was equal, even if only equally poor. In reality, the sort of life implied by the iron rice bowl, always full and never broken, was seldom so easy, but it certainly was more predictable than that of the mercurial present.

For most Chinese people, urban or rural, current politics, economic opportunities, and social norms and views are startlingly different from what passed for convention in their lives only thirty-three years ago. Many have benefited from China's changes, others have lost considerably more than income, but all have had to change their attitudes and ideals as the reforms roll on. The past three decades have witnessed greater changes of all types on more people and at a faster pace than perhaps any time in China's long history. In fact, it may be the pace of this change, the startling rapidity of China's newest "revolution," that has fostered so much interest in the country in recent years.

Manifestations of these changes are apparent at all levels of Chinese society, from the individual to the corporate, from the urban to the rural, and from the provincial to the national. As might be expected, these social, economic, cultural, and political changes have not evolved in homogeneous fashion throughout China's many regions and provinces. Studies indicate growing disparities in income and per capita gross domestic product between wealthy and poorer places (Buckley 2010; Fan 1995, 1997; Gipouloux 2000; Lyons 2000; Wei 1999). Beyond income or productivity, however, there are also important differences across regions and provinces with respect to access to hospitals and health care professionals, social services, dedicated education funds, and transportation infrastructure (National Bureau of Statistics 2009). The changes have resulted in an increasingly uneven landscape—socially, culturally, and economically. Social inequality is a growing concern, and debates about the causes and solutions are carried out in the public arena and via public media as never before since the founding of new China in 1949 (Wang 2008).

Readers are certainly aware of China's growing impact on the world economy. Originally stimulated by the landmark economic reforms promulgated by Deng Xiaoping and boosted by subsequent and arguably greater adjustments, the great efforts of the Chinese people have transformed a once-backward economy into one of the world's largest. In 2009, China's reported gross national income of US$4.78 trillion ranked third in the world, although on a per capita basis the nation ranks 124th of the 213 nations that participated in the study. Today, China is the world's second-largest economy based on the aggregated value of goods and services produced, having passed Japan in 2010 (Bloomberg News 2010). Some studies project that China's economy will surpass that of the United States by 2030 or even earlier (Wikipedia 2010).

Ongoing reforms have also transformed the fundamental structure of China's economy from one based largely on agriculture and heavy industry to one that is much more balanced—now including considerable light manufacturing and a fast-growing service sector. Most economists believe that despite some significant domestic economic problems, China's global economic role will continue to grow. American homes and those of the people of virtually every industrial nation in the world are full of Chinese products—testimony to the benefits of low-cost labor, flexible production, and increased global trade. China's entry into international markets for consumer durables, electronics, tools, household items, sporting goods, clothing, and even some agricultural products, such as crayfish, fruit juice, honey, and essential oils, has already radically altered global trade in these items (Studwell 2002; Veeck 2006; Walcott 2002).

Manufacturing and exports continue to burgeon. In 2009, for the ninth year in a row, the largest trade deficit of the United States ($226.9 billion on $365.9 billion total trade with China) was recorded with the People's Republic of China (PRC; U.S. Census Bureau 2009). Most nations of the European Union are also running large trade deficits with China, attesting to the massive impact of China's reforms on the global economy. With such favorable terms of trade and trade balances, it is no surprise that China had accumulated approximately US$2.5 trillion in foreign reserves by 2010. China then invests some of these foreign reserves in the treasury bonds of countries like the United States and in that way increases its influence and power in global capital markets but also keeps interest rates low for American consumers as well.

China's role in global politics has also expanded, not only with its vital participation in the potential resolution of regional issues such as North Korean instability and nuclear proliferation but also more broadly as China takes a more active role in global politics via international organizations such as the United Nations and the UN Security Council, the World Health Organization, and the World Trade Organization (Zhao 2000a, 2000b). There is a new world order emerging, and while it is most difficult to anticipate the China of fifty years hence, it is certain that China will play an increasingly important role in global affairs in the coming decades.

Internal Forces of Change and Spatial Rearrangement

Less observable from afar and less reported in the popular press of Western nations are the continuous changes and challenges that have emerged in Chinese society, culture, domestic politics, and environment. In tandem with economic and international political change, Chinese society and cultural values are also being reformed and redefined at an astounding pace. Most would argue that there is a trend toward growing differences across China or that, in other words, China is becoming more heterogeneous. The pace of change and the impacts of this change now vary considerably from place to place, sector to sector, and even household to household (Fan 1995; Wei 1999). These growing differences throughout the nation currently influence some of China's most significant and confounding development policy debates. Beyond the income disparities mentioned above and discussed in several of the chapters to follow, unequal economic growth and development have created many other new social, cultural, and political problems that vary significantly by location and socioeconomic status. Yet there are many clear winners, especially in coastal areas, whose lives are appreciably better in many ways than they were prior to economic reform (Dwyer 1994; Fan 1995). In short, it is increasingly inappropriate to make broad generalizations about China or the Chinese people.

Generation gaps, economic gaps between rich and poor, regional gaps in economy and opportunity, new environmental challenges, and tensions between modernity and tradition abound in this large, ancient, and great nation. There are spatial patterns associated with all of these issues. In the chapters that follow, many of these issues are explored with the goal of making sense of contemporary China. There is what might be called a spatial "geography" to China's economic and social changes, challenges, and outcomes. Of course, different places in China have different resources, variable infrastructures, and different social problems, and they also have developed different resolutions to these problems over time. In the reform era, more of these solutions (and, unfortunately, problems) are of local character and/or origin. Exploring these locational differences in issues, resources, and solutions allows us to move beyond national generalizations to gain a more comprehensive and accurate understanding of contemporary China's true complexity as well as of the diversity of issues confronting China's people and leaders today.

Goals and Conceptual Approaches

A major goal of this volume is to elucidate and analyze the changes and challenges facing China in the present and those that it has faced in the past. At the same time, we want to summarize and explain China from a geographic perspective. It is certainly true that there have been many books originating from many disciplines that have presented summaries of modern China. Many of these are cited in our own chapters. The impetus for this specific volume, however, is to explore modern China in terms of two geographic themes: (1) China's spatial organization, including its rapidly grow-ing links to the changing global system, and (2) the human occupancy and use of the many types of environments (physical and cultural) found within the country's current borders.

These themes are common in geographic research but are sometimes overlooked in studies originating in less integrative disciplines. Geography, in parallel with history but unlike most social science disciplines, can be distinguished by its approach—the ideal of a genuine integration or synthesis of information and explanation from many disciplines dedicated to the exploration of change across space and time—rather than by any claim to unique subject matter. In this book, we apply this holistic approach, weaving research from many disciplines beyond geography into a tapestry or frame-work that allows useful generalization while at the same time representing the diversity that is modern China. As geographers, we believe that an effective assessment of China can best be made through simultaneous appreciation of the complex forces caught up in what might be viewed as the interactions between axes of time and space. The geographer's art is to explore the nuances and implications of location not only in terms of changing environmental and cultural landscapes but also as these landscapes and their uses (in this case, by the people of China) change over time (Hart 1982; Williams 2002, 223). Like those of any country, the history and geography of China are always interconnected, with contemporary society, culture, economy, and politics emerging as distillations of both past and present conditions in any specific location.

Our challenge is to present this mosaic in a manageable volume free from jargon and excessive oversimplifications. The challenge for our readers is to constantly recall that the chapters that follow provide, at best, a series of insightful explanations and snapshot summaries of a nation that is reinventing itself in all aspects at astonishing speed. At the outset, it is of paramount importance that readers recognize the vast implications not only of the changes that have come to China during the past several decades of reform but also of the very pace of these changes and their implications for the world. Rapid change makes policy development and implementation more difficult because the major issues associated with seminal problems change constantly, just as the facets of a gemstone variably refract light and color throughout the course of a day. Recognition of the pace of change as an almost independent factor is vital to any understanding of contemporary China.

It could be argued that during the Maoist era (1949–1976), strong Chinese Communist Party (CCP) control of China's economic, political, and social infrastructure and ideology resulted, at least briefly and somewhat superficially, in the homogenization of not only national goals and aspirations but, to some extent, of many aspects of Chinese culture and society as well (Blecher 2000; Lippit 2000; *Rural Cadre Handbook* 1981). Under Mao, central planning directed not only the use of resources and capital but also the spatial organization of the nation and its ideological views. This resulted in relatively uniform strategies for the nation as a whole as well as a strong inward orientation and self-imposed isolation—a plan and policy direction that were not altogether successful. Admittedly, the impact and implications of policy formed at the national level were buffered by regional and local conditions, but prior to the 1978 reforms, there was far greater consensus, at least among those making decisions, regarding how best to proceed.

This is not to imply that after 1978 market forces were given free rein. Indeed, while the future role of the state with respect to directing the economy is not clear, its control over many important aspects of everyday life in China remains a general feature of "socialism with Chinese characteristics" (Liu and Wu 1986). In many sectors and aspects of the economy, the central government continues to play a critical role in planning and guiding the direction of development (Duckett 1998; Garnaut 2001, 15–17; Hinton 2000). As we discuss at length in many of the following chapters, the consensus viewpoint of the Mao years has to some extent been lost in contemporary China—local conditions matter more now than at any time in the past fifty years. Research from many places in China reflects the growing importance of varying levels of regionalism by which provinces, prefectures, counties, and cities increasingly manage their own affairs, resolve disputes, and plan their future in step with increasing ties to and interactions with the global system (Chu and Yeung 2000; Lin 1997; Ma and Cui 2002; Shieh 2000).

The fundamental American debate represented by the contrasting views of Thomas Jefferson and Alexander Hamilton regarding the role of the central government is as important in contemporary China as in most industrial nations. What is the role of the central government in promoting economic growth and social equity? How should revenues be raised and distributed?

What should be taught in the schools, and how should schools be supported? How should health care and social services be provided, and how should they be paid for? In the reform era in China, it is clear that many more decisions, once determined by central government fiat, will now be made by local governments and local entrepreneurs as new sources of revenue are required for the rapid urban expansion (Lin 1997, 2009). The periphery is growing in power on the back of variable economic growth, differential revenue streams, and changing political realities and government policies. Yes, the CCP is still in charge, but as the party is slowly separated from the day-to-day functions of local government, the increasing number of democratically elected local officials attests to a deep-seated desire to reform the political system in many ways. Often in rural China, village or township elections have multiple candidates, and sometimes non-CCP candidates win.

An explosion of scholarship, in and out of China, coupled with the rapid pace of change noted above, make any summation of contemporary China both easier and more difficult. China is so large and the many histories of the people and places now found within its national borders are so complex and different that generalizations will always be problematic. Still, without any generalization, there can be no comprehension of the whole. We hope to achieve balance while presenting this remarkable nation from a variety of perspectives.

The remaining sections of this chapter provide additional background information and introduce our geographic approach to the material in the topical chapters that follow.

China's Land and People: Implication for State Organization and Control

As is discussed in considerable detail in chapter 2, an understanding of the vastness and grand scale of China and of the implications of this scale with respect to China's varied environments and human activities is essential. Consider the size of several individual provinces. Many of China's provinces and autonomous regions are individually bigger than a number of the larger and most important countries of Europe (see figure 1.3). Sichuan, Guangdong, Henan, and Shandong provinces, for example, are larger than any European country except Russia, each with more than 94 million people in 2008. In fact, the provinces of Shandong, Guangdong, Sichuan, and Henan all have more people than Germany—Europe's second most populous nation to Russia (National Bureau of Statistics 2009, 96; *Encyclopaedia Britannica* 2003, 480).

Figure 1.3. China's first-order (provincial-level) administrative regions

China's population of 1.34 billion is very unevenly distributed. Seven of the eleven cities with populations over 4 million within their urban districts are located within East China. Shanghai and Beijing are each home to more than 17 million (National Bureau of Statistics 2009, 362). Of China's largest cities, only Xian, Wuhan, and Chongqing are located in the interior center or West. As elsewhere, political and economic power tends to concentrate in the most-populated regions. Places with greater populations generate more revenues and have better transport, more universities, and more political representation. All of this confounds the effect on West China of its spatial isolation and higher transport costs, with a parallel increase in costs associated with manufacturing and industrial production—activities that have higher net returns than traditional activities, such as agriculture, forestry, or mining.

Equally significant for China's past and present is the country's location in Asia. China, like the United States, is a middle-latitude land, and its eastern and southern peripheries open onto the Pacific Ocean and adjacent seas. The effect of this is to give the eastern half of the country a high index of accessibility to ocean shipping and international air transportation (see figure 1.1). Middle-latitude location and proximity to the ocean and seas help provide a source of moisture and moderate temperatures, and both accessibility and middle-latitude location are great advantages when compared with the relative isolation and extreme climatic conditions of China's northern neighbor, Russia. These advantages resemble, in part, similar conditions found in the eastern United States.

China, however, has the disadvantage of a West that is closed off by a high and extremely dry region of mountains and basins. China's West suffers from inaccessibility and the rigors of a harsh, dry climate. West China, past and present, is sparsely peopled, and much of it remains poorly developed. A disproportionate number of China's chronically poor are found within its interior borderland provinces, this being partly a reflection of both their location and their environment but also a testament to their lack of political and economic power. In many of the chapters that follow, contrasts are made between coastal and interior provinces as new regions emerge and local and regional economies and societies rearrange themselves. Differences in infrastructure, educational opportunities, transport, and central government patronage underlie the current coastal East versus interior West debate. Such palpable differences and inequities no doubt stimulated the contemporary central government policy "Develop the West," which began in 2000 but continues to the present and is designed to foster greater regional equity while slowing migration to urban areas and coastal provinces. Regional equity is an important issue in many of our chapters and, as noted earlier, an issue of increasing concern to China's leadership as well. It is important to keep in mind the implications of China's great size and the eastern concentration of its population because these factors relate to the past and present development efforts that are summarized in many of the chapters as well.

China's Political and Cultural Permanence

China's great size and large number of people have been cited as two of the outstanding features of the country and proof of its significance in the panorama of contemporary world events. Another aspect of China that illuminates its importance in world history is its long record of political and cultural tenure. The Chinese people today occupy the original core area (as well as additional territory) that was occupied at the beginning of Chinese recorded history in 1700 BC (the beginning of the Shang dynasty). The Zhou (1122–221 BC) and Qin (221–206 BC) dynasties followed and signaled an age of cultural development and unification in China. Qin Shi Huang, the first emperor of the Qin credited with unifying China (see chapter 3), faced many of the problems that, in somewhat altered form, remain today. Imagine the problems facing China's leaders over time as they sought to manage culturally distinct regions so immense and varied in natural resources. How should the central state deal with and overcome the "friction of distance" among places and regions while establishing effective political control? How and where should investments be made for the construction of public works that would best serve the nation's interest by improving production? These were some of the tasks that confronted Qin Shi Huang in his attempt to unify China and provide an overarching sense of nationhood to *Zhongguo* (the Central Kingdom), or "center of the known universe," as the Chinese refer to their own country.

To achieve his goals, Qin Shi Huang created an imperial dynastic system that exhibited a strong degree of centralized authority. He established a bureaucratic mechanism that used a standardized written language to communicate as it sought to maintain administrative control over the national territory. He standardized currency to promote interregional trade. He also expanded the canal system and road network, investments that aided commerce and were critical for political control of key areas in the early years of the dynastic era. Such accomplishments laid the foundation on which the ensuing Chinese Empire was built.

China expanded its frontiers in subsequent centuries, relentlessly pushing southward and absorbing a great many local tribal peoples and states into the powerful framework of the Chinese cultural system. This was done despite conflict and struggle. By the beginning of the Christian era, China had carved for itself a distinctive national territory, administered by a central political and bureaucratic system and underpinned by a powerful set of shared values and culture. That functioning cultural and spatial system lasted more than two thousand years, until the twentieth century when the traditional dynastic form of governance and administration was overturned by the Republic of China (1911–1949). This twentieth-century revolution later resulted in the establishment of a socialist and communist system of government and national management, the PRC.

Martin Jacques (2009) has argued that China is best described as a civilizational state, what the political scientist Lucian Pye (1992) described "as a civilization masquerading as a nation state." This is an interesting and provocative viewpoint that suggests some fundamental differences about China from other modern or developing nations. Such a view also may help us understand better and interpret the enduring idea the Chinese people have of their own culture and way of life and of the manner in which the governing system of the central state should operate. Consequently, it may be argued, there is no other appropriate national analogue for us to use in comparing or analyzing China's current development trajectory.

Geography, Globalization, and China's Path to the Future

Answers to the why and how of China's political and cultural stability and permanence are historically interesting and are valuable for the insights they can provide with respect to contemporary visions and plans in China for the future economic development, environmental management, and spatial organization of the nation. These answers and questions are geographic because environment and space (territory), two of the key aspects of the subject matter of geography, are heavily involved in these debates (Tuan 1969). It is for this reason that we have selected the two topics mentioned above (spatial organization and human occupancy and use of the environment) to follow as basic themes in unfolding and helping to explain the "geography" of modern China.

Parallel to the internal development of China is and has been an accompanying set of forces that are typically identified as global. Yet as Dicken (2007) has forcefully reminded us, these forces are often misunderstood and misinterpreted. He goes on to point out that global forces can best be understood in the context of a series of networks of related production and distribution that operate on a variety of levels or scales from the local to the global. His analysis suggests that power relations vary in different production processes and networks depending on the role of the various agents involved: the multinational corporation, the central state, and the local enterprise and state.

This observation is particularly appropriate to an analysis of China, for, as was noted above, the reforms in China have increasingly promoted decentralized power in economic decision making and have led to spatial rearrangements and the emergence of new economic regions based on rapidly changing production activities and behaviors, themselves the results of regional and global economic forces of investment, manufacturing, distribution, and trade. These forces, based on reform and restructuring policies of the central state in place since 1978 and further liberalized after Deng Xiaoping's southern trip in 1992, have been of great benefit to China's economic growth and growing power. Until 2000, the spatial outcomes of these shifts were largely in favor of the coastal regions that have been in the vanguard of the new economy, but as China's leadership tries to reduce regional inequality, there is clear evidence that the western and central provinces are receiving increasing investment and attention.

Yet the economic gains have not been matched by political shifts that might lead to a more open and democratic system of governance and to rearrangements in the manner in which political power and regions are constituted. China remains a Leninist state in which the 75 million members of the CCP make the decisions that direct the social, political, and cultural directions of the state as well as playing a key role in overseeing the general direction of economic growth and change. In this sense, the forces of globalization are buffered and muted and are not allowed to intrude in ways that might challenge the authority or power of the party and the central state.

The large territory and varied natural environments of China require considerable attention if we are to achieve a real understanding of the manner in which the contemporary Chinese are setting about their goals of economic, social, and political reform. Understanding the interaction of the Chinese and their environment is fundamental to comprehending contemporary China and the steps taken by its leaders to promote development and the well-being of its citizens. At the same time, coming to terms with China as a giant spatial and human system is another intellectual and theoretical challenge that offers us an opportunity to bring the special tools and conceptual approaches of geographic analysis to readers of this volume. Finally, using these themes and approaches will help inform our understanding of how China is increasing its global role as its economy expands rapidly and as the nation's political, military, and security roles grow correspondingly.

Questions for Discussion

1. What are some of the indicators that the authors use to report China's economic success since 1978?
2. What are some differences between the Maoist era and the Reform era?
3. When did the first unification of China occur, and what were some of the measures taken by Qin Shi Huang to accomplish this?

4. How would you describe a "geographic approach" to the study of China? (What is different about this approach from other disciplines? What is the same?)

References Cited

Blecher, Marc. 2000. The Dengist period: The triumphs and crises of structural reform, 1979 to the present. Chap. 2 in *The China Handbook: Prospects onto the 21st Century*, ed. Christopher Hudson. Chicago: Glenlake, 19–38.

Bloomberg News. 2010. China overtakes Japan as world's second-biggest economy. August 16. http://www.bloomberg.com/news/2010-08-16/china-economy-passes-japan-s-in-second-quarter.

Buckley, Peter J. 2010. *Foreign Direct Investment, China and the World Economy.* New York: Palgrave Macmillan.

Chu, David K. Y., and Y. M. Yeung. 2000. Developing the "Development Corridor." Chap. 13 in *Fujian: A Coastal Province in Transition and Transformation*, ed. Y. M. Yeung and David K. Y. Chu. Hong Kong: Chinese University Press, 305–26.

Dicken, Peter. 2007. *Global Shift: Mapping the Changing Contours of the World Economy* (5th ed.). New York: Guilford.

Duckett, Jane. 1998. *The Entrepreneurial State in China.* London: Routledge.

Dwyer, Denis, ed. 1994. *China: The Next Decades.* Essex: Longman Scientific and Technical.

Encyclopaedia Britannica. 2003. New York: Encyclopaedia Britannica Educational.

Fan, C. Cindy. 1995. Of belts and ladders: State policy and uneven regional development in post-Mao China. *Annals of the Association of American Geographers* 85, no 3: 421–49.

———. 1997. Uneven development and beyond: Regional development theory in post-Mao China. *International Journal of Urban and Regional Research* 21, no. 4: 620–39.

Garnaut, Ross. 2001. Twenty years of economic reform and structural change in the Chinese economy. Chap. 1 in *Growth without Miracles: Readings on the Chinese Economy in the Era of Reform*, ed. Ross Garnaut and Yiping Huang. Oxford: Oxford University Press, 1–18.

Gipouloux, Francois. 2000. Declining trend and uneven spatial development of FDI in China. In *China Review 2000*, ed. Chung-ming Lau and Jianfa Shen. Hong Kong: Chinese University Press, 285–305.

Hart, John Fraser. 1982. The highest form of the geographer's art. *Annals of the Association of American Geographers* 72: 1–29.

Hinton, Peter. 2000. Where nothing is as it seems: Between Southeast China and mainland Southeast Asia in the post-socialist era. In *Where China Meets Southeast Asia: Social and Cultural Change in the Border Regions*, ed. Grant Evans, Christopher Hutton, and Kuah Khun Eng. New York: St. Martin's, 7–27.

Jacques, Martin, 2009. *When China Rules the World: The End of the Western World and the Birth of a New Global Order.* New York: Penguin.

Jefferson, Gary H., and Thomas G. Rawski. 1999. China's industrial innovation model: A model of endogenous reform. Chap. 3 in *Enterprise Reform in China: Ownership, Transition, and Performance.* Washington, DC: World Bank.

Lin, George C. S. 1997. *Red Capitalism in South China: Growth and Development of the Pearl River Delta.* Vancouver: UBC Press.

———. 2009. *Developing China: Land, Politics and Social Conditions.* New York: Routledge.

Lippit, Victor D. 2000. The Maoist period, 1949–1978: Mobilizational collectivism, primitive accumulation, and industrialization. Chap. 1 in *The China Handbook: Prospects onto the 21st Century*, ed. Christopher Hudson. Chicago: Glenlake, 3–18.

Liu Suinan, and Wu Qungan, eds. 1986. *China's Socialist Economy—An Outline History (1949–1984)*. Beijing: Beijing Review.

Lyons, Thomas P. 2000. Regional inequality. Chap. 14 in *Fujian: A Coastal Province in Transition and Transformation*, ed. Y. M. Yeung and David K. Y. Chu. Hong Kong: Chinese University Press, 327–51.

Ma, Laurence J. C., and Gonghao Cui. 2002. Economic transition at the local level: Diverse forms of town development in China. *Eurasian Geography and Economics* 43, no. 2: 79–103.

National Bureau of Statistics. 2009. *Zhongguo tongji nianjian 2009* [China statistical yearbook 2009]. Beijing: China Statistics Press.

Pye, Lucian, 1992. *The Spirit of Chinese Politics*. Cambridge, MA: Harvard University Press.

Rural Cadre Handbook. 1981. The correct handling of love, marriage, and family problems. Reading no. 81 in *Chinese Civilization and Society: A Sourcebook*. New York: Free Press, 371–77.

Shieh, Shawn. 2000. Centre, province, and locality in Fujian's reforms. Chap. 4 in *Fujian: A Coastal Province in Transition and Transformation*, ed. Y. M. Yeung and David K. Y. Chu. Hong Kong: Chinese University Press, 83–117.

Studwell, Joe. 2002. *The China Dream: The Quest for the Last Great Untapped Market on Earth*. New York: Atlantic Monthly Press.

Tuan, Yi-fu. 1969. *China*. Chicago: Aldine.

U.S. Census Bureau. 2009. Foreign trade statistics. http://www.census.gov/foreign-trade/balance/c5700.html#2009 (accessed November 3, 2010).

Veeck, Gregory. 2006. Post-WTO agriculture in East Asia: A case study of apple products. *Asian Geographer* 25, no. 1 and 2: 173–91.

Walcott, Susan M. 2002. Chinese industrial and science parks: Bridging the gap. *Professional Geographer* 54, no. 3: 249–364.

Wang, Feng. 2008. *Boundaries and Categories: Rising Inequality in Post-Socialist Urban China*. Stanford, CA: Stanford University Press.

Wei, Dennis Yehua. 1999. Regional inequality in China. *Progress in Human Geography* 23, no. 1: 48–58.

Wikipedia. 2010. Economy of the People's Republic of China. http://en.wikipedia.org/wiki/Economy of the People's Republic of China

Williams, Jack F. 2002. Geographers and China. *Issues and Studies* 38, no. 4; 39, no. 1: 217–47.

World Bank. 2009. *World Development Indicators*. Washington, DC: International Bank for Reconstruction and Development.

Zhao, Quansheng. 2000a. China in East Asia: Changing relations with Japan and Korea. Chap. 5 in *The China Handbook: Prospects onto the 21st Century*, ed. Christopher Hudson. Chicago: Glenlake, 69–80.

———. 2000b. The China-Japan-US triangle and East Asian international relations. In *China Review 2000*, ed. Chung-ming Lau and Jianfa Shen. Hong Kong: Chinese University Press, 77–103.

CHAPTER 2

China's Natural Environments

Territory and Location

China's territory encompasses approximately 9,600,000 km² and is roughly the same size as the United States and smaller than only Russia and Canada. Among these very large nations, China is exceptional in that its land and resources must support and provide for more than 1.34 billion people—approximately 19.7 percent of the world's population—on only 6.5 percent of the earth's land area (China PopIN 2009). The sheer size of the nation has many implications, and China exhibits numerous similarities to the United States. For example, both countries are situated in the middle latitudes of the Northern Hemisphere, and both have extensive coastlines fronting on middle-latitude oceans and seas. Approximately 98 percent of China's land is located between latitudes 18°N and 50°N, and the densely settled eastern half of the country is mainly a temperate and subtropical land. Physically, the southeastern United States and southeastern China are very similar in their soils, climate, and topography. Although China has approximately 18,000 km of coastline and is easily accessible by water from the east and south as is the United States, the great northern and western portions of China are enclosed and isolated by massive mountain systems, great deserts and basins, and high plateaus (figure 2.1).

The Implications of China's Physical Landscape

The barriers represented by the rugged terrain of China's western regions were historically very significant, isolating China from neighboring nations and cultures. Even at present, the formidable terrain and great distances of its West hamper China's westward connections with the remainder of Asia. Unlike Europe, whose mountains are clustered in its center, China has mountains that encircle the country and encourage development within the more accessible eastern alluvial and coastal plains. Historically China's developmental energy has been focused on overcoming the challenge of its internal physical geography, and through the centuries, China's governments have

16

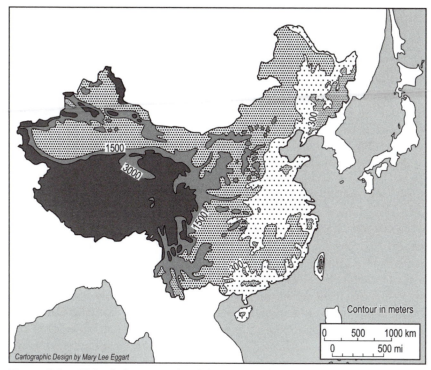

Contour in meters

0   500   1000 km

0   500 mi

Cartographic Design by Mary Lee Eggart

Figure 2.1. China's topography (Wu Yuanli 1973)

expended considerable time and energy melding the country's various cultures and regions together rather than promoting its external expansion. In this respect, China's experience is in marked contrast to that of Europe, where maritime colonial expansion beyond traditional borders made small nations such as England and the Netherlands, during their heyday, among the most powerful in the world.

Western China generally is high and mountainous, although the center of the Turfan Basin in Xinjiang is a remarkable 505 m below sea level, and parts of the Tarim Basin are only 1,000 m above sea level (ASL). The vast remainder of West China, however, is well over 2,500 m ASL, and the massive Tibetan-Qinghai Plateau is mostly over 4,000 m ASL. Because of its interior location within the great Eurasian landmass and the orographic effect of these great mountain systems that block out moisture, West China tends to be extremely dry. For three millennia, this western region has remained sparsely populated, despite ever-growing population pressures in East and Southeast China. Spatial isolation, coupled with an inhospitable climate, has given this region a very distinct developmental and cultural history when compared with the remainder of China.

This isolation has created problems of national integration. Traditionally, West China has been viewed by most Han Chinese (the ethnic group comprising 92 percent of China's population) as a remote and desolate frontier region. All of China's provincial-level autonomous regions (offering, in theory, some measure of political autonomy for minorities) are in the West and Southwest. To some extent, the stereotypes of the early American frontier parallel those held by the majority of Chinese regarding China's Far West. As in America, the frontier in China projects a range of cultural images from the romantic and free to the wild, unsettled, and remote. Further, like the American Wild West, these regions are contested space where local ethnic-minority traditions, growing Han Chinese intrusions, and conflicting perceptions of modernity have yet to reach a balance. Millennia-old conflicts between West China's indigenous groups and the growing numbers of Han Chinese settlers or migrants arriving to exploit its great natural resources are on the increase. This is but the first example of many in this book of how demands placed on China's environment have shifted dramatically in response to China's changing economy and society.

A country with such a large area characteristically has a large resource base. Although precise identification and surveys of China's mineral resources remain incomplete in a few difficult cases such as petroleum, natural gas, and precious metals, it is clear that China has vast reserves of most major minerals, including tungsten, coal, antimony, tin, zinc, iron sulfide, vanadium, molybdenum, titanium, mercury, salt, fluorspar, and magnesite (Zhao 1994). China also has substantial reserves of iron, manganese, aluminum, limestone, and petroleum. Its reserves of tungsten and antimony are believed to be the world's largest. Only in nickel and aluminum, among the most commonly used and requisite minerals for modern industrialization, is China seriously deficient. China's supply, production, and use of mineral resources are examined in greater detail in chapter 9, but from the outset, it is important to realize that China's mountainous topography is both a blessing and a curse. The western mountains and plateaus have restricted economic development, transportation, and national integration, but they are also rich in the mineral resources that are vital for China's continued economic development.

Another consequence of China's size and location is the great diversity of its regional and local climates. In West China, *continentality* is the most important climatic factor because the dominant westerly winds from Central Asia render the northern and western interior regions cold and very dry throughout winter. With the concept of continentality, meteorologists recognize the fact that land areas experience greater temperature variations than water bodies, and larger land areas experience greater ranges than smaller land areas. Continental climates tend to be drier than oceanic or coastal climates with greater temperature ranges through the year. Continentality influences both temperature and moisture, the major drivers in regional climate conditions. As a consequence, precipitation in North and West China is as limited as that of the Great Plains and basins of the western United States.

In contrast to the north and west, Central and southern China are heavily influenced by the great high-energy summer *monsoon systems* originating in the Pacific. These systems result in onshore winds laden with precipitation, and the climate of these regions ranges from temperate to tropical, with typically 1,000 mm or more of rainfall per year (see figure 2.2). Monsoon influences are also found throughout eastern China, although northern China receives significantly less precipitation than does the South.

A third driver of regional climates is commonly referred to as the *orographic effect.* Under proper conditions, mountains serve as barriers or impediments to the progres-

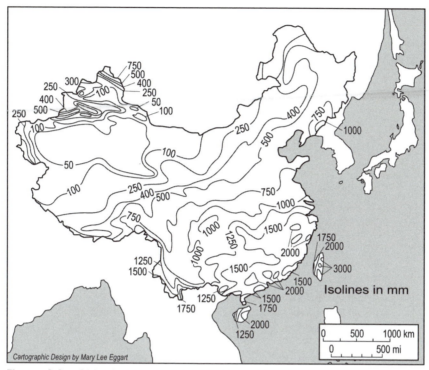

Figure 2.2. China's annual precipitation (*Zhonghua renmin gongheguo dituce* (Atlas of China), 1979 (in Chinese))

sion of air masses across landscapes. As air masses are driven into the mountains or elevated as they move over plateaus, the air cools as it rises, producing rainfall on the windward side of the mountain system and significantly drier conditions on the lee-ward side. A classic example of this orographic effect may be observed south and north of the Qinling Mountains, where a major environmental distinction may be made between humid, subtropical China south of the mountains and drier, more continen-tal China north of the mountains. In part, this division follows the east–west trend of the Qinling Shan (Qinling Mountains), which extend eastward to the East China Sea along the Huai River. Thus, this mountain chain divides China into regions of water surplus and deficit—a critical issue for agriculture and settlement (see figure 2.3). Based on such a division, southern China and the southern parts of the North China Plain and the Northeastern Plain (Dongbei Pingyuan) are regions where rainfall gener-ally refreshes annual groundwater supply. In sharp contrast, all of western China and the remainder of China north of the Qinling Shan are water-deficit regions—where groundwater is but inconsistently replaced by precipitation. Not surprisingly, some of China's worst environmental problems have emerged in the dry, ecologically unstable regions of the North and West.

As this survey of the spatial organization of China's economic activities and popu-lation distribution unfolds in subsequent chapters, the environmental advantages or constraints of any given region take on great significance—location matters in so many

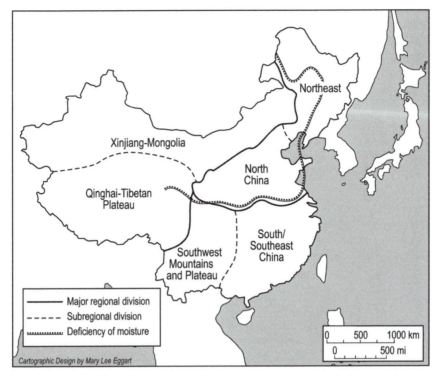

Figure 2.3. China's major environmental regions (after Pannell and Ma 1983, 16)

ways. The two major themes of this text—spatial organization and human–environment relations—provide a geographic explanation of the challenges of modernization, development, and cultural and economic change in contemporary China as these vary across the diverse landscapes that make up this great nation.

For much of its long history, China has been a nation of farmers. China's great environmental diversity permits the use of a broad range of agricultural environments, resulting in a comprehensive array of agricultural products. This environmental diversity also offers some measure of insurance against local or regional natural catastrophes because these can be absorbed by the country's larger environmental and economic systems. The summer floods on many of China's rivers in 2010 were some of the worst in fifty years, yet the national summer grain crop that year was one of the largest on record because conventionally dry, interior grain areas recorded unusually high yields. Protected in this way by its sheer size, China is able to meet upwards of 90 percent of its citizens' food requirements while producing an ever-expanding variety of different commodities—grains, fruits, vegetables, commercial and industrial crops, and livestock—for export as well as domestic consumption (see chapter 8 for more on food security).

In all large nations, some regions are more favorably endowed than others. Typically, inhabitants of resource-rich regions become wealthier than their counterparts in poorly endowed or isolated regions. For the successful and ecologically sound development of resources, capital is also a prerequisite. As a consequence of limited capital, many parts of West China remain relatively poor despite significant subsurface resources. On the other hand, degraded environments in China, as elsewhere in the world, are frequently associated with poverty. Regional inequities in quality of life and income levels result from environmental degradation, and social problems of many types generally follow quickly on the heels of uneven economic growth. Often the poor make improper environmental decisions concerning land and water resources. Just as poverty results from degraded environments, degraded environments often are a result of poverty. Recent high-level government decisions to protect China's beleaguered environment offer hope, but the past fifty years of economic expansion have caused myriad environmental problems that will take many decades to reverse (McElroy, Nielson, and Lydon 1998; Smil 1993). The familiar list of environmental problems in most nations, rich and poor, hold for China as well, but at times, because of the nation's size, some problems are more intense than for other nations.

China's environmental challenges should be broadly classified into two categories: (1) pollution problems resulting from point-source pollution (e.g., industrial effluents, airborne releases from factories, and power plants) and (2) ecological problems, such as deforestation, desertification, salinization, land and air pollution from fertilizers and farm chemicals, and erosion brought about by uninformed and avaricious land-use practices and changes in regional precipitation patterns (Day 2005; Edmonds 1994). Point-source pollution problems once originated largely in urban areas because, up until the reform era that began in 1978, major cities (over 500,000 persons) claimed the lion's share of industrial production and coal and oil consumption. After the reforms, however, the problems associated with point-source pollution spread throughout most of eastern China and into many areas of the West (Economy 2004; Smil 1984). As Chinese and international firms sought lower labor and land costs by shifting manufacturing and raw material processing to the western regions, the health effects from point-source pollution, once only found in the East, are now a fact of life facing many areas of West China as well. Township and village industrialization now accounts for more than 32 percent of rural employment. All too often, local government officials and agencies favor growth-at-any-cost policies that generate higher wages and ensure their reelection while ignoring the severe pollution and the environmental degradation that results from unregulated factories, mines, pulp and paper mills, and agroprocessing facilities. Water quality in most areas of rural China has deteriorated rapidly in the past thirty years because of the dumping of industrial chemicals, wastewater, salts, bleach, and other toxins into rural China's rivers, ponds, and canals (Day 2005; Economy 2004). In some areas, riots and mass demonstrations regularly occur as citizens decry the greed that has left a legacy of birth defects and health problems in many areas. Ecological problems associated with agriculture, husbandry, and forestry have also impacted much of China, but the most pressing challenges are found in the central and western regions that are at once drier and more isolated. Here the use of rangeland for farming, unregulated logging, and the careless disposal of mine tailings associated with mineral extraction have resulted in many tragic local ecological disasters that are only now being identified. Farm-chemical abuse throughout the nation has resulted in polluted groundwater and surface water, causing many types of illnesses in vulnerable populations. Looking to the next decade, China's curbing and mitigating of its environmental problems are among the most difficult challenges facing the nation.

Surface Structure and Geomorphology

The surface structure and landforms of China are very complex. If viewed in the context of continental drift, the tectonic structure of China has resulted from the gigantic eastward- and southward-moving Eurasian continental plate colliding with the westward-moving Pacific plate. In addition, southwestern China came into contact with the Indian Ocean plate which is moving northward. The intersection of these gigantic plates, or what is called the *geotectonic structural platform* of China, explains the major structural surface features of China's diverse contemporary landscapes. On the macroscale, the progressive decrease in elevation from western to eastern China is most important for the purposes of this text. Elevations of the higher mountain systems in China, such as the Kunlun and the Himalayas, average over 6,500 m ASL. These elevations in the Far West decline to about 5,000 m on the eastern face of the Tibetan Plateau and eventually drop to sea level near the coast of the Yellow Sea.

The Tibetan Plateau, a huge and high *massif*, or primary mountain mass, is rimmed on the north and east by the remainder of China. The eastward descent from this western landmass to China's eastern coast is much like coming down three steps of a gigantic staircase: (1) the Qinghai-Tibet Plateau, (2) the Central Mountains and Plateaux, and (3) the Eastern Lowlands.

In sum, the country may be divided into these three macroregions: West, Center, and East. This is true not only for physiographic comparison but also for the economies and cultures that emerge from these environments. In all three parts, the landforms are composed of several giant *geosynclines* and related *geanticlinal* features that are the result of the tectonic movement described above. China's synclinal structure is made up of alternating ridges and plains. In the East, the *strike* (or directional axis) of these is generally northeast to southwest. Moving from east to west, the most conspicuous features of this structure are the Changbai Mountains of northeastern China that form the country's border with Korea, the Northeast and North China Plains, the Taihang Mountains, the Inner Mongolian Plateau, the Loess Plateau, the Sichuan Basin, and the Yunnan-Guizhou Plateau.

Cutting across these features almost at right angles and on an east–west strike in the East are the Qinling Mountains, the Dongnan Hills (Nanling Mountains), and many other smaller systems (see figure 2.4). China's topography has often been described by geographers as a "giant checkerboard" with north–south and east–west striking mountains intersecting constantly, often disrupting the flow not only of rivers but of people and goods as well.

In the Far West, the strike of most of the major geomorphologic features is again east to west. Here are the truly great mountains and basins of China. Indeed, these are some of the most extensive physical features on earth—the Himalayas, the Tibetan Plateau, the Kunlun Mountains, the Tarim and Junggar basins, the Tian Shan

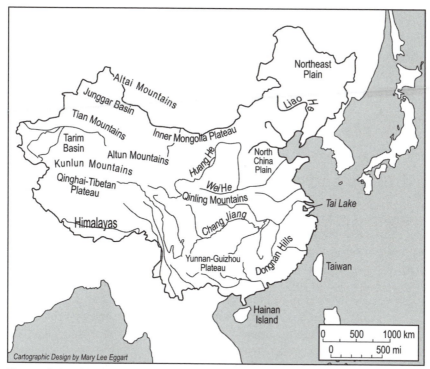

Figure 2.4. China's physical features (Zhao Songqiao 1986; C. S. Hammond and Co., New York, sheet map number 47)

Mountains, the Junggar (Dzungarian) Basin, and the Altai Mountains (see figure 2.4). Massive in scale, these features combine to isolate East and Southeast China from the remainder of Asia. Almost 58 percent of China's territory lies 1,000 m or more above sea level, and only 25.2 percent is below 500 m (Zhao 1994, 17).

Geologically, much of China's territory was formerly below a great sea, although three ancient granitic massifs (one in southeastern China, one in Tibet, and one in Mongolia) formed huge stable islands of ancient Precambrian rock (prior to 620 million years ago) within this sea. Movements against these massifs created the various series of parallel ridges, basins, and plains described above. In addition, a subsequent submergence of troughs and basins during the Paleozoic era resulted in the formation of vast quantities of coal and natural gas, which are of critical importance with respect to meeting the nation's equally large energy needs.

Variations in the geomorphological structure of China's regions and the resultant surface landforms have shaped and molded the distinctive and complex drainage system of China. At the same time, this drainage system has attracted but also challenged the energy and imagination of China's people as they have gone about settling and exploiting the land's many diverse environments.

It is important to realize that the complex patterns and processes of China's contemporary river and drainage systems are not merely a result of China's physical geography. Rather, current surface conditions are also the result of thousands of years of complex interactions between these landscapes and the people who have occupied them. These "feedback" systems have always yielded both sweet and bitter fruit: greater prosperity but more environmental degradation. From this long-term perspective, the Three Gorges Dam project that went into full operation with twenty-six giant power generators in 2009 is simply the latest of efforts among those taken by countless generations in China to massively transform the landscapes on which they live. Discussed further in chapter 8, no landscapes on earth have been as extensively altered by human use as those in China.

Hydrology and River Systems

One geographer estimated that China has over fifty thousand rivers with drainage basins of over 100 km² (Ren, Yang, and Bao 1985). Thirty-six percent of China's territory has *interior drainage* (rivers having no outlet to the sea), and the remainder drains into the oceans and seas around China and Southeast Asia. China has a number of very large and famous rivers within its boundaries, as well as the headwaters of several other major rivers of the world (see table 2.1 and figure 2.5). Most of China's major rivers originate on or near the Qinghai-Tibetan Plateau and flow south or east, emptying into different seas associated with the Indian or Pacific oceans. China's three major self-contained rivers that flow generally from west to east are the Huang He (Yellow River), the Chang Jiang (Yangtze River), and the Xi Jiang (West River). Of these, the Chang Jiang (see photo 2.1) is easily the largest in terms of length, area of drainage basin, and volume of discharge (see table 2.1). The Huang He is next in length and area of drainage basin, although its discharge is modest compared to the Chang and the Xi because of the dry lands through which it flows. The Heilong Jiang (Black Dragon River, or the Amur in Russia), along with its largest tributary, the Songhua, is the most important river in Northeast China and forms the boundary between China and Russia for part of its course. Perhaps because the Northeast was integrated

Table 2.1.   Drainage, Length, and Flow of China's Major Rivers

River	Drainage Area (sq. mi.)	Length (km)	Annual Flow (100 million cu. m)	Drainage Area (% of China total)
Chang (Yangtze)	1,808,500	6,300	9,513	18.92
Huang (Yellow)	752,443	5,464	661	7.87
Songhua	557,180	2,308	762	12.15[1]
Liao	228,960	1,390	148	3.61[2]
Zhuhe (Pearl) and other southeastern rivers, including Xi	453,690	2,214	3,338	6.07[2]
Hai	263,631	1,090	228	3.33[2]
Huai	269,283	1,000	622	3.44[2]

[1]This figure includes other rivers in Northeast (Dongbei) China.
[2]Includes major tributaries.
Source: National Bureau of Statistics (2009, 382); Zhao (1986, 47; 1994, 38).

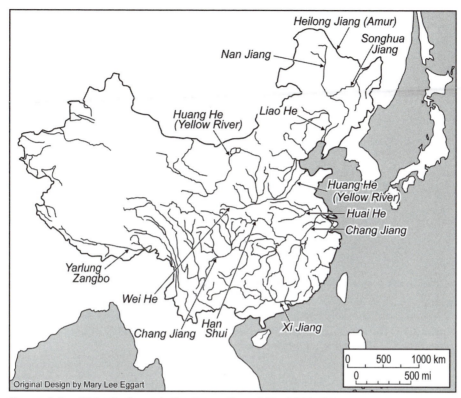

Figure 2.5. China's rivers (after Pannell and Ma 1983, 26)

into China only during the Manchu-controlled Qing dynasty (AD 1644–1911; see chapter 3), the Heilong figures less in the history and cultural evolution of China and is often overlooked. Given that the Northeast is one of the most important new agricultural and industrial regions of China, the ecological health and well-being of this river, along with that of the Songhua, the Nen, and the Liao (its major tributaries), have become of critical concern. International issues complicate environmental protection of the Heilong because it flows through Russian territory to eventually drain into the Sea of Okhotsk just west of northern Sakhalin Island, so water flow and quality are international issues of concern not only to China but to Russia as well.

The Huang He: China's Sorrow

The Huang He and some of its tributaries, such as the Wei, Fen, and Luo, are of special interest because of their role in the emergence and evolution of Chinese civilization. The middle part of the drainage system of the Huang He cuts through the Loess Plateau, as do a number of the river's important tributaries. It is the enormous quantities of yellow loessal silt eroded from the plateau that give the river its name and characteristic color (*loess* being the German term for loamy, yellow deposits of windblown soil). Uncontrolled logging over the centuries has resulted in extreme

Photo 2.1. The Chang Jiang River near Nanjing, the capital of Jiangsu province, taken in the spring of 1987 (Gregory Veeck)

erosion in many parts of China, but the problem is particularly severe within the Loess Plateau. As the Huang He descends onto the North China Plain, its gradient is reduced sharply, and river velocity slows—limiting its ability to transport the massive silt load generated by erosion in suspension. The result is gradual deposition along the riverbed and the accumulation of large quantities of loess on the bottom of the river and, of course, on the river's natural levees when it floods. The *aggradation* (building up) of the riverbed in turn has led to a perennial threat of floods that through the centuries were ameliorated only by building the levees higher and higher. As the riverbed rose in any given location, so did the height of the levees built by farmers to protect their land and homes. In many areas, as a result of centuries of levee construction, the bed of the river is now high above the surrounding agricultural plains (see figure 2.6). Whenever the levees breach, the river pours down onto the plains, causing massive property damage and loss of life—thus the river's other ancient name, "China's Sorrow."

Beginning in the 1970s, the Chinese government launched a major program to tame the Huang He and utilize the river more effectively, but, just as with the more publicized conflicts over the Three Gorges project, these projects remain highly controversial. The large dam and reservoir at San Men Xia (Three Gates Gorge) near the Tongguan elbow of the river began to silt up almost immediately after construction, and technical problems limit its power generation. Overall, the river has been contained but at high cost. There is something akin to squaring the circle in all these efforts. In order to obtain funding from Beijing, every water-control project must serve

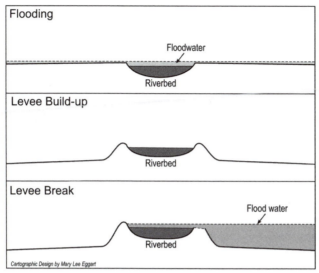

Figure 2.6. Flooding and levee construction on the Yellow River (original drawing by Mary Lee Eggart)

many masters. Water and waterways are scarce in North China, and so each project must have multiple goals to receive funding. Ideally, a successful project will control flooding; generate hydroelectricity; expand irrigated area; provide industry and urban consumers with clean, sediment-free water; and offer improved navigation. A compromise dam is often not very successful in meeting any of these goals, particularly one that carries the amount of silt found in the Huang He. Water quality must be improved before it gets to the river through the promotion of sound farming practices, pasture protection programs, and reforestation.

Another environmental problem of growing political concern is the high consumption rates of water from the Yellow and the Chang Jiang's upper reaches by cities and farmers in western provinces. Eastern provinces are demanding new agreements amid charges that water is wasted in the western arid environments through the development of increasingly popular "water parks" in cities such as Xian or the production of inappropriate crops such as irrigated rice in provinces such as Gansu, Ningxia Hui, and Xinjiang.

In 1999, a series of logging bans and mandatory reforestation projects went into effect on the upper and middle stretches of the Huang He, and these will eventually reduce erosion and improve water quality and flooding issues. Environmentalists see control of deforestation and erosion on the Huang He as one of the major policy battles that illustrate the importance of environmental policy and regulation in China's continued economic development. There must be better enforcement of and compliance with these environmental regulations. This means that more compliance officers are needed, officers who will be able to assess steep fines and provide other coercive measures for noncompliance that will convince local officials to protect the environment. Passing a law is the easy part; enforcing it is much more difficult.

The Chang Jiang: China's Mightiest River

China's largest river is the Chang Jiang (Long River), which drains over 1.8 million km² of Central China. Not surprisingly, given its name, it is also China's longest river (6,300 km). Rising in the rugged mountains of the Tibetan Plateau, the river flows generally eastward to empty into the East China Sea at Shanghai. The importance of the river cannot be overstated. By draining the central part of China, it has provided water for irrigation to generations of farmers and has served as the nation's major west–east transit route for three thousand years. The river has, by far, the largest annual discharge, with an average discharge at its mouth of 31,055 m³/sec. (Zhao 1994, 18). This compares with an average through-flow for the Mississippi River near Vicksburg, Mississippi, of 18,000 m³/sec. or with 1,822 m³/sec. at the mouth of the Huang He. The average discharge rate of the Chang Jiang near its mouth, then, is about seventeen times greater than that of the Huang He.

As with the Huang He, deforestation in the upper reaches of the Chang Jiang has exacerbated the dangers of flooding on the river. The floods of the summer of 2010 saw the river's water level at its highest since 1954. In August 2010, the water level in the Three Gorges Reservoir was measured at over 160 m—about 15 m above the planned maximum. Reports are that the dam has performed well in this respect, although siltation in the reservoir remains a major issue given the river's role in the national transport system (see chapter 9). Unlike the silt-laden and often shallow Huang He, the Chang Jiang is the main corridor for maritime transportation from the great economic engines of the east coast such as Shanghai, Nanjing, and Wuhan to the heart of Central and Southwest China. As Shanghai reasserts its role as the dominant port in central coastal China, better management of the river and improving water quality is vital.

For many years, the Chinese have discussed diverting water from the Chang Jiang and routing it north to the water-deficient areas of the North China Plain and especially to Beijing and Tianjin. Mao Zedong mentioned the idea in a 1952 speech (South to North Water Diversion Project, China 2008). Three routes are generally cited, but none has been fully implemented—due as much to funding and jurisdictional issues as to the environmental concerns and engineering challenges these projects present. Estimates for completion of all three routes range from $52 billion to over $70 billion—three to four times the cost of the famous "Three Gorges Dam." These projects, collectively called the *Nan Shui Bei Dao*, literally "southern water, northern route" but more commonly referred to as the *South to North Water Diversion Project*, continue to be debated extensively in China, and while construction is essentially complete on an eastern route, it is still not fully operational. Construction on the central route started in 2004, and the first 10,000 persons– of an expected 320,000—to be relocated were moved to new villages in August of that year. The central route will link the Danjiangkou Reservoir on the Han River via new canals to Beijing and Tianjin—a diversion of 1,267 km. The western route will not even start until 2010 and is expected to take forty years to complete. Low water levels on the Huang He, which limited irrigation in Shandong since the mid-1990s, encouraged supporters of the eastern route, so it was started much earlier, in 2002. This project not only uses a series of pipelines to pump Chang Jiang water north through the Grand Canal but also uses four lakes (including shallow Hongze lake in Jiangsu) as portions of

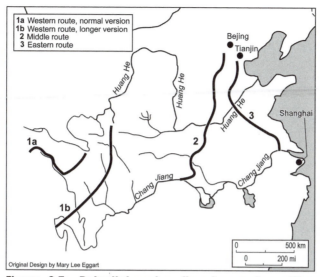

Figure 2.7. Potential water diversion routes, either hypothesized or under construction. (author unknown, "Nan Shui Beidao: China's South-North Water Transfer Project," after a map presented at the East Asia Research Conference, University of Sheffield, July 1999)

the system. A telling stalemate, however, between the provincial governments of Jiangsu and Shandong provinces has frustrated national planners who are unused to taking "no" for an answer. The massive task of constructing a viaduct underneath the Huang He was a significant engineering challenge, but in China as elsewhere, technical difficulties are often overcome faster than political disputes. As a measure of the increasing power of China's provincial governments in relation to Beijing, Jiangsu province refuses to allow the project to begin operation unless more water is allocated to its northern portion. Farther north, Beijing and Shandong province claim that their water needs are more critical, particularly for urban and industrial uses. As more and more of China's grain is produced in North China, arguments for water transfers shift from regional political conflicts to the national stage. Northern grain production cannot keep up with southern China's consumption trends without the water transfer schemes (Ma et al. 2006). Potential western and central routes that will come online later will be much more expensive but would traverse less populated areas and possibly encountering less political resistance. On the other hand, all three transfer programs have slowed in 2009 faced with legal and procedural challenges by environmentalists as well as some provinces wishing to protect their resources and environments (*People's Daily* 2004; Zhang et al. 2009) (see figure 2.7).

The Three Gorges Dam

The impoundment of the Chang Jiang at Three Gorges (San Xia) is among the largest and most controversial construction projects ever built—currently the world's largest dam. The massive amount of hydroelectric power should, in time, allow more effective economic development of all of Southwest China, an area currently among China's poorest. This region has been neglected despite its considerable contributions to the country, and the dam represents a political and economic commitment to the region's future as well as to its integration into a more prosperous future China. Low-cost power is seen as a critical input if this once-remote area with its lower wages and tax revenues is to attract manufacturing. Signs in 2010 suggest that this is occurring as many firms from East China, as well as international firms such as Ford Motor Company, Toyota, Hewlett Packard, and Cummins, have recently located in Chongqing, the megacity of 30 million on the western edge of the massive Three Gorges Reservoir. The public controversies that developed over this project included the project's construction cost, its displacement of farmers, the destruction of hundreds of historic and prehistoric artifacts, and the many potential environmental impacts (Nickum 1998). As many as 1.2 million people will eventually be relocated, and more than 50,000 ha of agricultural land have been inundated. Resettlement, mostly complete, has not gone smoothly, and to this day citizens demonstrate regularly over inefficiencies and corruption associated with the mandatory relocations. Many residents, removed to new villages higher upslope but lacking good soils, feel disenfranchised and powerless. Funding for relocation and new village and town construction has still not met expectations—two years after the dam became fully operational.

The Xi Jiang

Of China's three major river systems, the Xi Jiang (West River) is the smallest in terms of both length and size of drainage basin. However, its position on China's southern coast, where the summer monsoon season generates tremendous rainstorms, means that the Xi has a greater annual discharge than the Huang He. Great efforts are under way to capture and store an increasing amount of this discharge for the rapidly growing cities of the Pearl River Delta, including Dongguan, Foshan, Guangzhou, Hong Kong, Jiangmen, Macau, Shenzhen, Zhongshan, and Zhuhai. The Xi Jiang and its tributaries drain 448,000 km², an area that is home to more than 115 million people. The Pearl River Delta remains China's fastest-growing urban and industrial region, and conflicts over water use and water quality, pitting farmers and consumers against urban and industrial concerns, are increasingly common. A project led by the Chinese Academy of Science and partly funded by the World Bank that is currently under way seeks to close polluting factories, prosecute violators of environmental regulations, and dramatically improve water quality by 2020.

Climatological Patterns and Resources

Climate is as much a part of China's resource base as its land and subsurface resources. Moisture and radiation patterns are quite varied across the nation. Northern areas tend to be dry, with distinct seasons much like the temperate regions of the United States.

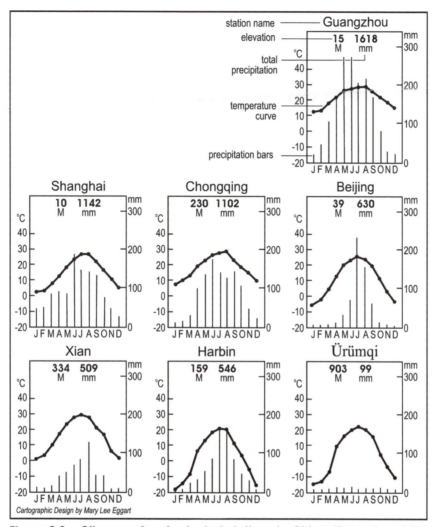

Figure 2.8. Climagraphs of selected stations in China (Pannell and Ma 1983, 21; Zhang and Lin 1992, chap. 4)

The more humid southern parts of China are generally monsoonal in character. Precipitation tends to concentrate in the spring and summer months. Figure 2.8 provides typical rainfall and precipitation rates for a selection of China's major cities.

Many parts of southern China, represented by Shanghai and Guangzhou (Canton) in figure 2.8, record around 1,500 mm (~60 inches) of rain annually. In contrast, 500 to 650 mm (~20–26 inches) of precipitation is common throughout much of North and Northeast China (e.g., Shenyang and Beijing; Zhang and Lin 1992). Given these volumes and the common incidence of drought in these areas, without irrigation, much of North and Northeast China is of marginal utility for intensive agriculture. At present, water shortages are a serious environmental handicap in the improvement and continued development of sustainable agriculture systems throughout the North and

West. A report published by the World Bank (*Economist* 2005) notes that one in three rural Chinese lacks access to safe drinking water and that more than one hundred large cities are "severely" short of water. The solution of the past fifty years, ever-expanding irrigation systems and deeper and deeper well drilling, has brought the water crisis to a head. In 2005, per capita water volume in Beijing was a very low 300 m³/person. Further, as discussed in chapter 8, irrigated rice production in the Northeast (Dongbei) has increased faster than in any part of China. Defying ecological logic, northernmost Heilongjiang is now a major rice-producing province despite potential long-term declines in water tables. The use of irrigation has also expanded on the North China Plain, and the water table there is also falling dramatically as consumption rapidly outpaces replacement through precipitation. In part, support for the water-transfer projects discussed earlier is due to these shortages and to the looming threat of extensive arable land loss due to widespread salinization. Efforts to promote water-saving practices in agriculture have continued for many years, but only since 2006 has there been a significant investment in technology for this purpose.

Western China has an even more severe annual moisture deficit, with annual rainfall in some areas averaging a mere 100 mm/year (<4 inches) (see figure 2.8). Initially, most of the settlements in far western China developed around sources of reliable groundwater. At present, efforts to capitalize on Xinjiang's long growing season and the intense radiation associated with cloud-free skies (no moisture) has led a drive to expand irrigation through meltwater transfers and tube-well irrigation. Cotton and fruit production, especially grapes, has expanded dramatically, and there is considerable concern among China's ecologists that groundwater reserves are being depleted

Photo 2.2.   Desert encroaching on stand of trees in the northern Taklamakan Desert in Xinjiang Autonomous Region, 2002 (Gregory Veeck)

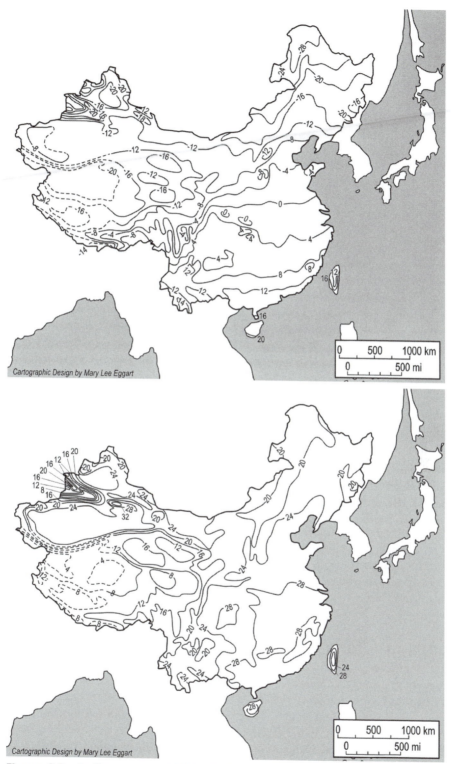

Figure 2.9. Isoline maps of China's January and July mean temperature (Zhang and Lin 1992, 38, 47)

Photo 2.3. Canal feeding into the Grand Canal in Suzhou, 2002 (Gregory Veeck)

too quickly throughout the Northwest (Zhang and Lin 1992). As oil production and refining has also increased in Xinjiang, so have water shortages (see photo 2.2).

Temperature regimes in China typically vary by north–south location, although there are some deviations from this pattern south of the 16°C isotherm (figure 2.9). South of this isotherm, which runs roughly along an east–west axis, average annual temperature is high, with generally mild winters and sufficient precipitation (see photo 2.3). The middle and lower Chang Jiang Basin region has similarly hot summers and mild winters (see figure 2.8, Shanghai). South of the Chang Jiang, average temperatures increase gradually as one moves south into the tropical coastal areas. North of the 16°C isotherm, the drop in average annual temperature is steeper, and the effects of a continental climate are more prominent (see table 2.2, climate data for Beijing, Xi'an, and Harbin).

Soils of China

Soils are an extremely important, if overlooked, resource for a major agricultural country such as China, and greater efforts must be made in China to protect these vital resources. Vegetative and climatic factors are most important in soil formation. Soil formation is dependent on the nature of the underlying rock type or *parent material,* seasonal temperatures, the amount of precipitation and drainage, and the character of the relief on which the soils are formed. The unusual effects of more than twenty-five centuries of cultivation on some of China's soils must also be considered. In China,

Table 2.2. Temperature and Precipitation for Selected Chinese Cities, 2008

City	January Temperature (°C)	July Temperature (°C)	Annual Average Temperature (°C)	January Precipitation (mm)	July Precipitation (mm)	Annual Total Precipitation (mm)
Beijing	−3.0	27.2	13.4	0.2	79.3	626.3
Harbin	−17.6	24.4	6.6	0	94.8	439.0
Shanghai	4.2	30.2	17.2	90.9	105.8	1,086.5
Wuhan	1.2	29.3	17.6	72.4	148.1	1,266.8
Guangzhou	12.8	29.1	22.4	98.0	170.3	2,284.0
Chongqing	6.2	29.1	18.5	16.2	55.1	962.7
Xi'an	−1.7	27.1	14.9	19.1	83.7	525.2
Urumqi	−15.6	25.2	8.7	3.0	20.9	171.8

Source: National Bureau of Statistics (2009, 389–91).

more than in any other nation, *anthropogenic* (human-induced) effects associated with wet-rice cultivation and terrace construction have formed distinct and extensive groups of soils that are rare in most other temperate nations.

China is fortunate in having a large and diverse stock of soils that are suitable for intensive agricultural development. Equally significant is China's long peasant tradition of manipulating and improving the available soil in order to increase agricultural output. In large part, it is this tradition of intensive cultivation on paddy and terrace that led the Chinese to develop their own classification system, a system that allows greater differentiation among types of anthropogenic soils, or *Anthrosols*. Given the unique characteristics of the order of Anthrosols and the great role these soils play in the history of China, some mention of their classification in the Chinese system, more refined than the systems used in other nations, is required because these soils have been of vital importance to the Chinese both in the past and at the present time.

Anthrosols can be created *from almost any type of parent material and soil.* Anthrosols can be broken down into two suborders: *Hydragic Anthrosols* and *Dryagic Anthrosols.* The suborder Hydragic includes soils that are deliberately submerged in water and heavily manured by farmers for a significant portion of each year (e.g., for rice production). The period of annual inundation and the length of cultivation impact the resulting soils differently according to their parent material and the volume of compost that has been applied. Annual field inundation periods vary by available radiation, water supply, and the frost-free period. Variations result in very different types of anthropogenic soils with different concentration of metals in the *gley* layer (a level of soil or *pan* below the surface cultivation zone that ultimately retains the water because of its high clay content; Xu et al. 1980). For example, farmers in the Lake Tai region of southern Jiangsu and northern Zhejiang provinces have used an anaerobic, waterlogged composting system (*oufei*) since before the Song dynasty (AD 960). Denied oxygen by a more than two-hundred-day submergence in water, composted material is reduced to ammonium—the ideal fertilizer for paddy rice. Sadly, as farmers in this emerging industrial area concentrate on factory work, composting rates have declined along with the natural fertility of the soils, and the excessive amounts of inorganic fertilizers used extensively in their stead result in polluted runoff and eutrophication in lakes, rivers, and canals (Zhao 1986).

The second suborder, the Dryagic (Orthic) Anthrosols, is less intuitive. There are four groups within this suborder, designed to incorporate dryland soils (or periodically irrigated dry crops) that have been terraced or subjected to heavy applications of manure, compost, and crop residue over long periods of time. These soils, which receive less attention in the academic literature than (wet) paddy soils, tend to be located in the old farming regions of North China or in hilly areas in the South and Southwest. Over the decades, the high level of composting in these dryland soils has also altered their soil structure and chemistry to the extent that they no longer fit into traditional classifications of natural soils. If these *anthrosols* had not been developed by China's farmers, it is safe to say that Chinese history and civilization would be radically different, as it is the high yields that result from these soils that allowed for specialization of labor and the rise of the complex state.

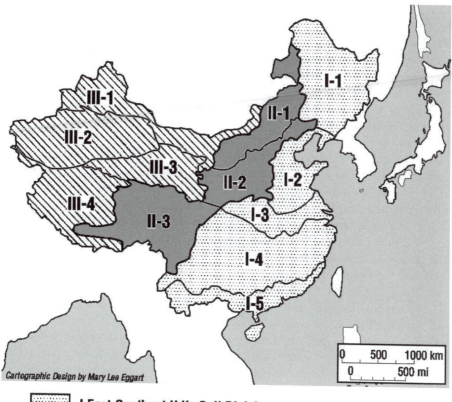

I East-Southeat Udic Soil Division
    I-1 Boric Luvisols and Udic Isohumisols dominant
    I-2 Udic Luvisols and Aquic Cambisols dominant
    I-3 Hydragic Anthrosols and Udic Luvisols
    I-4 Udic Ferrisols and Perudic Cambisols dominant
    I-5 Udic Ferrisols and Ferralisols dominant

II Central Ustic Soil Division
    II-1 Ustic Isohumisols and Ustic Sandic Entisols dominant
    II-2 Loessal or Orthic Entisols, Ustic Luvisols
    II-3 Ustic Isohumisols and Ustic Cambisols

III Northwest Aridic Soil Division
    III-1 Cyric Cambisols and Halosols dominant
    III-2 Mixed Aridisols and Halosols dominant
    III-3 Orthic Aridisols dominant
    III-4 Cyric Aridisols dominant

Figure 2.10. A generalized map of China's major soils and their distribution (simplified from Gong Zitong 1999, 860–70)

Setting aside the *anthrosols*, China's soil groups can generally be differentiated in a manner that correlates to the country's different climates and vegetation. That is, soils can be assigned to three macroclasses based on region: (1) the high or dry western and northwestern regions, (2) the humid central and southeastern regions, and (3) a zone of transition between the first two (see figure 2.10). In general, this basic regionalization follows a fundamental division in soil classification based on processes of soil evolution and chemistry (Gong 1999). One process concerns soil formation with high calcium and magnesium carbonate concentrations (pedocal formation) in dry areas. The other process involves leaching and eluviation of aluminum and iron oxides from the upper part of the soil and the deposition and concentration of these oxides (pedalfer formation) in the subsoil of soils in high precipitation areas. This latter process is common in more humid areas of southern, central, and eastern China. There are many further refinements and complicating factors that become important in considering subregions within each zone, but for our purposes, this general classification is sufficient. Gong (1999) divides China into three macroregions: the Southeast Udic (moist) Soil Division (*Dongnan Bu Shirun Turang Quhua*), the Central Ustic (transitional with limited moisture) Soil Division (*Zhongbu Ganrun Turang Quhua*), and the Northwest Aridic Soil Division (*Xibei Bu Ganhan Turang Quhua*; see figure 2.10). The Udic division (41.6 percent of national area) typically includes soils that occur in climates with well-distributed rainfall or areas where summer rainfall plus stored moisture equals or exceeds evaporation. The Ustic transitional zone (22.7 percent of national area) typically features soils formed under circumstances drier than for most of the soils found in the Ustic division but with sufficient moisture for the regular cultivation of crops in most years. Finally, the Aridic division (35.7 percent of national area) incorporates the driest areas of West and Northwest China where soil formation is severely hampered by limited moisture. Soils are often omitted from discussions of environmental and ecological issues, but soils damaged by compaction from overgrazing; made toxic from water- and airborne toxins including farm chemicals, industrial pollutants, and mining wastes; or salinized from excessive irrigation are a principal reason for large-scale human migrations, including those in China. When soils fail, only tragedy follows.

The Natural Vegetation of China

China, as befits a large country that stretches from 18°N to 53°N latitude, has a great range of natural vegetation. Included in this range are most of the vegetative types native to the Northern Hemisphere except for some varieties found only in arctic regions. For example, in Hainan Island and along the southern coastal littoral of the country are found tropical rain forests and other plants indigenous to the tropics. In the high mountains of western China and Tibet, alpine and subalpine plant communities may be found. The rest of the country holds vegetation communities common to deserts, steppes, savannas, prairie meadows, or coniferous evergreen and deciduous forests (see figure 2.11).

Generally, the classification of vegetation in China can be first grossly divided between a humid South and East China and a dry North and West China, exactly along the lines for rainfall surplus and deficiency discussed in previous sections of the chapter. The eastern humid region of the country must be further subdivided into a tropical southern area that supports a tropical rain forest; a subtropical southern and central area of broad-leaved evergreens, pines, and many varieties of bamboo; and a humid but cooler northern and northeastern area where evergreen conifers and other northern deciduous species such as birch are common.

In discussing natural vegetation in China, it is useful to remember that in the densely populated eastern part of the country, which has been settled for thousands of years, it is now really quite difficult to identify "native" vegetation with any certainty. China's farmers have been hard at work for several millennia cutting and burning trees, terracing hillsides, clearing shrubs and grasses, and farming virgin fields. As a consequence, identifying "natural" (original) vegetation is quite difficult. Intensive land use for agriculture, forestry, and husbandry over a period of this duration has often removed all evidence. In addition, new species from other portions of China and Asia have been introduced extensively for slope stabilization and forestry as well as for cultivation (grain crops, woody shrubs for medicines, fruit, and vegetables). Fuel for cooking and heat has long been scarce, especially in North China, and local farmers and other rural residents have further devastated forests and grasslands in their search for both fuel, medicines, and construction materials (for an excellent case study, see Coggins 2003).

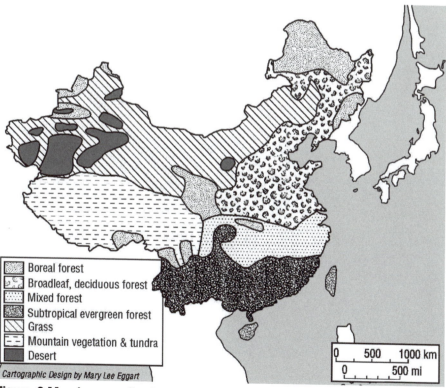

Boreal forest
Broadleaf, deciduous forest
Mixed forest
Subtropical evergreen forest
Grass
Mountain vegetation & tundra
Desert

Cartographic Design by Mary Lee Eggart

0   500   1000 km
0   500 mi

Figure 2.11. A generalized map of China's major vegetation types (Pannell and Ma 1983, 34)

Great pressure has also been placed on forest resources, altering original vegetation in a fashion similar to agricultural and husbandry areas. Most remaining old-growth forests are in the Northeast, in the Daxing'an and Xiaoxing'an mountain chains, but there are scattered reserves, such as in the Wuyi Mountains of Fujian in Southeast China, as well. The commercial forests first extensively exploited after 1949 are also found in the Northeast, but in the past twenty years, China's forest products industry has undergone major changes in both location and sophistication. Modern commercial timber operations have gradually shifted to the southern portions of the nation, where longer growing seasons with faster-growing trees increase annual production by a factor of two or three. Thus, it is hoped that the remaining forests of the Northeast will serve largely as biological preserves and ecological research centers. There are also a few smaller such reserves scattered throughout the Southeast and larger semitropical preserves in southwestern China.

International support and funding for biopreserves has played a critical role in the protection of these areas. Fantastic stands of conifers, birch, ash, and other northern species are still found on the picturesque slopes of the mountains that rim the Northeast Plain (see photo 2.4). Logging bans, increasingly stringent in the past decade, may protect the forests for posterity, but unemployment in the region is as high as in any rural area in China. Conservation, while a good idea, creates a hard and bitter life for many of the unemployed loggers and related workers in cities such as Yichun, the once-heralded "Wood City" sited within the foothills of the Xiaoxing'an Mountains. There are also fairly extensive areas of old-growth forests in Tibet, western Sichuan, and Yunnan. It is no accident that most of these areas may be characterized as poor

Photo 2.4. Fenglin Forest Preserve in northern Heilongjiang province, 1998 (Gregory Veeck)

and isolated. This isolation, although a major impediment to improving the living standards and economies of such places, has also protected the forests to an extent not possible in more accessible areas, where the primordial forests were cut long ago.

In eastern China, Chinese researchers have identified four main vegetational zones from south to north: (1) tropical forests and savanna, (2) subtropical evergreen forests extending north to the Chang Jiang and west to the Qinghai-Tibet Plateau, (3) mainly deciduous forests extending from the Chang Jiang far northward to the Song-Nen Plain, and (4) taiga in the Heilong Jiang Basin. In general, plant communities change with their location from south to north as well as with elevation. However, tropical plant communities are found as far north as 28°N in the river valleys of Yunnan province and southeastern Tibet. Likewise, the Sichuan Basin, characterized by a humid, mild climate and associated thermophilic plant species, does not exhibit the vegetation regimes suggested by its latitude.

The dry, western part of the country is similar to Central Asia, a land of very dry deserts, grasslands, and steppe, again mostly at great elevation. Its vegetation is monotonous and is composed predominantly of several species of grasses and hardy shrubs. The composition of its flora and the character of its vegetation along with its ecological conditions permit dividing this vast region into three major vegetation zones: (1) Mongolia, (2) Dzungaria and the Tarim Basin, and (3) Tibet.

Within these three zones, the flora varies according to local ecological conditions and altitude rather than as a result of shifts in location north and south. Thus, extensive forests are established where precipitation is greatest and average annual temperatures permit their growth. For example, extensive forests of spruce, birch, ash, and aspen may be found on the windward northern slopes of the Tian Shan, which rise to more than 7,000 m in height. There is, however, great variation in the extent and type of tree cover because of differences in terrain and climate. At higher elevations and in drier steppe locations, grasses and sedges are all that will become established in these harsh arid locations. These conditions are common throughout northwestern China. Ecologically, the most severe region in western China is the western part of the Tibetan upland, a high, cold desert over 4,500 m ASL.

Regional Challenges to Sustainable Development and Resource Use

EASTERN CHINA

An interesting and significant feature of China's physical and cultural landscape is that most of the country's people are concentrated in the eastern third of the country, a spatial imbalance that is discussed at length in chapter 5. Briefly, however, this concentration puts much more pressure on land—particularly precious farmland in this most populous and industrialized region. Early on, the East China economy benefited from capital flows derived from agricultural surplus and a high return on transportation investments in the alluvial plains precisely because the land there is flat, the roads are cheaply constructed, and much of the soil is fertile because of many thousands of years of flooding. The alluvial plains of the Huang He, Huai, Chang, and Xi rivers and their tributaries constitute China's greatest agricultural resource but ironically are now the locations of the greatest conflicts over land use as the eastern seaboard becomes increasingly crowded and more and more land is required for housing, transportation, and industrial uses. Most of China's greatest cities are located in eastern China and are within these rich agricultural areas. As these cities and their associated transport systems grow rapidly in the modern era, however, much high-productivity cropland is being lost. National environmental protection laws enacted in the late 1980s help to stem this loss, but, as noted earlier, local enforcement is often limited and politically difficult when jobs are at risk (Economy 2004).

SOUTHWEST CHINA

In contrast to eastern China and the Sichuan Basin, many other areas of China are not especially suitable for intensive settlement and development. The latter includes ex-

Photo 2.5. Karst (Limestone) landscape in Guangxi Autonomous Region (near Liuzhou City) in Southwest China, 1996 (Gregory Veeck)

tensive areas of karstic limestone found in southwestern China (see photo 2.5). Other locations throughout the world possess similar features, but nowhere is the tower karst as extensive, nor is the impact of humans on these landscapes as long established, far reaching, and striking. Covering much of the area of the three southern provinces of Guangxi, Guizhou, and Yunnan are massive beds of limestone, or *karst*, created when the entire region was part of a huge ancient southern ocean. After water levels fell and the ocean receded, these "exposed" karstic landscapes were dissolved and eroded by very high annual temperatures and frequently extreme precipitation events. The result is some of the most beautiful and bizarre karstic landscape found anywhere in the world. Limestone towers and domes abound, with some rising 200 m above base level. These are the major terrain features of the Southwest. Other specialized karst features, such as *dolines* (solution valleys), sinkholes, caves, and underground rivers, are also present. The scenery is spectacular, but, as is often the case, these remote and beautiful rugged environments also are associated with considerable poverty. To be farmed, the steep terrain must be terraced, or, alternately, farming is confined to the narrow stream valleys between the karst towers and domes. The limestone-rich soils produced through erosion of the karst are geologically new and infertile, with low levels of organic material and nitrogen, except in the river bottoms. Unfortunately, even where the landscape is level or near level, it is difficult to keep water near the surface, where it can be accessed by crops. Because of the solubility of the limestone, the water continuously percolates downward. Enough level land amid the mountain systems within the karstic region is available nonetheless to support a modest level of agricultural development and population density. Particularly since 1990, tourism has become a very important mainstay of many local economies here, but the long-term environmental impact of mass tourism on the region is only now being systematically studied.

NORTHEAST CHINA

Although the Han Chinese have occupied and farmed most of Northeast China (Manchuria in the history books) for less than one hundred years, agricultural and industrial development and population growth in the area have been very rapid during the reform era, especially in the two southern provinces of Liaoning and Jilin. The Northeast's major port city, Dalian, has increased in size many times over as capital from nearby Korea and Japan has poured into the city for industrial joint ventures. Dalian is now one of the largest ports in China. Increased industrial production has taken people from the land, and at the same time the region's farms have increased in size and efficiency. The state farms that mostly disappeared from the rest of China with the end of the commune era in 1978 remain quite important in the Northeast.

Wheat, corn, barley, millet, sorghum, soybeans, and even rice are grown on farms of thousands of hectares on the calcium- and humus-rich *Isohumisols* and *Luvisols* (the *chernozems* in the Russian system). In the northern province of Heilongjiang, a very active land reclamation program (particularly in the Sanjiang or Three Rivers Plain) has recently clashed with a growing ecological protection agenda that has increasing domestic support both within and beyond government agencies. Certain areas have already benefited from designation as national biopreserves, and in 1991 UNESCO recognized the 18,000-ha Fenglin Forest Preserve as such a site. As ecological research increases, it is hoped that sustainable agricultural developments can coexist with nearby protected environments. While it is vital that China preserve its scarce natural land for future generations, it is also important to be sure that those persons removed from protected areas be helped to reestablish themselves elsewhere. This has not always been the case, and many people relocated to new areas after being forced to move from ecologically protected areas are disillusioned with government efforts to help them.

WESTERN CHINA: XINJIANG, QINGHAI, AND TIBET

The western Chinese regions of Xinjiang, Qinghai, and the Tibetan highlands account for about half of China's area. Western China is a land of extremes—of very high mountains and plateaus, extensive basins, and huge arid deserts. These harsh landscapes, however, offer extremely serious environmental problems for human occupancy and sustained agricultural development. A large portion of this region, including the Tarim, Junggar, and Qaidam basins as well as much of northern and western Tibet, is extremely dry. Seasonal rivers fueled briefly each year from snowmelt and the occasional intense rain drain to the centers of these basins. The headwaters of several great rivers—the Huang He, Chang Jiang, Mekong, Salween, Brahmaputra, and Indus—originate on the eastern and southern faces of the Tibetan Plateau.

Xinjiang's most imposing physical feature is the Tarim Basin, one of the driest places on earth. Rimmed by tall mountains, most of the basin is uninhabited desert (*Takla Makan*) with very little vegetation. Only along the foothills of the bordering mountains is there opportunity for human occupancy. This is due to the presence of mountain streams fed by snowmelt and also because of the easily worked alluvial soils eroded from the mountains that fan out into the basins for short distances. North of the Tarim Basin, the Tian Shan (Tian Mountains) rise to elevations over 7,000 m ASL. Noteworthy among the growth centers of this region has been Ürümqi (Urumchi), the capital of Xinjiang Autonomous Region and a major industrial city of 2.6 million now refining much of the oil in the region and surrounded by a thriving, irrigated-oasis farming system. Based on improved control and management of streams that empty from the adjacent mountains, extensive land reclamation and cropping have proceeded throughout the southern and western margins of the Junggar Basin (for an excellent summary of Xinjiang, see Toops 2004).

The Tibetan highlands (including the Tibet Autonomous Region and Qinghai province) represents more than one-quarter of China's area. Three main physical divisions make up these highlands. The first of these, the large Qianzang Region in the North and West, is an area of mountain ranges, valleys, and plains. Because of its great elevation, coldness, and aridity, most of this region has only sparse vegetation and human habitation. Second are the alluvial river valleys of the South and Southeast, generally with elevations below 4,000 m that contain most of the developed farmland and are home to most of the Tibetan population. Here, along the Yarlung

Zangbo (Brahmaputra) River and its tributaries, is the heartland of Tibet and its capital, Lhasa. The third division comprises the older districts of Tibet that are now located outside the boundaries of the autonomous region and include the grasslands in Qinghai province and the rugged mountain systems in western Sichuan and north-western Yunnan. Although much capital has been invested in building roads in the southwestern mountains, this has been aimed at territorial integration through spatial linkages rather than at extensive agricultural development. The Tibetan Highlands are likely to remain marginally developed and sparsely populated. Isolation and the region's rugged environment offer little incentive for significant investment and de-velopmental commitment. Programs directed at poverty alleviation have centered on improvements to infrastructure, including massive road construction projects and the new rail link that connects Lhasa to Xining (1,956 km), which started service in July 2006. Although only the first phase of a three-phase development program, it is hoped that rail service and better roads will allow increases in both tourism and mining throughout the region with the ultimate goal of raising incomes. There are numerous indigenous-rights groups that decry the loss of local autonomy that comes with these massive government investments despite the undeniable benefits to the economy. As discussed in subsequent chapters, balancing local interests, including those of China's minority ethnic groups, with state plans for economic development and growth is a major challenge moving into the century's second decade.

Conclusion

China is an enormous nation, and its physical environments are exceptionally di-verse—covering virtually all types of terrestrial biomes. On the other hand, much of China is ecologically fragile and/or difficult to access. So, while the nation is approxi-mately the same size as the United States and has 4.35 times (as of September 2009) the population of the United States, the vast majority of these people are living within the eastern lowlands, while most of the west—excepting the great river valleys of the Huang and Chang and their tributaries—is very lightly populated. Current challenges to long-term national development can be understood only if this marked difference in population distribution and the ecological fragility of much of the nation are taken to heart. This is not environmental determinism but rather recognition of the simple fact that proper stewardship of the land will establish a sound foundation for all that follows. In less populated, land-extensive nations such as Australia, Canada, and the United States, significantly lower population pressure limits the use of the most mar-ginal ecological regions, so much of these are given over to national parks or preserves. In contrast, despite the dangers of continued exploitation of these regions, China will probably press ahead simply because there are few alternatives. Efforts to develop these peripheral lands have many implications, and many environmentalists bemoan the seemingly relentless onslaught of development in these fragile environments.

However, there is more reason for optimism at the present time than at any time in the past five decades. There is a growing awareness worldwide that resources must be used more effectively, and many manufacturing processes and agricultural technologies (including forestry and rangeland management) have been transformed with an eye toward lower costs and promoting greater efficiency in resource use. There is a growing awareness among China's leaders that further growth must be sustainable. Of course, the future is uncertain, but at the end of the first decade of the new millennium, the costs of previous mistakes and growth-at-any-cost policies are clearly apparent. At the very highest levels of government, China's leaders realize that China cannot afford to continue without greater attention toward environmental protection. Currently, the costs of reversing environmental damage from decades of careless actions now often exceed the benefits of these shortsighted activities. Many of China's leaders recognize this simple but powerful economic fact and support environmental protection. Despite its current problems, China's recent efforts in environmental protection should not be dismissed out of hand. Rapid economic development can be more blessing than curse if greater prosperity results in greater awareness and in greater expenditures directed at protecting the environment and replacing dated, polluting, industrial technologies. On the other hand, massive public works such as the Three Gorges project or the North-South Water Transfer Project raise parallel questions about China's long-term priorities. A prosperity based on environmental exploitation in the absence of sustainable practices will be short-lived.

Questions for Discussion

1. What is the relationship between China's population and the three great "topographic steps"? That is, where are most people located, and why?
2. Why is the Chinese mainland described as a "checkerboard," and what are the implications of this condition?
3. What are the benefits of the Three Gorges Dam? What are the drawbacks associated with the dam?
4. What is the South-North Water Transfer Project, and why is it being built?
5. Anthropogenic soils are vital to China's food supply. What are anthropogenic soils, and how are they created?
6. What are some important environmental problems facing China that are discussed in this chapter?

References Cited

China PopIN. 2009. 中国人口信息 [Zhongguo Renkou Xinxi] (China population information). http://www.cpirc.org.cn/index.asp (accessed September 23, 2009).

Coggins, Chris. 2003. *The Tiger and the Pangolin: Nature, Culture, and Conservation in China.* Honolulu: University of Hawai'i Press.

Day, K. A., ed. 2005. *China's Environment and the Challenge of Sustainable Development.* Armonk, NY: Sharpe.

Economist. 2005. China and water: Drying up. May 21, 46.

Economy, E. 2004. *The River Runs Black: The Environmental Challenges to China's Future.* Ithaca, NY: Cornell University Press.

Edmonds, R. L. 1994. *Patterns of China's Lost Harmony: A Survey of the Country's Environmental Degradation and Protection.* New York: Routledge.

Gong Zitong, ed. 1999. *Zhongguo turang xitong fenlei* [Chinese soil classification system]. Beijing: Science Press.

Ma, J., A. J. Hoekstra, H. Wang, A. K. Chapagain, and D. X. Wang. 2006. Virtual versus real water transfers within China. *Philosophical Transactions of the Royal Society* 361: 835–42.

McElroy, Michael, Chris P. Nielson, and Peter Lydon, eds. 1998. *Energizing China: Reconciling Environmental Protection and Economic Growth.* Cambridge, MA: Harvard University Committee on Environment.

National Bureau of Statistics. 2009. *Zhongguo tongji nianjian 2009* [China statistical yearbook 2009]. Beijing: China Statistics Press.

Nickum, James E. 1998. Is China living on the water margin? *China Quarterly* 156 (December): 880–98.

Pannell, Clifton W., and Laurence J. C. Ma. 1983. *China: The Geography of Development and Modernization.* New York: Halsted.

People's Daily. 2004. Key section of south-north diversion project ready for construction. November 26. http://english.peopledaily.com.cn/200411/26/print20041126_165217.html (accessed September 24, 2009).

Ren, Mei'e, Renzhang Yang, and Haosheng Bao, eds. 1985. *An Outline of China's Physical Geography.* Beijing: Foreign Language Press.

Smil, Vaclav. 1984. *The Bad Earth: Environmental Degradation in China.* Armonk, NY: Sharpe.

———. 1993. *China's Environment: An Inquiry into the Limits of National Development.* Armonk, NY: Sharpe.

South to North Water Diversion Project, China. 2008. http://www.water-technology.net/projects/south_north (accessed September 24, 2009).

Toops, Stanley. 2004. Demographics and development in Xinjiang after 1949. Working paper. Washington, DC: East-West Center.

Wu Yuanli. 1973. *China: A Handbook.* New York: Praeger.

Xu, Qi, Yanchun Lu, Yuanchang Lu, and Hongguan Zhu. 1980. *The Paddy Soil of Tai-hu Region in China.* Shanghai: Shanghai Scientific and Technical.

Zhang, Jiacheng, and Zhiguang Lin. 1992. *Climate of China.* Translated by Ding Tan. New York: Wiley.

Zhang, Q. F., Z. F. Xu, Z. H. Shen, S. Y. Li, and S. S. Wang. 2009. The Han River watershed management initiative for the South-to-North Water Transfer project (Middle Route) of China. *Environmental Monitoring and Assessment* 148: 369–77.

Zhao, Songqiao. 1986. *Physical Geography of China.* New York: Wiley.

———. 1994. *Geography of China: Environment, Resources, Population, and Development.* New York: Wiley.

Zhonghua renmin gongheguo dituce [Atlas of China]. 1979. Beijing: Cartographic Publishing House. (in Chinese).

CHAPTER 3

Ancient Roots and Binding Traditions

Historical Geographies of China

In considering China's economic and cultural development from a geographic perspective, it is important to start with the simple fact that contemporary China is based on all that came before in the nation's long and complex history, as these events played out on the diverse landscapes of this extensive nation. This chapter reviews China's past in light of the two interrelated, principal themes outlined in the introduction: (1) the spatial organization of people and activities in China and (2) the implications of human–environment interactions throughout China's long history. Through the centuries, the complex tapestry woven from these interactions and implications has influenced how society and polity were organized, how both functioned over time, how environments were exploited differently across both space and time, and how the production of goods and services took place. China's history extends from a time when the territory was shared by hundreds of kingdoms and states to that when it was an empire, then a nation, and finally to today with the country's present status as an emerging global power.

Early Humans in China

The origins of China and the Chinese people are not as clear as one might think given that China's civilization has long been studied and venerated for its age, sophistication, and achievements. The territory that is now China was home to at least three distinct Neolithic cultures and probably more yet to be discovered. The archaeological record remains surprisingly incomplete for many regions. While many questions remain, it is clear from archaeological finds and related evidence in *palynology* (the study and analysis of pollen samples) that prehistoric humans, initially *Homo erectus* and later *Homo sapiens sapiens*, lived across extensive areas of contemporary China for at least 1.5 million years (Wu and Poirier 1995). Most archaeologists feel that additional archaeological sites will be discovered within Chinese territory as more time and resources are

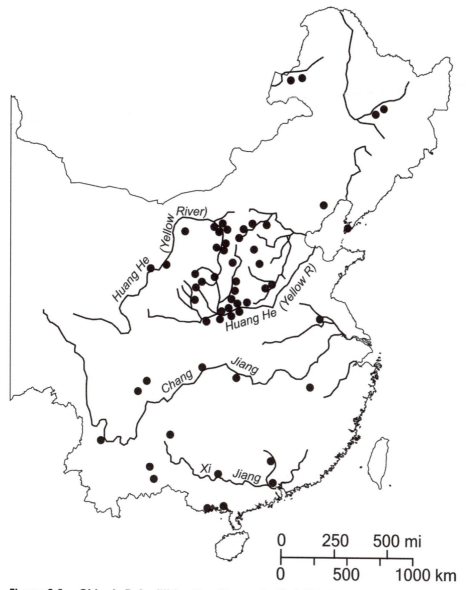

Figure 3.1.   China's Paleolithic sites (Ikawa-Smith 1978; Zheng 1997)

devoted to larger-scale systematic excavations. China's prehistory, then, is still a story in progress. Pleistocene hominids of one type or another have been found in twenty-three of China's thirty provinces, or first-order administrative regions. The earliest recognized *Homo erectus* fossils found in the Yuanmou Basin in Yunnan province in Southwest China are dated to 1.7 million years ago. A considerable number of other Paleolithic sites have also been found in southern Yunnan and the surrounding provinces of Sichuan and Guizhou, but there are also important sites in northern China. Major finds, with skeletal evidence in combination with worked stone tools, have been located in Yunxian county, Hubei province; Lantian county, Shaanxi province; Hexian county, Anhui province; Nanjing City (Tangshan Zhen), Jiangsu province; and, perhaps most famously, the caves of Zhoukoudian, which are only 50 km from Beijing (see figure 3.1). It is clear that within what is now currently held Chinese territory, there were a number of diverse cultures evolving simultaneously. Less clear are the nature and distribution of common ancestors among these peoples, the times of their separations, and their relationships (genetic and cultural).

A common mistake is to assume the Chinese and Chinese culture emerged from some single ethnic stock, the origins of which could be linked to a particular place and time. But rather than attesting to such a common source, the evidence points to a pattern in which, over time, many distinct groups were blended and absorbed into the composite that we now think of as the people of the Han—named for the dynasty that constituted one of China's golden eras. This blending and amalgamation would be far less complete in the absence of the dynastic system for which China is famous. Military conquest and territorial expansion, however, were but the most visible forces leading to this homogenization. Commerce, exploration, and even religious pilgrimages that fostered the spread of Buddhism all worked to integrate the peoples of China.

Considering China's many environments, the diverse use of plants and animals by early humans throughout China should come as no surprise. Throughout the dynastic era, when new crops, agricultural techniques, methods of manufacture, medicines, architectural designs, or livestock breeds were developed in one region, they were quickly dispersed throughout the empire by imperial bulletins, ambitious court officials, and traders seeking profits. In this sense, it was not until the march of the dynasties that Chinese culture began to approximate the present. And the very foundations of this flexible system of human–environment relations came from the diverse practices of the ancient peoples of the Pleistocene.

As it was in Europe and North America, the Pleistocene epoch in China was a time of dramatic change in many of the country's natural environments. Major natural changes included tectonic shifts and mountain building, changes in climatic cycles with several associated periods of extensive glaciations, and major cycles in fluvial activity with distinct large-scale erosion and sedimentary periods. The Chinese Pleistocene was characterized by several periods of climatic fluctuation accompanied by related changes in landforms, flora, and fauna. Four major cold-moist stages were involved, and these correlated to periods of glacial advance in all the highland areas of northern China.

Although there is some debate about the conditions following the final Pleistocene stage, available palynological evidence indicates a gradual warming trend matched with a considerable increase in precipitation. More rainfall resulted in more erosion, deposition, and the initiation of sedimentary cycles that dramatically altered the topography. For example, during this period, much of the North China Plain—now one of the most densely populated regions of modern China—was composed of marshes and deltaic swamps (see figure 3.2). The Shandong Highlands at this time formed a series of islands separated from the Chinese mainland either by the sea or by a series of intermittent saline marshes and lakes. This *ecotone* (the margin between two distinct ecological zones) was ideal for settlement because it provided a wide diversity of useful

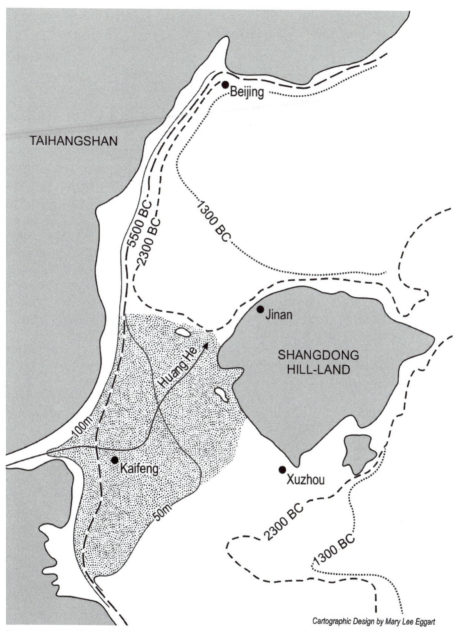

Figure 3.2. Deposition and geological formation of the North China Plain (after Pannell and Ma 1983, 46)

plants and animals to the early settlers of the region and the areas immediately westward that became the cradle of Chinese civilization.

China's Neolithic Period

The transition from gathering and hunting to the cultivation of plants and the domestication of animals is a turning point in many ancient human societies. The domestication of animals and the adoption of sedentary farming practices and permanent settlement increased the reliability of food supplies and led to food surpluses, allowing for the specialization of labor. Such specialization was crucial for human progress, and it permitted human groups to turn their energy and creativity to other activities, including the construction of cities and states.

Initially, some scholars were convinced that China did not have an indigenous civilization based on the independent development of agriculture (Smartt and Simmonds 1995). For some years, it was incorrectly believed that diffusion from western Asia had brought agriculture comparatively late to North China (Fairbank and Goldman 1998, 29). Enough new archaeological evidence has been assembled from North China to establish at least one independent center of Chinese agrarian

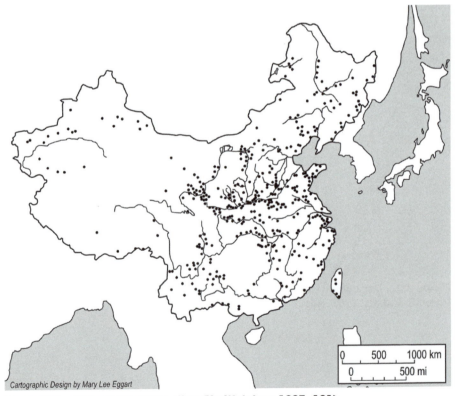

Cartographic Design by Mary Lee Eggart

Figure 3.3. China's Neolithic sites (Yo Weichao 1997, 101)

civilization in the region known as Zhongyuan (Central Plain) in the middle Huang He and Wei He basins. More recently, two other areas of China (first, the southern coastal area of Taiwan, Fujian, Guangdong, and Guangxi and, second, the Pacific coastal area stretching from Shandong south to Zhejiang) have been identified as probable loci of other early independent agrarian civilizations (see figure 3.3). As the distances between these areas of China might suggest, the prevailing view in Chinese archaeology and prehistory is that the Neolithic farming revolution and its related cultures developed in these three regions independently and at approximately the same time. What remains to be determined are the kinds of interrelationships and linkages that may have existed among parallel groupings of Neolithic development in China. In addition to Zhongyuan, other important Neolithic sites discovered in China include sites related to the Qingliangang culture of the Pacific seaboard and the Dapenkeng culture of the Huai River Basin of Shandong and northern Jiangsu provinces.

Of the three ancient Neolithic cultures in China, the best known and most important with respect to tracing the earliest threads of Chinese culture to the present is the Yangshao culture of the middle Huang He and Wei He basins (the Zhongyuan region). Based on archaeological records, Yangshao culture appeared by 5000 BC and may have existed considerably earlier. Yangshao culture has traditionally been regarded as the source of much of what eventually would come to be known as Chinese civilization. In support of this view, it was from this region that the first recorded dynasties sprang and grew: the apocryphal Xia (ca. 2200–1766 BC), the Shang (1766–1122 BC), and the Zhou (1122–221 BC; see table 3.1).

Yangshao culture was characterized by sedentary farming and village settlements scattered along the loess terraces found in the river valleys of the drainage basin of the middle Huang He. The main crop of the Yangshao cultivators was foxtail millet (*Setaria italica*). A variety of stone tools and implements were employed by Yangshao peoples, the most important of which were hoes, spades, digging sticks, and stone disks, probably for hammering. Knives, axes, adzes, and chisels were also important as well as distinctive pottery jars of many different types and designs that were used for storing grain, a crucial function in times of agricultural surplus. The pottery was handmade, and no evidence of potter's wheels has been found for this early period. Research on Yangshao artifacts indicates that much of the pottery was fired at temperatures approximating 1,000°C. The ability to control and maintain such temperatures within the kilns represents a significant technology.

The sedentary agriculture of Yangshao cultivators was heavily supplemented by wild-grain collecting, hunting, and fishing. Many settlements seem to have been discontinuously occupied, which would be in accordance with the slash-and-burn style of cultivation (resulting in plots known as *swidden*) that is evident at the village of Banpo near modern Xi'an, Shaanxi province (Roberts 1996, 3).

The Chinese Neolithic period extended from approximately 6000 BC to 3000 BC, although the exact period varied from locale to locale. But it was at the next stage of development, commencing at approximately 3200 BC, that broadly common cultural types emerged. These types are conventionally called *Longshanoid*, and they represent the forerunners of true Chinese civilization.

Table 3.1.   Population Estimates of Major Chinese Dynasties and Periods

Xia	Possibly 2206–1766	Population in Millions[1]
Shang	1766/1480–1122 BC[2]	Not estimated
Zhou	1122–221 BC	Not estimated
Spring and Autumn	722–481 BC	Not estimated
Warring States	452–221 BC	Not estimated
Qin	221–206 BC	Not estimated
Han	206 BC–AD 220	59.6 (AD 2)
Wei, Jin, and Northern and Southern dynasties	AD 220–581	16.2 (AD 280)
Sui	AD 581–618	46.0 (AD 606)
Tang	AD 618–907	52.9 (AD 755)
Five Dynasties and Ten Kingdoms	AD 907–959	Not estimated
Northern Song	AD 960–1126	46.7 (AD 1110)
Southern Song	AD 1126–1279	Not estimated
Yuan (Mongol)	AD 1271–1368	58.9 (AD 1290)
Ming	AD 1368–1644	59.9 (AD 1381) 63.7 (AD 1562)
Qing	AD 1644–1911	102.8 (AD 1753) 377.6 (AD 1887)
Republic of China	AD 1912–1949	474.8 (AD 1928)
People's Republic of China	AD 1949–	541.4 (AD 1949) 1.33 billion (2009)

[1]Estimates prior to 1949 are often widely debated. These are from Zhou (1994).
[2]Estimates for the beginning of Shang vary widely. The earliest estimate is Zhao (1994), and the later is Sivin (1988).
Sources: National Bureau of Statistics (2002, 2009), Sivin (1988), Zhou (1994).

Longshan and Longshanoid Culture

The Neolithic archaeology of China becomes increasingly complex as it approaches and blends into recorded Chinese historical civilization, which began during the Shang dynasty (1766–1122 BC). The painted pottery of Yangshao culture—and the cultures that this distinctive pottery represented—was gradually replaced starting around 3200 BC by Longshanoid black pottery produced on a potter's wheel. There are close linkages between newer Longshanoid or Longshan-type culture and features of the later Shang period, specifically in symbolism on pottery, the extensive use of oracle bones, and house and village-wall construction techniques.

Features of Longshanoid Culture

Longshanoid culture assemblages have been located in Shandong, Shanxi, and Shaanxi provinces and along the lower reach of the Huai and Chang (Yangtze) rivers as well as in several southeastern provinces (see figure 3.4). Although these sites may vary in particulars, hence the distinction between "true" Longshan sites and those locations

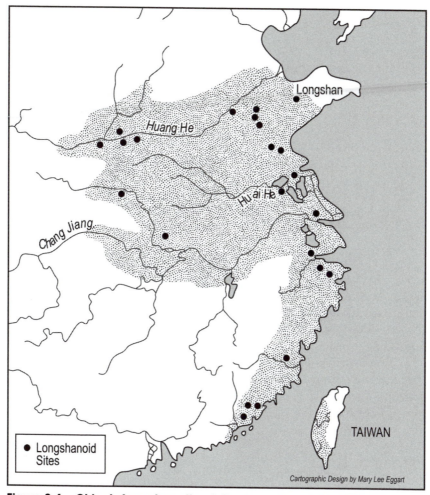

Figure 3.4. China's Longshan sites (after Pannell and Ma 1983, 52)

with only some Longshan culture artifacts (Longshan type), they share certain characteristics and features that have allowed their classification as Longshanoid assemblages.

First, all these Longshanoid cultures were based on subsistence agriculture still supplemented to varying degrees by hunting, fishing, and gathering. Possible proof of rice cultivation has been found in areas of the central and southern region, but current evidence from Longshanoid and Yangshao culture sites suggests that rice cultivation was not important. Yangshao farming, specifically, was based mainly on millet cultivation.

Second, all Longshanoid sites include distinctive types of stone implements useful for carpentry and agriculture. In addition, very distinctive pottery remains have been uncovered that included such diagnostic forms as the *ding* (a tripod vessel for use over an open flame), the *dou* (a vessel with a distinctive ring-foot design), and the *gui* (a tripod jar with handles; see figure 3.5). As noted earlier, the potter's wheel is believed to have come into use in the Longshanoid period.

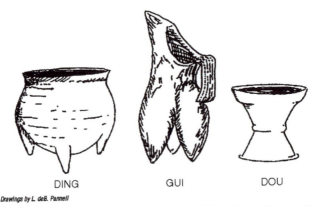

DING   GUI   DOU

Drawings by L. deB. Pannell

Figure 3.5. Examples of Longshanoid pottery (Pannell and Ma 1983, 57)

Finally, the various Longshanoid or Longshan-type cultures occupied distinctive and clear chronological niches that followed the Yangshao and blended into later Longshanoid assemblages. Thus, a Longshan-type culture might have been more advanced at a specific time in one place or another, but a steady evolutionary chronology would have developed for each of the Longshanoid cultures. Longshan-type cultures appear to be the main and true prototypes for Chinese civilization as it emerged in the first historic period of the Shang dynasty. The occurrences of Longshanoid sites are frequent and common enough and share a sufficient number of characteristics to indicate that a substantial population sharing this culture occupied a large area of the eastern part of China.

Historic China

The transformation from fragmented, localized, mixed economies based on rudimentary horticulture, hunting, and gathering to a formalized, clearly stratified social system with marked specialization of labor and an administratively dominant urban center apparently took place in China sometime between 3000 BC and 2000 BC.

A legendary dynasty mentioned in records written long after its supposed existence, the Xia (ca. 2200–1766 BC) has been much discussed by historians and remains within our idealized lineage of Chinese dynasties despite the lack of definitive archaeological evidence to support its existence as a true dynasty. Based on the presence of several large palace-style buildings of extraordinary size, scholars have suggested that the extensive ErliTou site on the middle reaches of the Huang He might have been the site of the capital of the Xia state. Until proper archaeological evidence is discovered, however, the Xia dynasty must remain a legend with undefined boundaries and uncertain tenancy.

Legend and archaeology converge in the first archaeologically supported Chinese dynasty, the Shang. Recorded Chinese history begins ca. 1800 BC with the appearance of divination records in the form of oracle bones made from cattle and sheep scapula

Figure 3.6. Shang-influenced areas (Minneapolis In-stitute of the Arts, http://www.artsmia.ORG/ARTS-OF-ASIA/CHINA/MAPS; Cheng Te-k'un 1960, 14–16)

(shoulder blades) or tortoise plastrons (under-shells). It is on these artifacts that the earliest Chinese writing is found. As of 1996, more than 107,000 oracle bone pieces in eighty collections had been assigned to the Shang period (Roberts 1996, 8).

The Shang dynasty (1766–1122 BC; see table 3.1) was composed of an established lineage of rulers whose subjects occupied a definite territory (centered in northern Henan). Contemporary written records from a variety of sites support this fact, most importantly the records found at the city of Yin that served as the capital for the last twelve kings of the Shang. Other sites displaying similar material culture to that found in Shang dynasty territory but spread across a much wider area of China that was clearly not under the control of the Shang kings (see figure 3.6) are described as "Shang-type."

Not only did Chinese civilization emerge, in full, from the bloom of the Shang, but a combination of archaeological and textual supporting evidence gives historians a better understanding of the Shang state than they have of the states and kingdoms that came before it. And nowhere were the strength and beauty of this civilization and culture more obvious than in the Shang urban capitals, which were, as always, the seats of authority, power, art, and dynastic economy (Wheatley 1971).

A number of Shang urban centers have been found, but three major sites stand out. Two lie along the Huang He in southern Hebei, and the third, Anyang, is north of the Huang He in northern Henan province (see figure 3.6). The oldest site is near Yanshi, which contains what is thought to be the oldest palace site in China and is believed to be the site of the first Shang capital. A somewhat larger and more impressive walled city, rectangular in shape and enclosing an area of 3.2 km², has been found east of Ershi near the present city of Zhengzhou. Artifacts that are characteristic of Shang culture include very high-quality bronze, fine pottery, and oracle bones covered with written records. The Shang bronzes not only are marvels of metallurgy and artistry but also reflect a benchmark in technology and organization for bronze manufacture. Some of the larger artifacts weigh over 85 or 90 kg. Casting required the orchestrated efforts of a workforce of over three hundred artisans and laborers working to simultaneously lift and pour up to seventy crucibles of molten metal. These early bronzes were some of the hardest bronzes manufactured in China during the entire dynastic era and again reflect a remarkable level of technical sophistication.

The most important Shang capital, serving the last three centuries of Shang rulers, was along the banks of a river near the present city of Anyang. Discoveries in the vicinity provide written evidence of the Shang and indeed provide a surprisingly detailed record of the dynasty's tenure. Thus, the importance of Anyang to early Chinese history and archaeology is unmatched. The stratified nature of its houses and neighborhoods also suggests a far more sophisticated settlement than those of earlier periods. Elite neighborhoods, centers for artisans, and slums for the poor are clearly reflected in the archaeological record.

The Origins of Chinese Cities

Researchers now recognize that North China, along with the ancient Middle East, the Indus River Basin, and two locations in Mesoamerica, were the sites of early and independently developed urban civilizations, with settlements of many sizes and purposes joined at times for political and economic purposes. Goods and people circulated easily among these settlements, with a major city overseeing the production and distribution of goods and the defense of the whole. It is now believed that a number of these local systems in China were joined together (i.e., conquered) in the early part of the Shang dynasty to form the core of the fledgling state.

Although the Shang states were allied to defend their northern frontier from barbarian incursions, translations of the oracle bones indicate that the states fought continuously among themselves. Territory and spoils of conquest moved continuously from state to state through the centuries. Eventually, these endless wars, both within and without, took their toll, and a rival group, the Zhou, established hegemony over the Shang dominions while also expanding southward.

Historical accounts of the demise of the Shang dynasty establish a pattern, perhaps also a process, of succession that remained through the dynastic era. The last Shang ruler, Zhou Xin, was thought to be profligate, wicked, and lacking in all the noble virtues desired of a ruler. This lack of moral leadership intensified the effects of rebellion and external conflicts, and the common people were exposed to constant warfare, high taxes, famine, and pestilence. As a consequence, "heaven" was thought to have withdrawn its mandate (support) from the Shang kings in favor of a more virtuous leader and lineage, thereby ending the Shang dynasty—a victim of antipathy in both heaven and on earth. A better-organized and presumably more chaste Zhou dynasty rose from the ashes of the Shang—initiated by the most virtuous King Wu, the first king of the Zhou.

Tradition holds that King Wu, as the first ruler establishing a new dynasty, was successful only because heaven provided support for the transition. This cycle of dynastic loss and gain, commonly called the "Mandate of Heaven," was replayed many times—more than the number of China's dynasties, in fact, because often there were two or more competing dynasties operating within what is now China, all claiming the support of heaven.

The Zhou dynasty has a special place in China's historical geography. Many of China's greatest classical thinkers and philosophers, such as Confucius, Mencius, and Laozi (Lao-tze), lived during this era. Although the central authority of the Zhou began to break down and become diffused in the middle and later periods of the dynasty, the Zhou dynasty constituted the first classical age for Chinese literature, philosophy, and the arts. The value system set forth in classical literature from the period not only would survive but also would become the philosophy under which China and Chinese culture would flourish for more than two millennia. The Zhou dynasty also saw the further consolidation of institutions and traditions that came to be seen as distinctively Chinese. After 700 BC, agriculture shifted from swidden agriculture (where burned fields were used for three to four years and then the group moved on to other fields over an extensive range) to more productive farming systems that were completely sedentary. Agriculture benefited from technical improvements, such as the fallowing of fields to let their soil regenerate and the use of organic fertilizers. Iron plowshares came into common use. Directives regarding the need for the preservation of arable land have also been found in dynastic texts of the Zhou, showing an understanding of previous agricultural problems and the need to protect land resources just as in contemporary China. These concerns also reflect the presence of the environmental problems related to degraded land that plague China to the present.

The Chinese Language

Indispensable to an understanding of China's recorded history and development is some knowledge of the nature and working of the Chinese language. Perhaps for no other nation at any time was a written language as critical for nation building as written Chinese was for Chinese civilization. Once standardized, written Chinese served to unify the many disparate groups of peoples and cultures (each with a distinct spoken language) within the territory that would become modern China. Written Chinese has been in use for at least 3,500 years. This tenure makes it the oldest written language continuously in use to the present. The use of written Chinese was critical for homogenizing the many disparate cultures of the Middle Kingdom. In a very real sense, it allowed China's rulers to hold the nation together.

The Chinese language differs from European languages in several significant ways. First, the Chinese language does not employ an alphabet but rather relies on the use of Chinese characters that are either *ideographs* (picture symbols) or *phonemic graphs* (sound symbols) that have developed over a very long period to depict objects, actions, and ideas. Each character is unique and varies in its complexity (the number of strokes needed to represent it), but all are composed of combinations taken from a set of standardized strokes. Characters representing similar concepts may often have a common portion, known as the character's *radical*. There are 227 radicals in the Chinese language, including one that represents "metal," one that represents "speech," one that represents "plant," and so on.

While every Chinese character has meaning in its own right (i.e., it is a *morpheme*), in practical terms, most often Chinese words are combinations of usually two or three characters. This propensity to combine characters gave written Chinese the flexibility it needed to remain viable to the present. Even so, there are over sixty thousand characters. Most of these, however, are seldom used (Mair 1997). To read a typical Chinese newspaper, for instance, requires knowledge of only two thousand to three thousand characters. As new concepts, materials, and ideas arise, new characters or combinations of Chinese characters are introduced to produce new words.

In 1956, in order to increase literacy, the Chinese government simplified the writing of a number of characters in common usage (*jianti zi*) by reducing the number of strokes required to write some radicals and complete characters. It is much easier to learn these simplified characters, and literacy in China has increased rapidly. In 2009, 91.6 percent of the population of the People's Republic of China (PRC) was officially classified by the Chinese government as literate (Central Intelligence Agency 2009). The figure increases to 95.7 percent for men, and women have a 2007 estimated literacy rate of only 87.6 percent because of unequal access to educational opportunities in some rural areas. Overseas Chinese and residents of Hong Kong and Taiwan still use traditional characters.

Another innovation aimed at simplifying the language has been the introduction of systems of standardized "romanization" (based on the Latin alphabet) for the transliteration of Chinese terms. The most recent of these systems, known as *pinyin*, was designed as an aid in the pronunciation of Chinese characters, particularly in the first and second grades when Chinese students are just beginning to learn to write characters. Pinyin has also been widely adopted as a means of romanizing Chinese outside China. The translation of written Chinese has always posed a problem for foreigners. Prior to 1972, the United States, as well as virtually all other Western nations, used the Wade-Giles system, which was created by Thomas Wade and Herbert Giles, two of the first professors of Chinese at Cambridge University in 1867. After President Richard Nixon's historic meetings with Chinese leaders Mao Zedong (Mao Tse-tung) and Zhou Enlai, the pinyin system was gradually adopted in the United States and most other Western nations. This shift in romanization systems has caused more than its fair share of confusion despite the fact that pinyin has been used in most cases for more than forty years. Under the Wade-Giles system, the capital of China was *Peking*. Using pinyin, this became *Beijing*. The Chinese characters that mean "Northern Capital" have not changed; the change is only in the letters of our alphabet that we westerners use to represent those characters.

Spoken Chinese

China's languages are tonal in nature, and words with the same sound but rendered in different tones have different meanings. This is what made a standardized writing system so critical for holding the Chinese nation together. At the same time, spoken

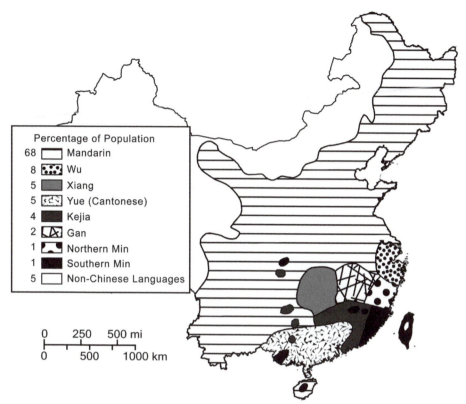

Figure 3.7. Chinese language distributions (after Pannell and Ma 1983, 65)

Chinese is represented by many different regional languages and dialects (see figure 3.7). That there should be so many Chinese languages is reasonable given China's great expanse, rugged topography, and complex history. These numerous and various regional languages (sometimes inaccurately called dialects), moreover, are often generally unintelligible to each other. The most commonly spoken language of the Chinese is Mandarin (named after the Qing court officials that westerners first heard speak it in the late 1700s). Mandarin has four tones and is spoken by the largest number of people in China. This official language of modern China, also known as *putonghua* (common speech), is the language of North China and is found in its standard, pristine form in and around the city of Beijing. *Putonghua*, or Mandarin, is used throughout North, Northeast, and Central China. Other major language groups are found in the East and South. Among these languages, Cantonese and southern Min (Fujian) are especially significant for overseas Chinese because most of the Chinese emigrating to other countries since the nineteenth century originated from these areas. Cantonese has nine tones and is usually considered more difficult to learn.

Because of the tones, Chinese is a difficult language to master. Still, the spoken language in some ways is quite simple in that the verbs have no conjugations and the nouns no declensions. Beginning in the early 1950s, everyone in the PRC was required to learn *putonghua* (Mandarin), and all television and radio broadcasts as well as movies were required to use it. Again, it is the official spoken language of China. However, in the past twenty years, there has been a tendency to return to regional dialects in local television and radio programs, particularly in the southern provinces of Guangdong and Guangxi. This return to local languages, particularly in the public arena, represents a rebirth of regional culture and pride.

Imperial China: The Dynastic Centuries

The divisions between competing states in the third century BC diminished the power of the last Zhou rulers and led to the emergence of a new and powerful state, the Qin, which through superior military tactics and equipment was able to consolidate power and reunite the country in 221 BC. The rise of the Qin was improbable. Initially, the Qin was the westernmost of the Zhou states. It started out as a minor royal domain responsible for the defense of the western borders and for the provision of horses to the Zhou army. The ruler of this new state, Zheng, after subduing his foes (the leaders of the other Zhou states), later proclaimed himself Qin Shi Huangdi (first emperor of the Qin) and established many of the structures and agencies that allowed China to become a great empire. Although the Qin dynasty was short-lived (221–206 BC), the

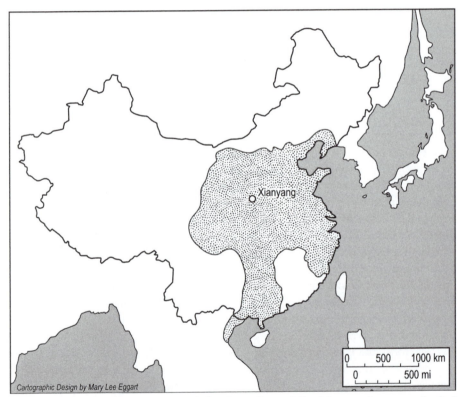

Figure 3.8.   Qin dynasty territory (after Leon Poon, http://www.chaos.umd.edu/history, accessed October 7, 2010)

unified, dynastic-based imperial system that it established lasted as a governing and administrative system without basic changes until 1911 when a republican form of government was introduced. The Qin conquests of many smaller, weaker states started China on its long path to the establishment of the boundaries of the modern Chinese state. The maps of the dynasties included in this section of the chapter, however, clearly show how Chinese territory regularly expanded and contracted over the 2,200 years of the dynastic era.

Exploding from the Qin kingdom based along the Wei River, Qin Shi Huang himself led the Qin armies to battle, first to the East and then more significantly to the rich farms of the South (see figure 3.8). Acquiring these rich southern lands and their associated food surplus was vital for fueling the massive public works and continued military expansions that marked the dynasty. The Qin dynasty was organized based on a legalist school of thought best represented by the writings of Yang Kungsun, chief minister of the Qin kingdom from 359 to 338 BC, and Han Fei (d. 233 BC). The legalists rejected Confucian visions of beneficent government and the moral suasion of citizens in favor of strict legal codes and regulations that addressed virtually all aspects of life.

Qin Shi Huang introduced many policies uniting and governing China, including the establishment of a national civil administration that divided the country into thirty-six commanderies, further divided into prefectures. This system of politically and militarily organized and united territories was a remarkable departure from the loose alliances and federations of the Shang or the Zhou dynasties. Money, weights, measures, and even the axle lengths of carts were officially standardized to facilitate trade and transport. Chinese writing was likewise made uniform through the development and promotion of official dictionaries.

The Spatial and Ecological Impact of Qin Shi Huang

Among the many accomplishments of the first emperor of the Qin dynasty in unifying China and in initiating the unified imperial dynastic system, three stand out because of their spatial and ecological ramifications: (1) the most ambitious rendition of the Great Wall to that date, (2) the extension of a national trunk-road network, and (3) the expansion and extension of the national canal network. The Great Wall was for security and boundary-delimiting purposes; the roads and canals improved transportation and communication. Most canals were also designed not only to improve transportation but to provide flood control and irrigation water for farming as well. All these innovations permitted Qin Shi Huang to acquire more territory and resources (food, lumber, horses, and trade goods). With a new type of chariot and better logistics, Qin dynasty armies were better supplied, better organized, and capable of faster movement than the opposing forces and national territory expanded quickly. But the Qin dynasty is regarded as one of the cruelest periods in China's long, storied history. The labor for its walls, canals, roads, food, and, above all, the recruits for its armies came from the weak and the conquered, and the reign was unpopular among all citizens.

China's Great Wall also came into being at this time. The Great Wall was an attempt to stabilize an ecological boundary between a zone of sedentary agriculturists, the Chinese, and a zone of pastoral herders, including the Xiongnu. The wall created a sharp and clear buffer separating the Chinese farmers of Inner China from pastoral nomads living in Outer China. It was also a symbolic, cultural boundary between the civilized Middle Kingdom in the South and the barbaric regions in the North and West.

Consolidation of the empire in 221 BC led immediately to the construction of new roads and highways designed to spatially integrate the political control center at Xianyang (immediately northwest of today's city of Xi'an) with the newly acquired regions. The empire's chief general, Meng Tian, was charged with building a main north–south highway, 800 km long, north across the Ordos Desert toward the present-day city of Baotou. Other new roads radiated out from Xianyang like the spokes of a wheel. Loewe (1986, 61) estimates that the Qin imperial highway network was approximately 6,800 km. This compares favorably with estimates of the Roman road system of 5,984 km at its greatest extension. This road network continued to expand during the Han period, but after the third century AD, the road network began to decline, possibly because of the increased importance of rivers and canals for transport. In South China, a short but significant canal was constructed in what is now Guangxi province, far to the south. The canal linked the drainage system of the Chang Jiang (Yangtze River) and Central China with tributaries of the Xi Jiang in South China. In this way, Central and South China were integrated more effectively for administrative, military, and economic purposes, and the southern territories were cemented further into Chinese territory.

Qin Shi Huang died in 210 BC, and the collapse of his legalist, authoritarian regime followed soon thereafter, but the tradition of imperial dynasties based on a royal family that changed with each dynasty continued for 2,200 years. The next dynasty, the Han (206 BC–AD 220) was established after a brief period of instability and fighting. The Han emperor reestablished aspects of a feudal landownership system that expanded the ruling class. The growing power of landed nobles during the Han dynasty, weakened under the Qin, led to a forced policy of subdividing land among all surviving sons. Unlike the European system of primogeniture in which land and titles passed to the oldest son, the Han policy assured the division of land into ever-smaller holdings, often resulting in holdings too small to support a family. This endless subdivision of land had consequences that continued right up to the twenty-first century.

Growth and Change in the Early Dynastic Era

The political and administrative traditions of China established under the Qin and the Han dynasties were durable enough to survive the 360 years of sporadic chaos that followed the collapse of the Han in AD 220 as well as the long periods of foreign occupation and domination of China by the Mongols (AD 1279–1368) and the Manchu (AD 1644–1911).

It is not that there were no changes in China's political system for 2,200 years; certainly, there were recurring periods of decisive change, progress, and setback. But while extensive coverage of the dynastic history of China would take volumes, there were some geographical and economic changes that are noteworthy (Huang 1997). Under the Qin and the Han dynasties, territorial expansion was largely to the East and South, but control of key commercial centers in the West along the Silk Road was of equal importance. Once the Chinese gained control over territories in Central Asia and established the overland trade routes that came to be collectively called the Silk Road, great prosperity quickly followed. If the granaries of China were to the South, then the West of China provided the lion's share of the more profitable commercial trade allowing access to the distant markets of Persia and Europe on which the fiscal system of the early dynasties was dependent.

Han Wudi, one of the most vigorous and famous of the Han emperors, sent multiple expeditions into the barbarian regions (Xiongnu federations) in the North and the West to pacify the populace, gain territory, and expand commerce. In 133 BC, a force of over 300,000 troops was used to establish four commanderies in what is now Gansu province in Northwest China. These troops—and the countless troops that followed—permitted the Han to trade their silk, tea, and porcelain to the great markets of Persia, India, Rome, and the remainder of Europe and Central Asia for centuries. In the Southeast, Han armies finally conquered the Yue and Min peoples of what is now southern Zhejiang and Fujian, giving the empire complete control of the Pacific seaboard (see figure 3.9).

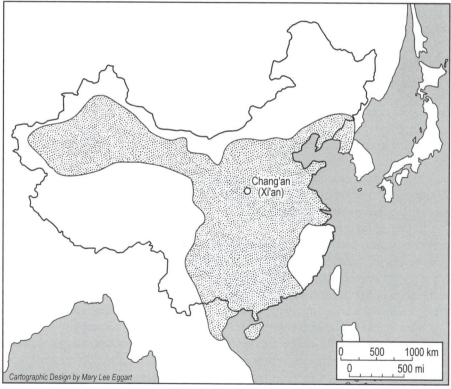

Figure 3.9. Han dynasty territory at greatest extension (after Leon Poon, http://www.chaos.umd.edu/history, accessed October 7, 2010)

The Han dynasty expansions, as the Qin before them, were costly in many ways. Regional conflicts and rebellions were the bitter fruits that sprang from the hardships sowed by court intrigue, bad government, forced conscription into the army, and the high taxes required to support the military. Eventually, the empire broke up into minor kingdoms and territories controlled by warlords powerful enough to maintain local control but incapable of pulling the empire together again.

It would be more than 350 years before China would once again be united under one ruler, with the establishment of the Sui dynasty in AD 581. Yang Jian, a Xianbei (non-Han ethnic group) ruler of the Northern Zhou dynasty that gained control of northern China in 577, established the Sui dynasty when his armies finally conquered the South. In the forty years of the dynasty's existence, Sui armies were almost constantly at war or brutally putting down internal rebellions. Appropriately, history treated the Sui and the Qin dynasties in similar fashion; both were periods of rapid expansion and unification, both were very short-lived, and both are now famous for the cruelty and hardship their leaders visited on the common people (Fairbank and Goldman 1998, 78).

The "Golden" Tang Dynasty

Just as the much-celebrated and relatively long Han dynasty followed the violent and brief Qin dynasty, the famous Tang dynasty followed that of the tumultuous Sui. Similarly, as the high costs of his Korean campaigns helped undermine support for Qin Shi Huang, unsuccessful Korean expeditions during the Sui dynasty so weakened the army that it was susceptible to the attacks of the eastern Turks. This chaos ultimately led to the establishment of the Tang dynasty (Roberts 1996). As with the Han, the key economic and cultural links of the Tang state to other nations and regions were westward. In the early, most vigorous years of the dynasty, Tang armies acquired or regained a remarkable amount of territory in Central Asia after campaigns against the Turks, the Tibetans, the Uyghur, and many other groups (see figure 3.10). In AD 635, the northern Silk Road route was regained. In AD 702, much of what is now Xinjiang province was pacified, and the southern Silk Road route was established as a more defensible option to the northern route.

In the eighth century AD, the Central Asian trade was so lucrative that Chinese prefectures were established in modern Afghanistan despite the very high costs of maintaining food supplies and troop garrisons. Trade along the Silk Road flourished, bringing not only wealth but also a remarkable array of foreigners, foreign religions, and foreign ideas into China. The Tang dynasty was far more cosmopolitan than any previous dynasty, and culture, high and low, benefited from foreign infusions.

It is important to note that during the Sui and Tang dynasties, the political orientation of the Chinese Empire and its economy remained directed toward the interior of Asia. Wealth and political control came with the maintenance and protection of interior trade routes to Central Asia and Europe. There were many other important and remarkable advances during the Tang dynasty, however. The legal system was completely overhauled, and many of these reforms remained largely in place for most

Figure 3.10. Tang dynasty territory (after Leon Poon, http://www.chaos.umd .edu/history, (accessed October 7, 2010)

of the remaining dynasties. The code of AD 653, which Ebrey (1996, 111) notes is the earliest comprehensive legal code to survive as a complete text, contains more than five hundred articles specifying specific crimes and associated punishments. Simultaneously, the government sought to equalize income by confiscating and redistributing landholdings. Equity was promoted not only for the sake of equity but also to minimize rebellions and uprisings, which in all times are bad for business and stability.

Ceramics, poetry, architecture, and the graphic arts all flourished, making lasting impressions on first the ambassadors and later the peoples of Korea, Japan, Vietnam, and even distant but wealthy Persia (modern Iran). The cohesion of the cultures of East Asia at the present time is in large part due to the great and pervasive influence of the Tang dynasty on the material and spiritual cultures of the other nations in the region. Buddhist sects flourished and spread throughout much of Southeast Asia during the Tang.

At its peak during the Tang dynasty, the capital city of Changan had more than 2 million inhabitants. The city walls enclosed more than 90 km². The food demands of this city and of many others of impressive size required an extensive canal system for food shipment and distribution. Under the Yuan dynasty (AD 1279–1368), the Grand Canal (*Da Yunhe*) had been extended farther north and east to the vicinity of Beijing, but it was during the Tang dynasty that engineering and excavation techniques

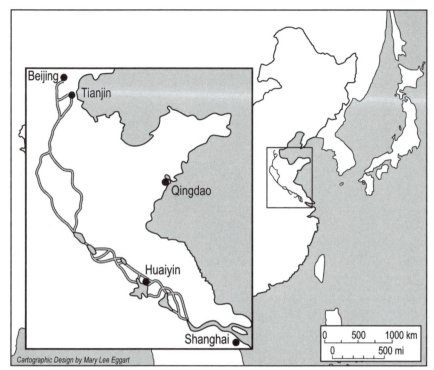

Figure 3.11. The Grand Canal routes, original (compiled from various maps, including the Atlas of China)

were perfected (see figure 3.11). The Grand Canal had two critical functions. It served a critical commercial purpose in fostering the exchange of goods and monies between North and South China, and it illustrated, to all, the great military power of the central government in the North and its ability to command grain shipments from South and Central China.

The Song Dynasty and a Turn to the Southern Seas

The western interior orientation (political and economic) characteristic of China from the early dynasties through the golden age of the Tang dynasty was abruptly changed during the Northern and Southern Song dynasties (AD 960–1279) that followed. During the more than three hundred years of the Song dynasty, the empire would never regain control over the vast territories held by earlier dynasties (see figure 3.12). It never acquired what is now Northeast China (Dongbei), where the Manchu Liao dynasty was established in AD 916 and remained unconquered until AD 1218. To the west, the Tibetan and the Tangut states threatened the frontiers and, along with the Uighurs, controlled the Central Asian caravan routes, thus removing the historically vital source of wealth for the new dynasty. Further, the Northern Song dynasty

Cartographic Design by Mary Lee Eggart

Figure 3.12.   The Song dynasty (after Leon Poon, http://www.chaos.umd.edu/ history, (accessed October 7, 2010)

never pacified the northern tribes that had breached the Great Wall. During the most prosperous years of the Song dynasty, much of the territory of North China was under the control of the rival Jin dynasty. The Loess Plateau and the North China Plain regions were lost in AD 1126 when tribes of the Jurchen (another nomadic group from Central Asia) sacked Kaifeng after an ill-fated alliance collapsed.

The territory firmly under Song control was but a fraction of that of Tang China, and the riches of the cross-Asia trade had fallen to other hands. One might think that the loss of half the national territory would have crippled the dynasty, but surprisingly the Southern Song dynasty was one of the most prosperous of the dynasties. Only the lands south of the Chang Jiang and eastward from the modern city of Chengdu in Sichuan were secure. In response to these territorial and economic losses, however, a new economic geography evolved through Chinese political and commercial expansions into the southern seas (Southeast Asia). In time, a rise in the very prosperous maritime trade with these regions more than compensated for the loss of the northern territories.

In terms of China's future economic geography, this shift to the seas represented a critical change in foreign diplomacy and economic policy. This is a very important departure from the past and requires special consideration. During the Southern Song dynasty, the economic orientation of China abruptly changed from a reliance on domestic overland trade and the Silk Road to one dependent on a Pacific orientation, supported by a large merchant marine and defended by a sophisticated imperial navy. Of course, as in Europe several centuries later, this maritime trade also transformed and expanded the domestic economy while shifting China's core economic region from its interior to the southern and eastern coasts. In measuring this commercial transformation, Ebrey (1996, 142) notes that "by 997 the Song government was minting 800 million coins a year, two and a half times the largest output of the Tang." As at the present time, international maritime trade transformed China in all respects.

The majority of maritime freight, by volume, during the Song dynasty was regional, integrating China with the countries of Southeast Asia, some Indian Ocean states, and many parts of the Indonesian Archipelago and the Philippines. During these centuries, China's influence over the economy and cultures of Southeast Asia was at its peak. There were also important long-distance routes to the Middle East, India, and even the coast of Africa. The most critical commodity in this Arab trade was porcelain, which, along with silk and tea, brought fantastic profits. Spices, precious woods, and tung oil for varnishes were important in the Indian Ocean trade.

Shipbuilding, fueled by the demands of coastal and canal trade, was far advanced in China relative to Europe at the same time. Levathes (1994, 40) reports that imperial revenues from overseas trade went from 500,000 strings of cash at the end of the eleventh century to more than 2 million by the middle of the twelfth. Major maritime ports of the Southern Song dynasty included Guangzhou, Quanzhou in Fujian Province, Xiamen, Fuzhou, and Hangzhou. Ironically, all these ports became "treaty ports" in the mid-nineteenth century, having been opened to international trade a second time by Western gunboats at that point in the century.

To protect its large merchant marine, the imperial Song navy grew to over six hundred ships. The largest merchant ships could carry three hundred tons of cargo (Levathes 1994, 43). As the maritime trade expanded, new technologies emerged to improve shipping and the naval presence required to defend Chinese interests on the seas. Song dynasty maritime innovations included stern-post rudders, multiple decks, compound composite wooden masts, and, perhaps most important, improved marine compasses.

In addition to new technologies represented by the flourishing maritime trade during the Song, there were also other technical innovations. Fairbank and Goldman (1998, 88–90) note that in AD 1078, northern China was producing more than 114,000 metric tons of pig iron per year, while seven hundred years later, England would be producing only half that amount. Other important Song dynasty commercial innovations in industrial technology included improved steel production through the use of coal-fired blast furnaces, glassmaking, and gunpowder. Papermaking, book printing, and production of ceramics all became established at this time as important commercial industries that relied on exports as well as domestic trade for their profits.

This remarkable boom in technology and innovation, what Elvin (1973) in his classic essay has called China's "medieval economic revolution," did not fuel long-term efforts at combining maritime trade with colonial expansion, however. Unlike Europe, China did not experience an age of global economic and political expansion (which would have been fueled, in part, by commercial applications of the inventions that would prove vital for European influence and control five hundred years later; Fairbank and Goldman 1998, 92).

The reasons for this are disputed, but all agree that on the eve of the twelfth century, China was the most technically advanced nation on earth (Needham 1962). For a variety of internal reasons, this economic and commercial golden age did not lead to the great territorial expansions that were characteristic of the early Han or Tang dynasties. The times were different, and China's borders during the Song dynasty were controlled not by the disorganized tribes of the past but by well-organized states (Khitans, Jurchen, Tibetans, and Tanguts) that demanded treatment as equals. Often, bribes of silk, silver, or gold were used to maintain the borders when Song armies, even with 1.25 million men under arms, proved incapable of the task. Just as its absolute territorial extent declined in relation to earlier dynasties, the internal organization of the territory under Song control also changed. Stimulated by increased food production and a growing specialization of labor, China's great cities experienced their own golden era. China now had dozens of cities with more than fifty thousand persons as the specialization of labor attained new levels on the back of increasing control by the central government (Skinner 1977).

Within China and especially southern China, farm output increased greatly. Several factors accounted for this, but most important was the widespread use of early-ripening Champa rice varieties acquired from northern Vietnam at the beginning of the eleventh century. There were also technical improvements in the cultivation and irrigation of wet rice, including new methods for leveling and smoothing land for paddy fields, constructing dams, and getting water to the fields by using new types of pumps.

Agricultural and technical primers also became popular in these times with the advent of commercial printing and the promotion of modern agricultural methods by the government. Perhaps the most renown of these primers is the classic *Jiming Yaoshu* (Essential Ways of Living for the Common People), written by Jia Suxie in approximately AD 533–544 and promoted by repeated printings and distribution throughout the Tang dynasty as a way to introduce better agricultural practices (Shih 1982, 2–5).

Fairbank and Goldman (1998, 108) note that the great paradox of the Song dynasty is that just as China reached its acme of civilization, it was conquered by outsiders. With the exception of the 276 years of the Ming dynasty (AD 1368–1644), the remainder of the Chinese dynastic era, from the fall of the Song dynasty in 1276 to the 1911 republican revolution, is a history of foreign dynasties.

The Mongols and the "Empire without End"

Mongol conquest of China was substantially linked to China's process of economic reorientation that began at the end of the Tang dynasty as the empire turned slowly from the lands and trade routes of Inner Asia to the southern seas and oceans. As Inner China became increasingly isolated and ignored, the old ties between the states and peoples of the West and the people of Inner China deteriorated. And as trade and mutual strategic interest declined, alliances were increasingly difficult to maintain amid contentious diplomacy that gave way to mutual contempt and conflict. With China's sphere of influence reduced, numerous local states emerged to fill the gap, only to fall to the Mongols. The Ruzhen (Jin), the Tangut, and other states could not match the Mongol's troop strength, strategy, or planning. Eventually, all of North China was in the hands of the Mongols. Ogodei, Ghengis Khan's third son, initiated the assault on Southern Song China. Victory led to the establishment of the Yuan (Mongol) dynasty in AD 1279 (see figure 3.13). The slaughter of civilians associated with these invasions is legendary, and many parts of China were seriously depopulated for more than one hundred years by the wanton violence of the Mongols. By way of example, after conquering Sichuan, Ogodei is said to have ordered the slaughter of Chengdu's more than 1 million inhabitants (Ebrey 1996, 169–70). The invasion of South China was completed by Kublai Khan, Ghengis Khan's grandson, but only after Kublai Khan had studied China by ruling a prefecture in Hebei for a number of years (Davis 1996, 211–18).

The Mongol conquest (and the resultant Yuan dynasty) represents a benchmark in China's history. While northern China had been under the control of foreigners (non-Han) a number of times, including during the Liao and Jin dynasties, the defeat of the Song dynasty marked the first time that southern China had ever come under foreign control. For the first time, under the Yuan dynasty, all of China was under foreign occupation. Ironically, this was also the first time that all of China had been united since the end of the Tang dynasty (AD 907). Mongol control of China was

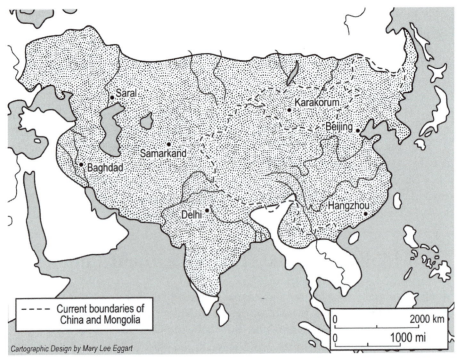

Figure 3.13. Yuan (Mongol) dynasty (Minneapolis Institute of the Arts, http://www.artsmia.ORG/ARTS-OF-ASIA/CHINA/MAPS, accessed October 7, 2010)

short-lived, though, because rebellions, stimulated in part by Song loyalists, erupted almost immediately in the South and later in the North (Davis 1996).

The Yuan dynasty had few long-term effects on China's economic geography because most Mongol innovations did not survive beyond the dynasty. In the end, for example, the Song dynasty administrative system remained intact, and Song systems of taxation, currency distribution, and the legal code were mostly unchanged. The Mongols did restore and improve the vaunted transportation systems of the Song and Tang dynasties by using massive amounts of corvée labor for clearing canals and repaving roads. A postal system with relays using 200,000 horses was also set up to further improve communications.

Perhaps the most influential aspects of the Yuan dynasty were cultural rather than economic or political in nature. Contrary to the popular conception of a closed and isolated China, the Yuan dynasty continued a cosmopolitan trend that had begun in the Tang and the Song dynasties. In 1275, Marco Polo arrived in Kublai Khan's summer residence, Shan-tu (this became Xanadu) in the Yuan capital of Cambaluc (later, Beijing), and found visitors, advisers, and ambassadors from many nations. Marco Polo marveled at the wealth and sophistication of China and remained in China for seventeen years, supposedly serving for three years during this time as a tax inspector in the city of Yangzhou in Jiangsu on the Grand Canal. Skeptics point out the absence of many common features of life in China that are missing from his accounts, though, and question if he ever actually lived in China or was just reporting the stories of others. This debate may never be resolved. But the cosmopolitan character of the Yuan dynasty was indubitable, as the following quotation well shows:

> Arabs, Venetians, and Russians engaged in business in Chinese ports, and one Russian took first place in the metropolitan examinations in 1321 and became a high official in Chekiang [Zhejiang] in 1341. Chinese and Mongols were penetrating Persia and Europe. Chinese engineers, for example, were used to improve irrigation in the Tigris-Euphrates basin; there were Chinese quarters in Novgorod, Moscow, and Tabriz. As envoy of the Mongol [leader], Rabban Sauma, a Nestorian born in Peking about 1225, visited Byzantium and Rome in 1287–88, saw the King of England in Gascony and Philip the Fair in Paris, and left a description of his visit to the Abby of Saint-Denis and Saint Chapelle, among other places. (Goodrich 1969, 178–79)

The Yuan dynasty collapsed as much as a result of the disintegration of the great Mongol Empire as of events in China proper. Still, a nascent Chinese nationalism next helped usher in the start of the Ming dynasty despite the dynasty's lowly origins.

The first emperor of the Ming dynasty, once an indigent peasant, led a rebel army that rose up from sheer anger, hunger, and frustration against the foreign Mongol rulers. Wishing to return China to glory, the ever-growing rebellion carried Nanjing in 1356 and finally captured Beijing from the Mongols in 1368. The Mongol court fled northward and regrouped. The Ming armies never conquered the Northeast (modern Dongbei) or the Far Northwest (Xinjiang), but the dynasty's early years were some of the most prosperous in China's long history. Oriented to the coasts as during the

Song dynasty, the Chinese economy under the Ming flourished, predicated in part on improvements to the national transportation system completed by the Yuan. The first several Ming emperors were just as ambitious, however, and quickly expanded public works to include massive urban construction projects that resulted in the contemporary layout of many cities, including Beijing and Nanjing. The Great Wall was completely renovated and reached its greatest extent under the Ming (see figure 3.14), and Beijing also became the permanent national capital after the death of the first Ming emperor in Nanjing.

A very brief return to long-distance maritime trade and exploration marked one of the most interesting periods of the Ming dynasty. This small sliver of China's long history (AD 1403–1433), when China built the greatest wooden ships ever constructed for trade and exploration, best represents China's early interest in the remainder of the world. Everything about the estimated seven major voyages of the Ming fleet was fantastic. The ships were massive, the distances were incredible, and the treasures collected rivaled those of any of history's greatest treasure troves. Predating Vasco da Gama's African explorations by eighty years, the fleet that set out in 1405 from Nanjing to explore the western oceans had 317 ships with a total crew of 27,000 men. The largest of these ships, the *baochuan* (treasure ships) were from 118 to 127 m long and 48 to

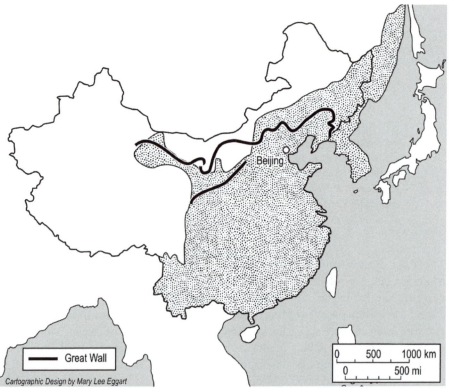

Figure 3.14. The Great Wall with Ming territory (Minneapolis Institute of the Arts, http://www.artsmia.ORG/ARTS-OF-ASIA/CHINA/MAPS, accessed October 7, 2010)

50 m wide and displaced around 3,000 tons. The largest ships measured more than four-and-one-half times the 26-m length of Columbus's largest ship on his 1492 voyage, the *Santa Maria* (Levathes 1994, 80–83). Led by the eunuch admiral Zheng He, a Muslim of modest means who rose because of his valor and ability, who was intent on visiting the Holy Land, the seven voyages were meant to be as much diplomatic and exploratory as mercantile. In the course of the voyages, Zheng He was to map all the northern shores of the Indian Ocean, the coast of Africa as far south as Madagascar, and much of the Indonesian Archipelago (see figure 3.15). From all the nations encountered, the great Ming fleet exacted tribute and exchanged court officials and, when possible, forged advantageous trading concessions, breaking the Arab lock on the cross–Indian Ocean trade. Giraffes, rhinoceroses, and many other animals, along with exotic products, plants, jewels, and curiosities, flowed back into China along with an ever-increasing knowledge of the world to the west. Given the tremendous capital required to construct and maintain such a fleet, it is surprising that in thirty years, the great Ming navy was allowed to disappear. There were many sadly mundane reasons for its demise, including court grumbling about its cost, increasing internal rebellion, and the completion of Grand Canal renovations, which made defense of the coastal trade from pirates (both Chinese and Japanese at this point) less necessary. The most important reason, however, was one that continues to influence China to the present: political rivalry between feuding leaders. Court conflicts between eunuchs and Confucians in the 1430s erupted as a struggle for control of the throne. Since the maritime trade was under the control of the eunuchs, the Confucians continuously advised

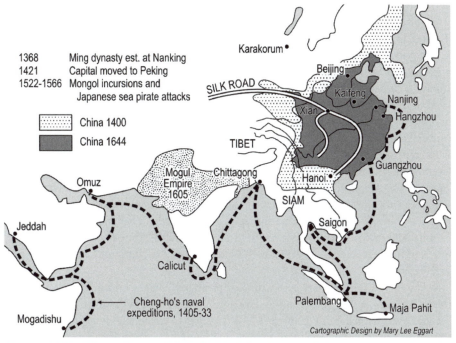

Figure 3.15. Ming China, 1368–1644. http://www.chinapage.com/zhenghe.html, (accessed October 7, 2010)

against investments in maintenance of the navy and the merchant marine. For these reasons, the Ming navy, at its peak consisting of three thousand ships, went into fast decline. Unprotected, much of China's vitally profitable coastal trade with Southeast Asia died along with the disappearance of its navy.

Although the Ming dynasty continued for more than two hundred years after the last of the great treasure ships was cut from its stays in Nanjing, its vigor was lost long before its final demise. The greatest days of the dynasty corresponded to periods of expanding global and regional trade. The collapse of the Ming dynasty was due as much to internal problems associated with more than eighty years of peasant rebellion, indifferent government, and profligate spending as to the invasion by the Manchu, who established the Qing dynasty.

In 1644 and after many years of war, the Manchu, based in Northwest China (hence the name *Manchuria*), overthrew the declining Ming dynasty. The Manchu originated as a Tungic tribe from Central Asia and did not have the maritime interests of the Song and the Ming dynasties. Crossley (1997, 81) estimates that the invading force of Manchu numbered only 120,000 to 150,000 banner men (soldiers organized into military divisions). This small force would not have been successful if Ming China had not been ripped apart by internal rebellions and famines occurring concurrently in many of the provinces. Just as the Mongols before them, the Manchu's small force, supported by foreign mercenaries but also including no small number of "Han" Chinese, was able to overcome Ming resistance. Again, following the methods of the Mongols, the Manchu used fear and stunning violence to acquire what their soldiers could not. After the siege of Yangzhou (in central Jiangsu province), over 800,000 corpses were cremated in mass graves outside the city walls.

For most of the 267 years that the Manchu ruled as the Qing dynasty (see figure 3.16), China's economy prospered in direct proportion to the expansion of its trade. Corresponding to the European mercantile era, the Qing dynasty supplied the world with tea, porcelain, silk, furniture of rare woods, pepper, and many other products. Qing manufactures were traditional, and few new technological innovations contributed to this prosperity. Agriculture grew more labor intensive, and there were critical adoptions of Western crops, such as corn and potatoes, but beyond this, few innovations were made. The Manchu successfully pacified the western tribes of Mongolia, Turkestan (Xinjiang), and Tibet, securing the greatest territory of China's long history. Commerce between the East and West continuously expanded from around 1680 to 1820, at which point the Europeans grew avaricious and the "illegal" opium trade grew out of control. Reestablishing Central Asian trade with the modern state of Russia, in addition to maritime trade with Southeast Asia (especially Thailand), the Philippines, Europe, and the Americas, also brought great wealth to the Qing court. Unfortunately, much of it was squandered. When the Western powers challenged, China was ill prepared to mount a viable defense.

The early Qing dynasty was a time of prosperity that would not continue given the lack of innovation and a rapidly growing population. The population doubled or perhaps even tripled from 1600 to 1911. This led to ever-increasing pressure on the land, and the internal rebellions that were spawned by environmental collapse and

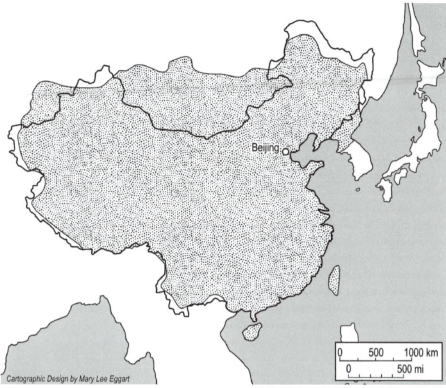

Figure 3.16. Qing dynasty China at greatest extent (Leon Poon, http://www .chaos.umd.edu/history, accessed October 7, 2010)

natural disasters at the end of the Qing dynasty clearly contributed to the overthrow of the dynastic system in 1911.

Early Modern China

China's modern era, as delineated by most historians, begins with the Treaty of Nanking in 1842 that ended the First Opium War between Qing China and Britain when China tried to stop the trade in opium. The lost war failed to regulate the trade in opium, and for the Chinese people, the Treaty of Nanking was the first of many unequal treaties, so many in fact that a common name for this period is the "unequal treaties period." These treaties, forced on the Manchu court by the foreign powers, opened up China to commercial exploitation and led to many significant geographic and economic consequences. The treaties provided extraterritorial privileges to foreigners, thereby making these foreigners immune to local Chinese law and custom while shifting control of the profitable international trade to foreign nations and companies. It was a time of great humiliation but also of economic deprivation as foreign firms and domestic collaborators made massive profits while wages stagnated and corruption destroyed the dynasty.

Signs of internal weakness had already been evident before the Opium Wars. Explosive uprisings as a result of poverty and excessive taxation occurred during the late eighteenth and early nineteenth centuries, such as the White Lotus Rebellion (1795–1804) and the Taiping Rebellion (1850–1864). This instability attests to the seriousness of the many domestic problems that developed during the waning years of the Qing dynasty. The Opium Wars and their denouement then set in motion a series of events that were cataclysmic and led to the overturn of China's entire political, social, and economic system.

The forces unleashed by the Treaty of Nanking on China after 1842 were social and political on the one hand—a basic challenge to the old order—and economic and technological on the other. These economic and technological forces had important ecological and spatial ramifications as well. In the nineteenth century, it was becoming obvious that the combination of an increasing population and the prevailing systems of farming and industrial development were unable to provide a satisfactory life and livelihood for all the Chinese people. It was also obvious that China's vast territory suffered from inadequate surface transportation. By the mid-nineteenth century, China had a poorly integrated spatial system that did not allow the development of a modern nation despite rich natural resources and low-cost labor.

The Treaty Ports and the Western Intrusion into China

One of the significant spatial consequences of Western intrusion in China was the emergence of what came to be called the "treaty ports." The Treaty of Nanking forced the weak Qing government to allow Western commercial interests to operate in select cities along China's coasts and major rivers, essentially without control. Tens of thousands of westerners established export and import businesses in coastal cities. Initially, five ports were opened. Eventually, there were more than one hundred, and a whole new stratum of commercial and industrial centers then overlapped the preexisting urban system of traditional and largely interior administrative and marketing centers. The treaty ports also introduced new technologies for manufacturing and developing tools, instruments, and weapons; new techniques for assembling, storing, processing, and distributing goods; and modern means of doing business and managing accounts and commercial affairs. These changed values would shake China to its very foundations before a new and revolutionary order gained control of China in 1949, but one of the most significant economic consequences of the establishment of the treaty ports was the spatial reordering and restructuring of the manufacturing sector toward export products. Development spread along the eastern coastal periphery of China, where the treaty ports attracted large numbers of industrial and service workers seeking to improve their lives and incomes. The transport system of China reoriented itself toward these new commercial centers, especially after the beginning of rail construction in the late nineteenth century (see chapter 7). American, German, and British firms built the first rail systems, followed by the Russians and Japanese. These systems were designed to expedite trade in areas nominally controlled by foreign commercial interests, and most of the routes ended at the coasts. As a consequence, new regional and local growth centers emerged, and the economic and political control mechanisms of the country gradually began to shift to incorporate these new forces and realities.

Ultimately, amid rebellion and nationalist sentiments, the Qing dynasty was overthrown, and a republican government emerged, based loosely on the democratic principles and ideals of Chinese nationalist Dr. Sun Yat-sen, partly educated in Hawaii. The Republic of China, established in 1911, was weak from the outset. A period of factionalism followed in which powerful regional warlords fought for territorial control and Western nations tended to recognize whoever controlled Beijing (then Peking). During this time of confusion and instability, two important events occurred. First, in 1921, the Chinese Communist Party (CCP) was founded in Shanghai. While it took a number of years for the CCP to grow, many urban intellectuals were drawn to its platform of renewed Chinese nationalism and independence from foreign control.

The second important event surrounded the rise to power of Chiang Kai-shek. During the early 1920s, the Nationalist Party (Kuomintang [KMT]; also Guomindang), originally founded by Sun Yat-sen and his associates to guide the republic, was reorganized with the help of Soviet advisers, fresh from the overthrow of czarist Russia. This reorganization was along the authoritarian lines of the Soviet Communist Party. Shortly thereafter, a new revolutionary army was created with Chiang Kai-shek as its head. Sun Yat-sen died in early 1925, and Chiang Kai-shek took over the reins of power of the KMT. Almost immediately, he launched military campaigns against local warlords but, more important, against the newly formed CCP. This conflict, so long ago, remains important because the split between Chiang Kai-shek's KMT and the CCP is at the root of the separation of Taiwan and mainland China today.

China's Communist Revolution

The story of the rise and development of the CCP and its famous leaders has been told often, and the details need not be repeated here. It will be sufficient to review briefly a few highlights and then explore the significance and effect of China's communist revolution on the country's physical environment and spatial organization and orientation. The CCP was initially conceived as an urban-based movement that sought support from the urban proletariat (factory workers and service workers). This was based on foreign theories of communism coming from the Soviet Union as well as from leading socialists in Europe. After a period of little success, some of its leaders turned to rural areas and focused their attention on the disenchanted peasant masses. The leading proponent of this rural-based strategy was a Hunanese cadre with a middle-class peasant upbringing, Mao Zedong. The man who eventually became "Chairman Mao" worked hard in his native Hunan and Jiangxi provinces in Central China to organize poor and middle-class peasants during the 1920s and early 1930s, achieving considerable success, especially in these poor and isolated regions. He and his cohorts became so successful at promoting a rural communist base of support that power and control in the party soon shifted away from the urban cadres to the charismatic and well-spoken Mao.

Eventually, competition between the nationalists and communists erupted into open conflict, and Chiang Kai-shek and his forces almost eradicated the communist forces in southern and Central China. In 1934–1935, Mao and his peasant forces (85,000 soldiers, 15,000 government and party officials, and 35 women, mostly wives of leaders) managed to escape through a long and harrowing migration to an isolated area of northern Shaanxi province—the epochal 12,500-km Long March. Only 8,000 survived the journey, arriving in Paoan County in northern Shaanxi and establishing a remote base area with its headquarters in the loess hills of remote Yan'an. From this isolated base, the CCP gradually expanded its influence. The survivors of the Long March proved to be the key leaders of China well into the 1990s.

The Sino-Japanese War and World War II provided the CCP with new opportunities to grow, and the communists exploited these cataclysmic events very effectively. Wisely, communist leaders focused their efforts on the patriotic struggle of the Chinese against the Japanese invaders rather than dissipating their strength against their domestic antagonists, the nationalists. By the end of World War II, support for the CCP and its military arm, the People's Liberation Army, had grown tremendously in North China. Efforts to mediate the political struggle between the nationalists and communists failed, and a civil war followed. The communists, with their strength in the rural areas, were able to dominate that struggle and isolate the nationalist forces easily and quickly in the North. By 1949, the nationalists had fled the mainland and set up an exile, rump government on Taiwan (see chapter 13).

On October 1, 1949, the PRC was established with Mao Zedong as its first chairman. This marked the end of a century of domestic turmoil and foreign intervention. And more than two thousand years of an imperial dynastic system had been replaced by a truly revolutionary system that was committed to changing China fundamentally and modernizing its society, economy, and politics. A strong, new central government with a firm and definite ideological foundation and position, supported by a powerful army and civil administrative apparatus, took over China. A new age had arrived. The CCP has led Chinese economic and social development to the present, and there is little reason to assume that the party will not remain the dominant decision-making political organization in China for many years to come.

In the following chapters, we discuss the kinds of changes that have resulted from the establishment of a communist system in China, with a special focus on the period after 1978. One of the goals of this geographic study of China is to provide at least partial answers to some of our questions concerning the spatial organization of this nation's politics, economy, people, and culture in the past, present, and future and to reflect on how the past has influenced the present.

Questions for Discussion

1. What is the relationship between early Paleolithic and Neolithic settlement in China and the environment?
2. Could you combine information from chapters 2 and 3 to summarize the types of environments or ecosystems where most of the early residents of China lived?
3. What were some of the accomplishments of the short-lived Qin dynasty, and why are they important for later dynasties?
4. What is *putonghua*, and why is it important that China have a national language?
5. What do the authors mean when they say that the Song dynasty had a different "spatial orientation" than the Tang or the other earlier dynasties?
6. What were the benefits of an extensive canal system to the growing Chinese state?
7. What do the authors mean when they say that the Yuan and Qing dynasties were "foreign" dynasties, and what are some implications with respect to economic development that might be related to these "occupations"?
8. What might be some of the implications of the Opium Wars and "unequal treaties period" with respect to modern China and how the Chinese view the rest of the world?

References Cited

Central Intelligence Agency. 2009. *The World Factbook* (China). https://www.cia.gov/library/publications/the-world-factbook/geos/ch.html (accessed August 17, 2010).

Cheng Te-k'un. 1960. *Shang China*. Vol. 2 of *Archaeology in China*. Toronto: University of Toronto Press, 14–16.

Crossley, Pamela Kyle. 1997. *The Manchus*. Cambridge, MA: Blackwell.

Davis, Richard L. 1996. *Wind against the Mountain: The Crisis of Politics and Culture in Thirteenth-Century China*. Cambridge, MA: Harvard University Press.

Ebrey, Patricia Buckley. 1996. *The Cambridge Illustrated History of China*. Cambridge: Cambridge University Press.

Elvin, Mark. 1973. *The Pattern of the Chinese Past*. Stanford, CA: Stanford University Press.

Fairbank, John K., and Merle Goldman. 1998. *China: A New History*. Cambridge, MA: Belknap Press.

Goodrich, L. Carrington. 1969. *A Short History of the Chinese People*. London: George Allen and Unwin.

Huang, R. 1997. *China: A Macro History*. Turn of the Century Edition. Armonk, NY: Sharpe.

Ikawa-Smith, Fumiko, ed. 1978. *Early Paleolithic Sites in South and East Asia*. The Hague: Mouton.

Levathes, Louise. 1994. *When China Ruled the Seas: The Treasure Fleet of the Dragon Throne, 1405–1433*. New York: Simon & Schuster.

Loewe, Michael. 1986. Introduction to the Ch'in and Han Empires. In *The Cambridge History of China*, ed. Dennis Twitchett and Michael Loewe. Vol. 1. Cambridge: Cambridge University Press, 2–19.

Mair, V. H. 1997. Script reform in China. In *China Global Studies*. 7th ed. Guilford, CT: McGraw-Hill Duskin, 175–79.

National Bureau of Statistics. 2002. *Zhongguo tongji nianjian 2002* [China statistical yearbook 2002]. Beijing: China Statistics Press.

———. 2009. *Zhongguo tongji nianjian 2009* [China statistical yearbook 2009]. Beijing: China Statistics Press.

Needham, Joseph, et al. 1962. *Science and Civilization in China*. Vol. 4, pt. 3. Cambridge: Cambridge University Press.

Pannell, Clifton W., and Laurence J. C. Ma. 1983. *China: The Geography of Development and Modernization*. New York: Halsted.

Roberts, J. A. G. 1996. *Prehistory to c. 1800*. Vol. 1 of *A History of China*. New York: St. Martin's.

Shih Sheng-han. 1982. *A Preliminary Survey of the Book Ch'i Min Yao Shu: An Agricultural Encyclopaedia of the 6th Century*. Beijing: Science Press.

Sivin, Nathan. 1988. *The Contemporary Atlas of China*. Boston: Houghton Mifflin.

Skinner, G. William, ed. 1977. *The City in Late Imperial China*. Stanford, CA: Stanford University Press.

Smartt, J., and N. W. Simmonds. 1995. *Evolution of Crop Plants*. 2nd ed. Essex: Longman Scientific and Technical.

Wheatley, Paul. 1971. *The Pivot of the Four Quarters: A Preliminary Enquiry into the Origins and Character of the Ancient Chinese City*. Chicago: Aldine.

Wu, X. Z., and Frank E. Poirier. 1995. *Human Evolution in China: A Metric Description of the Fossils and a Review of the Sites*. New York: Oxford University Press.

Yo Weichao, ed. 1997. *A Journey into China's Antiquity*. Vol. 1. Beijing: Morning Glory Publishers.

Zhao, Songqiao. 1994. *Geography of China: Environment, Resources, Population, and Development*. New York: Wiley.

Zheng Wenlei. 1997. *Paleolithic Age*. Vol. 1 of *A Journey into China's Antiquity*, ed. Weichao Yu. Beijing: Morning Glory Publishers.

CHAPTER 4

The Political Geography of Emerging China

A Political Geography for the Twenty-First Century

The political geography of a country as old, large, and diverse as China is both very broad and very complex. China has traditionally referred to itself as *Zhongguo*, meaning "Central Kingdom," a term that connotes an ethnocentric bias and indeed a superior attitude toward the rest of the world. Throughout much of its history, the country clearly perceived itself in a central, pivotal role wherein its relations with other countries, especially those on its immediate periphery, were typically those of a superior to an inferior. This was the self-view of Chinese culture and civilization even when its military forces were inferior. What does this mean for China's role in global politics and the security environment of the twenty-first century, and how does it affect China's behavior within the community of nations? A brief historical review will help inform our understanding.

In China's relations with other countries, there evolved over time political and diplomatic relationships that are best described as sovereign state to tributary state or patron to client. Until the nineteenth century, even when it was ruled by non-Han emperors as in the Yuan and Qing dynasties, China was always a great Asian power. Partly, or perhaps entirely, because of its ethnocentrism, China tended to be inward looking and self-satisfied with its position and role. Historically, this may have resulted from China's preoccupation with the internal consolidation of its national territory and its leaders' enduring struggle to keep their various regions tied and loyal to the center. Related to this was the symbolic role of the Chinese emperor (the "son of heaven" as he was known), who sat in the sacred center of this great empire. This locus of power, the place where the emperor was located, became the sacred center of the known universe for all Chinese (Wheatley 1971).

Yet there were times when China did in fact look beyond its land frontiers. As early as the eighth century AD, landless peasants, traders, fishermen, and pirates looked to the sea and places beyond the Chinese ecumene, and many set sail for distant lands to seek their fortunes and enhance their opportunities and livelihoods. Thus began an exodus and dispersion of the Chinese people first to Southeast Asia or the South Seas (the *Nanyang* as the Chinese refer to it) and later to more distant places, an exodus and spread that continues to this day (Wiens 1967).

During the Song dynasty, beginning in the tenth and certainly by the late twelfth century AD, China began to launch trading expeditions into the South China Sea. By the fifteenth century, this tradition had expanded to the sending out of the great Ming dynasty maritime expeditions headed by Zheng He, and a number of these reached the coast of Africa. An especially extensive network of trading ports was established during the thirteenth to fifteenth centuries in the South China Sea (Nan Hai) area. Yet, paradoxically, in late Ming times maritime policy was reversed, and China turned away from sailing and maritime trading activities just at the time when the age of great European exploration was getting into high gear. This change occurred, among other reasons, because of the high cost of fleet construction and also perhaps owing to a decline in the dynasty and its fiscal control and an increase in its instability. Whatever the causes, the results were an inward reorientation, rapid population growth, and stagnating per capita output.

From a twenty-first-century point of view, these early Chinese maritime forays were important events, for they showed the real and potential power of the empire despite their somewhat abrupt termination. Of great current interest and symbolic significance even today was a recent expansion of the Chinese navy and that navy's first circumnavigation of the globe, which a small fleet of warships made in 2002 by visiting a number of countries in a global showing of the flag of the People's Republic of China (PRC). Such an expedition brings to mind, for the first time since the voyages of Zheng He, the interest of the modern Chinese state in expanding its naval forces to a deep-water navy and its commitment to displaying its growing military and technological prowess to the world.

The Chinese Geopolitical View

In evaluating what this process of self-image formation has meant for China and the world, the traditional model depicts a dominant Sinocentric view when most Chinese knew little about the rest of the world. The exception to this included those peripheral neighbors with which China had traditionally sought to establish a superior–tributary relationship—a policy that served reasonably well for two millennia despite the vagaries of dynastic effectiveness. According to Ginsburg (1968) and Whitney (1970), the Chinese placed a different value on far-flung national territory. Therefore, the eighteen traditional provinces in the eastern half of the country represented a more valued national core, and the worthiness of China's remaining national territory declined as a direct function of its distance from this core (see figure 4.1). Parts of Southeast and Central Asia as well as Mongolia and parts of Siberia were inside the traditional frontiers of the area over which China had some kind of control or loose tributary relationship.

After the onset of European incursions and especially after the Opium Wars, this traditional view and approach was under serious challenge. China, perhaps because of

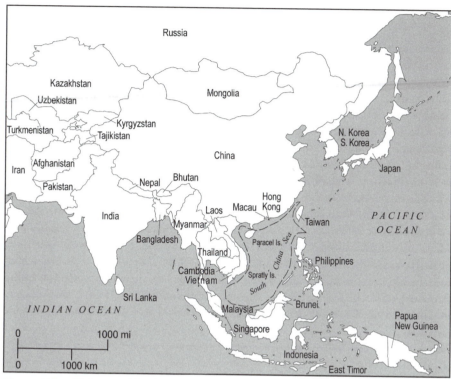

Figure 4.1. China and its Asian neighbors

external threats, began to assume much more interest in its peripheral territories and to issue more direct claims of territorial sovereignty to frontier regions. However, during the nineteenth century as the Qing dynasty became increasingly weak and ineffective, China had to yield considerable territory to other countries, such as Russia, Japan, and British-controlled South Asia.

As David Shambaugh (2000) recently noted, few issues in modern Chinese studies excite the imagination of scholars as much as the recent evolutionary trajectory of the Chinese state. As he goes on to point out, in the modern period this evolution proceeded from "imperial system to republican to revolutionary communist to modernizing socialist, and in Taiwan to democratic phases" (1). At the same time, he noted three enduring characteristic missions or goals of the Chinese state over the past century: "modernization of the economy, transformation of society, and defense of the nation against foreign aggression" (1). As we proceed with our discussion of China's development, it will be useful to consider the changes in how these missions were approached during the socialist period since 1949 and what methods or paths to their resolutions were followed.

One goal that China has clearly and tenaciously sought and partially succeeded in gaining is restoration to the state territory of those areas of Greater China that lay outside the territorial sovereignty of the PRC in 1950. The obvious territories were Hong Kong, Macao, and Taiwan. Hong Kong was returned to China in 1997 by the

British according to an agreement signed in 1984 that allowed the territory of Hong Kong to exist as a special administrative region (SAR) of China for fifty years, with many of its existing commercial practices and civil rights as well as its political democracy intact during that period (*Basic Law* 1991). The Portuguese and Chinese agreed to a similar arrangement for Macao, and it was restored to China in December 1999. Taiwan remains outside China, and its reintegration as a part of Greater China has become a particularly contentious and volatile issue.

Greater China (*da Zhonghua*) as defined here is simply those areas of the national territory peopled by Han Chinese or linked historically with it through tributary relations. In fact, the conceptual idea of a Greater China is slippery, complex, and subject to a variety of interpretations and designation (Harding 1993).

In the twentieth century after the establishment of the PRC in 1949, the country moved quickly to establish full and sovereign control over many of its outlying territories, such as Tibet, Xinjiang or Chinese Turkestan, Inner Mongolia, and all of Manchuria. In addition, serious border conflicts and fighting occurred with India (1964), the Soviet Union (1969), and Vietnam (1979). The PRC has also laid claim to virtually all of the archipelagos in the South China Sea, thereby essentially asserting that the South China Sea is in fact territorial water and an economic resource zone of China (Samuels 1982) (see figure 4.2). While these geopolitical claims are to some

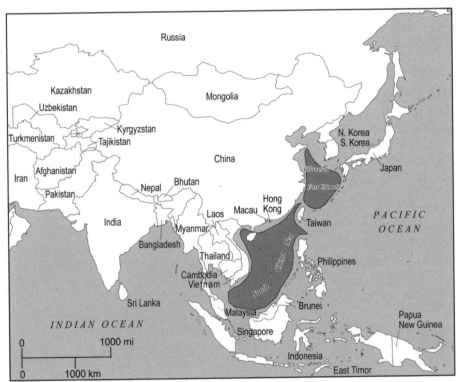

Figure 4.2.   China's claims in the South and East China seas (Office of the U.S. Secretary of Defense)

extent problematic, this activist and aggressive approach indicates that China has continued to shift from its early Qing dynasty inward-looking posture to a more outward policy that has sporadically been aggressive in asserting claims.

While there are many different interpretations and conclusions that may be offered about China's role in world affairs since 1950, it is clear that the country intends in the future to play a major role in East Asian and global affairs. As China's economic development progresses, its ability to develop a powerful military and political presence will continue to grow. To what extent this will be translated into challenges to those neighbors with whom it has territorial disagreements is difficult to predict. Such growing power could indeed presage a period of growing conflict. On the other hand, the presence of many ethnic nationalities in China's current frontier territories represents a threat to internal stability. Domestic stability, history suggests, will remain China's first priority. Further, China's historical record with its Tibetan and Central Asian minority nationalities is such that continuing conflict and internal instability here seem likely, and this will require the continuing attention and resources of the central state.

Empire or Nation-State: China's Goals for the Twenty-First Century

One of the more interesting and controversial questions often invoked as a result of China's rapid ascendancy toward global superpower status is how China should be viewed as a nation and country. Given the huge size of its territory and the diversity of its people, should it be considered an empire, reconstituted in a new, twenty-first-century political guise and form from the old Chinese imperial structure and system? Or should it more appropriately be considered a nation-state despite the many minority and ethnic groups that make up its population, including those various linguistic groups among the mainstream Han Chinese majority? Arguments can be made for both points of view, and in the discussion of minority nationalities that follows, the case for China's behaving like a latter-day empire, almost in the manner of the Soviet Union, can be seen.

The powerful force of Han Chinese assimilation as a mode of long-term nation building can also be seen at work in the manner in which groups on the margin of Han civilization have progressively been brought into the greater Han fold. One scholar, Martin Jacques (2009), has declared that China is in fact a "civilizational state" owing to its long record of political tenure in which its remarkable civilization has gradually simply absorbed and transformed most of those peripheral groups with whom it has come into contact. Over time, some of these groups, especially where *putonghua*, or Mandarin, is the first language, have come to be almost indistinguishable from Han Chinese despite self-identification (the Manchu or Hui, for example).[1]

Yet in reconsidering China's worldview and global ambitions for the twenty-first century, it is necessary to review China's domestic as well as international situation. Since the late 1970s, the goals of China's restructuring and reforming socialist system were to advance the economic growth and development of the nation while maintaining political and social control through the primacy of the Chinese Communist Party (CCP). Some political liberalization was allowed during the 1980s, but after this liberalization resulted in division and turmoil and the tragic events of June 1989, in which a number of students were massacred in the Tiananmen Square incident in Beijing, party and government emphasis was placed on maintaining order and control and avoiding what the Chinese perceive as chaos (*luan*) and unstable political conditions. The Japanese specialist on international politics Satoshi Amako (1997) has described as "unified authoritarianism" the political path or political model the Chinese leadership followed after the Tiananmen Square incident.

At this point, a lid was clamped over the domestic political scene, and little changed in Chinese politics for thirteen years until leadership succession was put in motion at the Sixteenth Communist Party Congress in the autumn of 2002. China then made a transition to a new generation of leaders, and President Jiang Zemin and Premier Li Peng yielded their leadership positions in favor of younger leaders Hu Jintao (president) and Wen Jiabao (premier). Yet after almost a decade of this new leadership cadre, it has become clear that little liberalization in the direction of political evolution in China has occurred. Certainly China's focus on economic growth has continued, and indeed the extraordinary success of China's economic growth and development are the keys to ensuring the legitimacy and success of the CCP as the sole arbiter of power and control in China.

Modern China since the communist takeover of the mainland in 1949 has been a state and nation that takes its basic direction and guidance for its development and growth from the CCP. Reaffirmed as recently as 2002 at the Sixteenth National Congress of the Communist Party of China, the goal of the CCP is

> to lead the people of all our ethnic groups in a concerted, self-reliant and pioneering effort to turn China into a prosperous, strong, democratic and culturally advanced modern socialist country by making economic development our central task while adhering to the Four Cardinal Principles and persevering in the reform and opening up. (*Xinhua News Agency* 2002)

The "Four Cardinal Principles" refer to maintaining socialism, the people's democratic dictatorship, and the primacy and leadership of the CCP based on Marxism-Leninism and "Mao Zedong thought." The key element in this is the preservation of socialism in whatever form it takes as it continues to evolve and emerge in China while maintaining the exclusive and primary leadership role of the CCP in the political control of the country.

The same document reaffirms and emphasizes the key role of economic development in leading the cause of socialism: "The Communist Party of China must persist in taking economic development as the central task, making all other work subordinated to and serve this central task" (*Xinhua News Agency* 2002). Such an emphasis suggests that the CCP and its top leadership recognize that domestic economic development and growth underlie their efforts to advance the country and that the goal of growing the economy is essential to maintaining a stable and productive society and polity in the future. Thus, continued success in political control would appear to hinge on effective efforts in economic growth.

Foreign Policy, International Relations, and the Taiwan Question

At the same time that the party-state must focus on domestic control and stability, it must study the forces of international politics carefully to ensure that China's security remains protected and that the country's ambitions as a growing regional and world power are taken seriously. In the twenty-first century, following the collapse of the Soviet Union and the emergence of the United States as the sole world superpower—and added to this the regional fighting in Afghanistan, South Asia, and Iraq following the events of September 11, 2001—China has sought to establish its proper role as a regional and emerging world power. But how does it deal with its neighbors, as well as the United States, and how does it pursue its long-held—almost sacred—objective of reuniting Taiwan with its motherland (Vogel 1997)?

China's foreign policy goals are straightforward and appear reasonable. An official Chinese website sums up the country's goals:

> The basic objectives of this policy are to safeguard the independence, sovereignty and territorial integrity of the country, strive to create a long-standing and favorable international environment for China's reform, opening-up and modernization drive, safeguard world peace and promote common development. (*China through a Lens* 2004)

These are certainly goals that appear reasonable and appropriate in the current global context. China goes further to affirm that it will follow an independent policy and avoid alliances while also opposing what it sees as "hegemonism" in any other state or group of states. Its other general goals are to develop good relations with all its neighbors, strengthen its relations with developing countries, be open to the outside world, and work toward world peace in the framework of the United Nations.

In China's stated goals of foreign policy and foreign relations, the only place where its current stated position appears aggressive and contentious is in its statements on Taiwan. Here China's policy is clear and unambiguous: Taiwan is part of the integral territory of the motherland, and China seeks Taiwan's reintegration although under terms that will allow Taiwan some degree of autonomy and time for special development. At the same time, while China calls for this reintegration into the national territory in a peaceful manner, it reserves the right to use force to bring about this territorial reintegration if necessary (Goldstein 2001).

It is the Taiwan issue, China's most contentious and serious goal, that has created ongoing tension, especially with the United States. China views Taiwan as a true and inalienable part of the motherland and will allow no variance from its view that it must reunite the island with the PRC. This is in line with the idea that China is a nation-state even though it contains 100 million people who are not Han Chinese.

At present however, despite a rapidly growing defense budget and much progress in military modernization, China does not possess sufficient military power to seize Taiwan. However, its recent accelerated development of military forces clearly appears focused on tactical amphibious, naval, and air weapons that are designed and suited for an invasion of Taiwan and to oppose the likely U.S. forces that would be sent to assist Taiwan in the event of an attack. Current insufficiency of forces, however, has not prevented China from issuing periodic threats and conducting military operations in the vicinity of Taiwan that appear to be intended to intimidate the citizens of Taiwan to the point where they will simply yield and capitulate to Chinese rule. China has proposed the idea of the "one nation, two systems" principle, wherein Taiwan would be allowed to maintain its own economy and political system for a limited period of, say, fifty years, similar to the model that is in place for Hong Kong. Yet the Taiwanese have resisted this model, and they reject it for now. With support from the United States, they have so far continued to maintain their autonomy from China. However, economic links between China and Taiwan are growing rapidly because many Taiwan industrialists have relocated their production operations to China, taking advantage of China's low-cost labor to keep their products competitively priced. A large number of Taiwan citizens also now live in the greater Shanghai region and are associated with these growing economic ties, ties that may someday help facilitate reconciliation of China and Taiwan based on the mutual benefits.

China and the United States have differed in their view of Taiwan, although both countries continue to support the "one China" policy and the idea that Taiwan is part of China. This was reaffirmed in the Shanghai Communiqué and in related communiqués between the United States and the PRC in the 1970s that set out the framework whereby the United States recognized the PRC as the legitimate government of China and rescinded its bilateral treaty with Taiwan. However, the United States also affirmed its position of rejecting the use of force in the return of Taiwan to China. Through enactment of the Taiwan Relations Act, the United States established the principle that it would provide Taiwan with defensive arms and military equipment to allow it to defend itself against attack from the PRC. While the PRC does not accept the validity of the Taiwan Relations Act and opposes strenuously the sale of U.S. weapons to Taiwan, it has not directly attacked Taiwan in an effort to bring about Taiwan's return. And because the island's population is divided over how to deal with China and how to proceed with Taiwan's political destiny, this matter continues to fester. Not surprisingly, it has soured relations between China and the United States. More discussion on this contentious and complex issue will follow in chapter 13, where we explore the political and social dynamics of Taiwan's people and their recent history.

China's relations with the United States continue to be characterized by uncertainty and difficulty as the two countries seek to engage one another constructively while also continuing to differ and squabble over Taiwan, human rights issues related to Tibet and minority peoples in China, and religious freedom. Military leaders in China remain very suspicious of U.S. goals and the perceived hegemony of the United States in world affairs, and they can be expected to oppose any efforts of the United States to establish or maintain outposts of U.S. power in the western Pacific or on the Asian mainland. Both the United States and Japan in turn view China's rapid increase in military spending and the modernization of China's military as a dangerous signal of China's long-term intention to establish itself as a dominant military and economic power in East Asia. Consequently, the trajectory of U.S.–China relations in the future must be managed very carefully by both sides to ensure a productive and peaceful outcome despite occasional disagreements and tensions.

China's Expanded Global Reach: The Search for Resources, Markets, Influence, and Growing Power

China's remarkable economic growth of the past twenty years has had a corresponding effect on its activities in searching for new resources for its industrial and growth machine as well as markets for its multifaceted products. This search has indeed been global in scope and has moved in parallel with an energetic diplomatic effort to win friends and influence among those with whom it does business. Nowhere is this more apparent than in Africa, where China has sent large numbers of commercial operators as well as contract workers and officials of the central state. The goals of this effort appear three-pronged: 1) find resources such as oil, strategic minerals, metals, and other raw materials for China's hungry industries; 2) establish markets for Chinese products and services, including the building of massive infrastructure projects such as dams, electrical grid systems, oil refineries, port facilities, railroads, highways, airports, bridges, municipal buildings, and housing estates; and 3) expand and solidify China's diplomatic influence and reach by providing economic aid assistance, cultural, medical, and educational support. China's leading source of foreign oil is now Angola, a country with a wealth of resources located on the southwest coast of Africa and where China has been very active in all three of the approaches noted above.

All this effort may be summed up as being of great value to China's rapid economic growth while also enhancing its global impact and influence in a manner that is described as the development of its "soft power," that is, bringing about desirable national goals without the application of direct military power or violence.

Recent analysis has focused on China's internal development and strategy for evolving its growing power in step with its economic growth. While much has been made of the rapid increases in the annual budgets for military expenditures, China's main focus continues to be on growing its economy owing to China's clear understanding of the relationship between a strong economy and the ability to develop a modern and strong military force. In the past two decades, its economic growth has permitted China to restructure its military and reduce the personnel while concentrating on modernizing the nature of its forces and weapons systems to keep pace with military modernization in other nations. These developments have advanced so quickly that they have raised serious concerns within the U.S. defense establishment (Office of the U.S. Secretary of Defense 2010).

One recent study has focused on a three-pronged Chinese approach to developing its national power in the early twenty-first century. David M. Lampton (2008) has

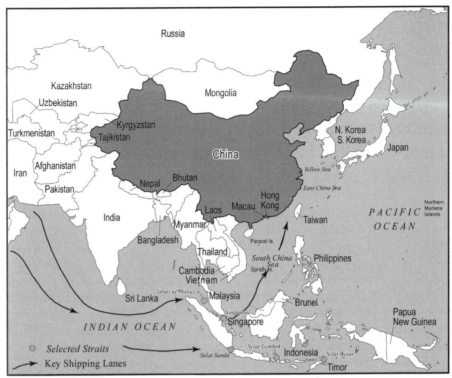

Figure 4.3. China's strategic sea-lanes in the Indian Ocean and South and Southeast Asia (Office of the U.S. Secretary of Defense)

identified what he calls the "three faces of Chinese power." These are money, minds, and might, and as he notes, China is moving ahead on all three fronts. Clearly, the money accumulation has been pursued aggressively with an export-oriented trade strategy that has allowed China to accumulate the largest foreign exchange reserves of any nation, a sum approaching US$3 trillion. The country has also invested heavily in improving its educational system with a particular focus on science and technology and research and development on a number of topics of strategic interest both to economic and commercial development and for security and military purposes. Finally, China has been careful in recent years in its application of coercive force. While it has reserved the right to use force against Taiwan as a means of ensuring the return of the island to the motherland, it has been careful to use restraint, especially where it may lead to armed conflict with the United States or in those cases where it is uncertain if its forces are sufficiently strong to guarantee victory. This appears consistent with traditional approaches to violent conflict as well as in keeping with the strategic approach of key leaders such as the late Deng Xiaoping.

This same strategy has been applied in other areas of the world such as Latin America, and China has made its search for strategic resources truly global in scope. China's economy is now the second largest in the world, having recently passed Japan, and the demand for resources is huge and continues to grow in step with economic growth. A consequence of this global search has prompted a much more aggressive

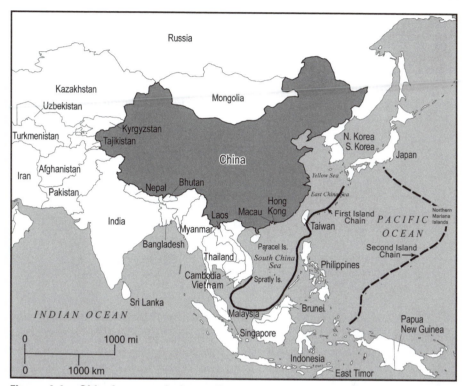

Figure 4.4. China's expanded security zone and the "second island chain" (Office of the U.S. Secretary of Defense)

foreign and security policy to ensure that China's strategic shipping lanes, particularly those that provide energy resources, such as oil from Africa and the Middle East, remain open. This has entailed massive new investment and development in naval power to create the capacity to extend China's defense and military reach beyond the traditional coastal and peripheral regions of the Middle Kingdom. Of particular concern is the expansion of the Chinese navy and strategic missile forces, which have created a much more aggressive and enhanced security zone in the western Pacific. China's expanded security zone now goes to the ocean waters beyond its focus on Taiwan and the South and East China Seas to a new so-called second island chain that extends to Japan and as far east as the island of Guam. Implicit also in this expanded new zone of security interest is the commitment to protect strategic sea-lanes associated with the Indian Ocean, such as the Strait of Malacca and those key choke points adjacent to the Indonesian Archipelago (see figures 4.3 and 4.4).

Internal Political Organization and Administration

One of the basic requirements of any country's political system is the organization of its territory for the most effective and efficient operation of governance. The functions of government and the control and support of the population must be managed effectively if the state is to rule as a credible and legitimate institution. The state must incorporate the territory of the country in a manner that allows that territory to contribute to the national well-being in a positive, economic, and political manner.

Traditionally in China, the main goal of the imperial authority was to establish central control, for the task of organizing China's large population and territory was gargantuan indeed. As Hsiao (1960) noted,

> The solution as it was worked out in China during the successive dynasties from Ch'in [Qin] to Ch'ing [Qing] consisted essentially in the development of an administrative apparatus which helped the emperors to assure obedience and forestall rebellion, partly by ministering to the basic material needs of the subjects so that few of them would be driven by unbearable hardships "to tread the dangerous path," partly by inculcating in their minds carefully chosen precepts (mostly from doctrines of the Confucian tradition) that tended to make them accept or acquiesce in the existing order, and partly by keeping constant surveillance over them so that "bad people" might be detected and dealt with in time. (3)

This type of administrative apparatus, supported by the army, permitted a dynasty to maintain control for long periods of time. Indeed, as we noted earlier, the basic system of control and governance in China lasted more than two thousand years. Our goal in analyzing China's internal administrative organization is not to recount the historical method of administrative control but rather to begin our discussion with the recognition of a significant change in that administrative system after 1949. Our specific goals are to describe the new administrative system and especially its territorial aspects, to explain and evaluate some of the functions and the effectiveness of that system, and, finally, to examine the nature of the system's organization of that territory not regarded as traditionally Chinese, that is, the outlying and associated territory peopled by China's ethnic minorities.

Geography of Administration

The administrative framework of the PRC has undergone significant changes since 1949. Under Nationalist (Kuomintang; also Guomindang) rule, China consisted of thirty-five provinces (including Taiwan), one autonomous region (Nei Mongol, or Inner Mongolia), one territory (Tibet or Xizang), one special district (Hainan Island), and twelve centrally governed municipalities. After 1949, major changes were effected.

First, the number of provinces and province-level cities decreased. In 1951, China reduced the number of provinces in the Northeast (Manchuria) from nine to six and again, in 1954, from six to three, the remaining provinces being Liaoning, Jilin, and Heilongjiang. During 1954–1955, the provinces of Rehe, Suiyuan, and Chahaer were eliminated, with most of their territories transferred to the Nei Mongol (Inner Mongolia) Autonomous Region (AR). In the 1950s too, the number of centrally governed cities was reduced from twelve to three.

Second, autonomous geographic regions in border areas inhabited by minority nationality groups were established. In addition to the Nei Mongol AR, established in 1947, four new province-level ARs were created in the decade from 1955 to 1965. Xinjiang (Sinkiang) province became Xinjiang Uygur AR in 1955. In 1958, Guangxi province was designated as Guangxi Zhuang AR, and Ningxia (Ninghsia) Hui AR was created. Finally, in 1965, Xizang territory (Tibet) was changed to Tibet AR (see figures 4.5 and 4.6). Meanwhile, lower-order minority districts, such as autonomous prefectures (*diqu*), counties (*xian*), leagues (*meng*), and banners (*qi*), were established as second- and third-order geographic areas. Minority peoples and areas, as a significant and sometimes contentious issue in the evolving political geography of China, are discussed in detail in the last section of this chapter.

Third, huge regional administrative units above the provincial level were created and then abolished. During the period from 1949 to 1954, when much of China's administrative power was decentralized, the country was divided into six Great Administrative Areas, plus Nei Mongol (Inner Mongolia) AR. Each area (except for Manchuria and North China) featured a Military and Administrative Committee that served as the highest organ of political control in charge of several provinces. These regional authorities initially had considerable power and enjoyed a significant degree of autonomy. But gradually, as the political power of the new regime was consolidated, Chinese leaders, perhaps motivated by the fear that the regional governments

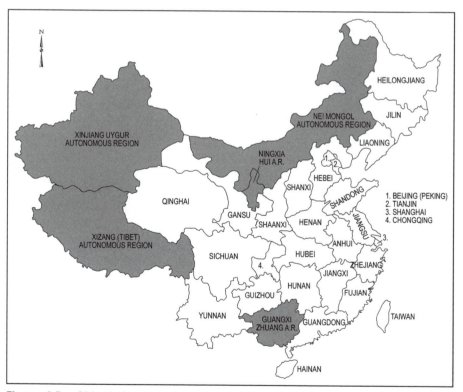

Figure 4.5. China's first-order administrative units

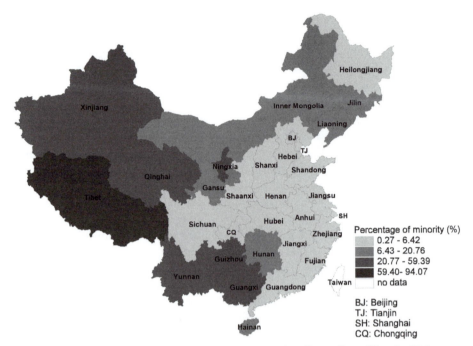

Figure 4.6. First-order administrative units for minority nationalities in China

could become too powerful for the central government to control, began to increase the power of the central government. In November 1952, the regional governments were stripped of most of their power. Two years later, the entire structure above the provincial level was abolished.

Several significant administrative changes have occurred in recent years, and the current administrative hierarchy is comprised of thirty-one first-order administrative regions plus the special administrative regions of Hong Kong and Macao. The thirty-one administrative regions include four national cities (Beijing, Shanghai, Tianjin, and, most recently, in 1997, Chongqing); five ARs (Xinjiang Uygur, Tibet, Ningxia Hui, Guangxi Zhuang, and Inner Mongolia) peopled with substantial ethnic minority populations; and twenty-two provinces of which the most recent, Hainan, was separated from Guangdong in 1988. Figure 4.5 indicates the considerable variation in the size of first-order administrative regions. This is paralleled by significant differences in their populations and population densities as well. Taiwan, as noted, is claimed by China but remains outside of China's direct control at this time.

Fourth, basic-level administrative units have been established. The most revolutionary changes in China's administrative structure have occurred at the grassroots level. Prior to the revolution, the central government and the basic-level administrative units below the county level were poorly integrated. Although there were administrative systems developed below the county level, such as the *baojia* and *lijia* systems, they were concerned chiefly with taxation, internal security, and corvée functions. Very little of the state's resources for economic development was allocated to villages, where most people lived, and there was virtually no systematic socioeconomic planning by the state for the masses at the village and urban neighborhood levels. Direct participation by the masses in the political and economic affairs of the state was hampered by a large administrative gap separating the central government from the villages. Members of the rural gentry class almost invariably dominated the affairs of the villages, often at the expense of peasant interests.

All this fundamentally changed after the communist revolution, and new basic-level administrative units were designed that proved to be highly effective. The entire administrative hierarchy in China is now tightly integrated. As a consequence, mass participation in and state control of the political, social, and economic affairs of the country at the grassroots level have been much more effective than before, and the authority of the state extends to the level of the household and even to the individual. At the same time, the effects of economic reform and the diminution of state control of the economy have begun to relax the hand of the state in economic activities. Shifting demands in the labor markets have led to a need for the increased migration of rural people to offer needed labor services in cities, and this in turn has led to reduced social controls in administering the *hukou* system of household registration and strict assignment of residential location.

It goes without saying that the CCP dictates urban and rural administration at all levels. Once a decision is made by the party, it is put into effect through the various levels of the administrative hierarchy of provinces, prefectures, cities, counties, townships, and villages and is then articulated through the spatial network of these various units.

Two major changes affected urban administration during the 1980s (Ma and Cui 1987). First, economic reforms were applied to cities in late 1984. While the reforms in urban areas did not have perhaps as great an effect as in the countryside, nevertheless some effort at creating a modest private-sector economy with vendors and petty service people had begun. Limited migration of rural people was also allowed to provide service and construction workers in the cities. These migrants were not afforded the conventional subsidies of housing, food, health, and educational services afforded to urban dwellers employed in the state sector of the economy. There are currently estimated to be more than 150 million of these migrants in China's cities, many of whom live in very crude and rough accommodations that are temporary (Chan 2010). Obviously, the government's commitment and capacity to provide administrative services and support for these migrants are different and less than that for long-term residents. Problems of social and political concern, such as increased crime, poor health conditions, limited educational opportunities, and crowded and poor housing, have increased.

The second major change was a set of political disturbances that occurred in many of China's large cities in the spring of 1989, disturbances known as the "democracy movement." This event had its roots in the general liberalization of both the economy and society that followed the Cultural Revolution and the demise of the Gang of Four in 1977. The violent suppression of this movement in Tiananmen Square in Beijing on June 3–4, 1989, ushered in a period of martial law in Beijing and a severe political crackdown throughout the country. The Department of Public Security (i.e., the *gonganju*, or police) initiated much more rigorous control measures, especially in cities that were seen as the focal points of trouble and dissatisfaction. This was done, in the eyes of the CCP, to prevent any full-blown assault on or threat to Chinese socialism and the integrity of the state. The immediate and direct consequence of these measures was to demonstrate that the CCP, buttressed by the army and its various security forces, was very much in control of China's cities. (For a full and insightful explanation of the Tiananmen Square incident, see Liang 2001.)

Retrospective View of China's Administrative Geography under Communism

As we noted previously, the twentieth century brought a fundamental change in the nature of China's political system with the overturn of its long-lived dynastic order. The CCP, which took over in 1949, focused especially on territorial organization and greatly strengthened the administration of territory at the local level. Among the most significant accomplishments of the CCP has been the extended control and involvement of local people at both urban neighborhood and township levels in the affairs of economy and politics. Transforming the economy has strengthened the role of local party cadres as they have embedded themselves more deeply in the expanding economy at local levels. Improvements in the economy have in turn offered greater legitimacy to the CCP at local, regional, and national levels. Yet political reforms have not accompanied these improvements in the economy.

New and younger leaders, such as President Hu Jintao and Premier Wen Jiabao, have assumed power, but they have done little to modify or liberalize the political evolution of China. Consequently, it remains uncertain as to if or when the CCP will allow any meaningful competition or democratic opposition to its rule. Until it does, China will remain a Leninist state as defined by the primacy and exclusivity of the CCP. There will be another Communist Party Congress in 2012, and this will bring in a new generation of leaders. While there is always much speculation about such leadership changes, only time will tell if leadership succession will bring about any meaningful political reform and liberalization.

At the same time, China's transition from socialism is leading to rapid although sometimes controversial growth in the private economy in which the role of local party cadres in facilitating and indeed in participating in the growing private economy is increasing. The line between socialist "public" and capitalist "private" is becoming less clear as cadres get more and more involved in business and China's dynamic market economic system (Wank 1999). At the same time, there is also discussion in the CCP about China's political destiny, and this has led to the study of limited experimentation with local democracy as well as with European-style socialist democracy.

Minority Nationalities

China is a state of many nationalities: fifty-six official ones. Approximately 92 percent of China's total population is composed of *Hanren*, meaning the Han people, an ethnic term that the Chinese have used since the Han dynasty (206 BC–AD 220) to distinguish themselves from non-Han minority nationalities.

A *nationality* is defined in China as a group of people of common origin living in a common area, with a common language, and having a sense of group identity in economic and social organization and behavior (Dreyer 1976). In 1982 in China, there were 67,233,254 people classified as minorities (see table 4.1) who belonged to fifty-five minority nationalities. By 2000, this population had grown to 106.43 million, 8.41 percent of China's total population (National Bureau of Statistics 2002). Fifteen minorities in 1982 had a population of more than 1 million each, and the number of groups with more than 1 million has increased to eighteen since then. Although China's minorities constitute only about 8 percent of the national total population, they inhabit about 60 percent of the nation's territory (see figure 4.6). While minority groups lag economically behind the Han Chinese majority, their significance is far greater than their numbers would indicate. They receive much attention from the Chinese government, which frequently claims that the minorities not only are politically equal to the Han majority but also deserve preferential treatment because of their small populations and less developed economic situations. Efforts have been made by the central government to ensure that the minorities are well represented at various national conferences. Certain national policies, such as family planning, that might hinder improvement in the socioeconomic level of minority nationalities are not enforced as rigorously in minority areas. Yet it is clear that fertility rates among minority women have diminished rapidly in recent years in parallel with those of the Han, although they have not yet declined to the same levels (Du 2000; Wu 1997).

Table 4.1. Population of China's Major Minority Nationalities in 2000

Zhuang	12,178,811
Manchu	10,682,262
Hui	9,816,805
Miao	8,940,116
Uygur	8,399,393
Tujia	8,028,133
Yi	7,762,272
Mongolian	5,813,947
Tibetan	5,416,021
Bouyei	2,971,460
Dong	2,960,293
Yao	2,637,421
Korean	1,923,842
Bai	1,858,063
Hani	1,439,673
Kazak	1,250,458
Li	1,247,814
Dai	1,158,989

Source: National Bureau of Statistics (2002, 44).

Several reasons account for the special treatment given to these minority nationalities. First, most of the minority groups occupy China's border areas, which are strategically important. Several groups, including the Shan, Koreans, Mongols, Uygurs, Yao, and Kazaks, are also found in neighboring Thailand, Burma, Korea, Mongolia, and the Central Asian republics. If they were hostile to the Chinese central government, these groups could significantly weaken China's border defenses and increase the threat of attack by foreign countries. A significant Tibetan exile group, focused around the Dalai Lama, has been based in India since 1959.

Second, most of China's minority regions are sparsely populated relative to the rest of the country but are richly endowed with natural resources. For example, there are extensive oil reserves in Heilongjiang and Xinjiang, tin and copper deposits in Yunnan and Guizhou, and uranium deposits along the Central Asian border in Xinjiang. The majority of China's forest land is also found in border regions, especially in the northeastern and southwestern provinces. Large numbers of livestock are raised in the arid and semiarid northwestern areas, where more than 80 percent of China's wool and animal skins are produced. In addition, certain minority localities have virgin land that can be reclaimed for settlement to alleviate population pressures in the densely populated regions of China. Extensive virgin land in Heilongjiang, Xinjiang, and other places has been developed by Chinese settlers in the past fifty years for agriculture.

Third is the matter of the political image of socialism. A contented, cooperative, and prosperous minority population is living proof of the superiority of socialism and enhances the political image of the central state. On the other hand, minority problems not only tarnish the image of the government but also generate domestic and international crises. China's record toward its minority nationalities, though, is mixed and has not always been clear. Moreover, government policy, despite its polite, upbeat rhetoric, has been ambivalent and at times harsh and repressive, as in Tibet and Xinjiang in the late 1980s and in the very recent past with major riots in Xinjiang in 2009.

Han Expansion

Historically, the Han Chinese were never isolated from the peoples who inhabited the areas surrounding the Chinese cultural realm. Interactions between the Chinese and the non-Han groups were already extensive as early as the Zhou dynasty (ca. 1122–221 BC). Frequent invasions of the Chinese territories by the non-Han peoples along China's northern borders necessitated the construction of defensive walls against these nomads by various states as early as the Zhou period. Although built primarily for defense against nomadic invasions from the north, the Great Wall was not always militarily effective. Its symbolic significance, however, never changed. For the Chinese, it symbolized a line of demarcation separating the steppe from the sown field, pastoral nomadism from sedentary agriculture, and barbarism from civilization (Fairbank 1992).

Chinese civilization spread primarily from the Huang He Basin toward the south, a region that offered the Chinese a favorable agricultural environment. Before the establishment of the first empire of the Qin in 221 BC, most of South China, including the middle Chang Jiang (Yangtze River) region, was inhabited by various Tai peoples, many of whom were displaced, absorbed, or acculturated in later centuries by Han colonizers (Moseley 1973).

The tempo of the southward spread of the Chinese varied considerably. During periods of nomadic invasions from the northern border regions, alien governments were established in North China, and large numbers of Chinese were pushed to the Chang Jiang valley areas. There were, of course, gradual shifts of population to the productive Chang Jiang Basin, where the alluvial soil has always been fertile and the amount of precipitation normally adequate. Further southward expansion took place along the tributary river valleys of the Chang Jiang, notably along the Gan, Xiang, and Yuan rivers. The highly fertile farming region of the Guangzhou (Canton) Delta began to grow rapidly after the eighth century.

With the exception of the Mongols during the Yuan dynasty (AD 1271–1368), virtually all the non-Han peoples who had established alien states on Chinese soil, including the Xianbei, the Jurchen, the Khitans, the Tanguts, and the Manchus, were in the end assimilated by the culturally more advanced and numerically greater Chinese. Over the centuries, intermarriage between the Han and other groups has made the population of China genetically heterogeneous. Genetically, there is no such thing as a Chinese race. Authoritative sources have indicated genetic differences, for example, between northern and southern Chinese populations (Cavalli-Sforza, Menozzi, and Piazza 1994). Alien groups, once they had been sinicized, were readily accepted by the Chinese. The term *Chinese* appears to be a cultural rather than a racial term.

The spread of Chinese culture was confined primarily to areas where agricultural cultivation based on lowland rice was possible, leaving the more hilly, arid, and non-productive border regions largely to non-Han peoples. Although some of the border areas, such as Xinjiang and Tibet, were nominally Chinese provinces, they existed largely as de facto kingdoms, frequently challenging Chinese control until after the communist revolution.

Gaubatz's (1996) description and analysis of Chinese colonization of frontier areas through the establishment and growth of urban garrisons that later grew to full-blown cities provides an insightful interpretation of the growth and spread of the Chinese ecumene and cultural area on the western frontier margins of the national territory. Continued rapid growth of the Han population in these peripheral areas indicates that this process of sinicization of frontier areas remains an active policy of the current government of the PRC.

Autonomous Areas

The special status of minority groups in the PRC is manifested in its administrative system. Parallel to the Chinese provinces are the five leading minority nationality ARs: Uygur, Mongol, Hui (Chinese Muslim), Tibetan, and Zhuang (see figure 4.6). In the ARs where minority groups concentrate, as pointed out earlier, there are second-order administrative units known as autonomous prefectures (*zhou*) and autonomous leagues (*meng*), which are the equivalents of Chinese prefectures. Third-order minority units are autonomous counties and banners (*qi*), which parallel the counties in Han areas. An AR usually carries the name of the largest minority group living there. For example, in Xinjiang Uygur AR, the Uygurs are the largest minority group. Where no single group dominates, an area is named after its two or three leading minorities, such as Jiangzheng Hani-Yi Autonomous County in Yunnan and Haixi Mongol–Tibetan–Kazak Autonomous Zhou in Qinghai. Not all minority nationalities have their own ARs, however, and no AR may be established for the Han, even in places where they are a minority (Schwarz 1971). Table 4.1 provides a list of the largest minority nationalities and their populations according to the 2000 census report.

According to the Chinese government, the purpose of this regional autonomy is to guarantee political equality for national minorities and to give special consideration to the development of minority areas. In these ways, the government indicates, the policies and principles of the CCP can be implemented more effectively. It would seem, then, that minority nationalities are entitled to the right of self-government in their own areas. In reality, however, political administration in such areas is identical to that in Han regions, and the minority nationalities must conform to the rules established by the CCP. Yet, although political autonomy exists in name only, there is a considerable degree of cultural autonomy. Comparatively greater freedom is given to peoples who work to retain their ethnic customs and beliefs and to use their native language to conduct such official business as public meetings and legal proceedings. Although the Chinese language (Mandarin or *putonghua*) is taught everywhere in China, serious efforts have been made by the government to create new scripts or to improve the written languages in minority areas. Of the fifty-six officially recognized nationalities, at least twenty-five now have their own written languages.

Xinjiang Uygur

Xinjiang, meaning "New Territory" in Chinese, used to be known as Chinese Turkestan. With more than 1,600,000 km² of territory, it is the largest among China's provinces and ARs. Aside from Tibet, Xinjiang is the only AR where minority nationalities outnumber the Han Chinese, although Han in-migration has been so rapid that the Han share of the population is about to overtake the Uygur and may soon form an absolute majority. Thirteen nationalities are found in Xinjiang: Uygur, Han, Kazak, Mongol, Hui, Kirgiz, Russian, Uzbek, Xibo, Tajik, Tartar, Daur, and Manchu. In the mid-1960s, there were about 4 million Uygurs and 500,000 Kazaks. By 1982, the Uygur population had grown to almost 6 million, and Kazaks numbered over 900,000. The total population of Xinjiang in 1987 exceeded 14 million. Less than two decades later in 2001, the Uygurs were estimated at approximately 8.4 million and the Kazaks at 1.3 million. The total population of Xinjiang in 2001 was 18.76 million. Han Chinese numbered 7.42 million in 2001 and had increased their share of Xinjiang's total population to 40 percent (National Bureau of Statistics 2002; Starr 2004).

The population of Xinjiang has been increasing steadily as a result of improved health care services. Smallpox has been eliminated, and leprosy, malaria, typhoid, venereal disease, and other infectious diseases have been brought under control. Infant mortality rates and postnatal deaths have been greatly reduced. Birth control is not enforced rigorously in ARs because their populations are generally much smaller than in other parts of China, but family planning guidance and assistance are readily available.

A key factor of population growth in Xinjiang has been the large-scale in-migration of Han Chinese. The majority of these migrants were demobilized troops and their families, young people recently graduated from secondary schools, groups sent out from large cities, and civil administrators. The rate of Han influx was particularly high during 1957–1958. By the end of 1958, some 556,000 Han Chinese had settled in Xinjiang, and, as noted above, the Han numbered over 7 million in the year 2001 and accounted for approximately 40 percent of the total population.

The Production and Construction Corps of the People's Liberation Army formed the basis of Chinese colonization of Xinjiang. The size of the corps increased from 100,000 in the early 1950s to more than 500,000. The troops reclaimed more than 700,000 ha (more than 1.73 million acres) of virgin land in Xinjiang for cultivation. In addition to reclaiming land, the army also developed mines, built railroads and factories, and operated more than one hundred military farms. The in-migration and growth of the Han Chinese population has been remarkable. Many of them are concentrated in industrial regions and cities such as Shihezi at the northern foot of the Tian Shan Mountains (Pannell and Ma 1997).

For many centuries, Chinese or East Turkestan, as Xinjiang was called, was a politically contested and unstable area. More recently, Xinjiang has been a politically sensitive area because of its proximity to the former Soviet Union, now the Central Asian republics. Several minority groups, including the Uygurs, Kazaks, Hui, Uzbeks, Tajiks, and the Kirgiz, inhabit both sides of this international border. Isolated rebellions by the Kazaks here against the Chinese government took place in the late 1950s and the early 1960s when China's relations with the Soviet Union became strained. Since then, Uygur separatists have created serious problems, and there have been bombings and other incidents involving Uygurs (Gladney 1990; Toops 2003) . The terrorist attacks of September 11, 2001, further heightened Chinese anxiety and resulted in the dispatch of increased numbers of Chinese troops to the region to counter the perceived threat of a link between Uygur separatists and forces from outside China, such as any members of al-Qaeda trained in Afghanistan. Chinese concerns were also raised over an increase in the military activity of the U.S. Army and Air Force in Afghanistan and nearby Uzbekistan.

In the summer of 2008 and the autumn of 2009, a series of bombings, riots, and attacks and counterattacks between Uygurs and Han caused major problems in the cities of Kashgar and Urumqi and other places. Some of these events (the bombings) seemed timed to coincide with the 2008 Olympic Games and aimed at embarrassing China during a major public event. The Chinese have placed blame for these outbursts on a terrorist group, the East Turkestan Islamic Front. The more recent events of the autumn of 2009 were blamed initially on labor discrimination against Uygur migrants in faraway Guangdong province. Yet these explosive riots and their death toll among both Uygur and Han tell us that ethnic minority issues in China remain a contentious problem despite the work of the central state to improve the status and condition of minority populations. Perhaps nowhere is this more serious than in the Tibet AR.

Tibet (Xizang)

Until the 1950s, Tibet was largely isolated from other parts of China and experienced only variable Chinese influence. The Tibetans are culturally distinct from other nationalities of China. Prior to the entry of the CCP, Tibet was a traditional theocracy of about 3 million people living in an area approximately twice the size of Texas. Tibetan politics were dominated by a small group of nobles and by the powerful Lamaist monasteries. The best agricultural land belonged to the monasteries, whose members in the upper clergy were drawn from the noble class. Approximately one-third of male Tibetans were monks, and political and religious powers were centralized in the hands of the Dalai Lama, traditionally a king, god, and high priest in one who was believed to be a reincarnation of the Buddhist deity of Avalokitasvara. When a Dalai Lama died, his successor was sought from among the children who were born shortly after his death. The new leader was then identified by a complicated religious ritual. The present Dalai Lama is the fourteenth in succession and, as noted, is in exile in India.

Much publicity was given by the Western press to Tibetan resistance against Chinese control in the 1950s. The relationship between China and Tibet in historic times remains confusing. In Tibet, there were periods of tributary relations with China and periods when China had little or no influence. In the eighteenth and nineteenth centuries, Chinese officials in Lhasa were known as Ambans, or Residents, and were appointed by the Chinese government. The Chinese government regarded the Residents as Chinese governors, a status the Tibetans never accepted. However, Tibetans periodically made gestures of accepting Chinese overlordship, mainly for the sake of furthering their independence. In the waning years of the Qing dynasty, Tibet almost completely broke away from China. After the downfall of the Qing, regional autonomy in Tibet existed for about four decades while China proper was torn by Chinese warlords and Japanese invaders and by the civil war between the nationalists and the communists. The nationalists never had an opportunity to deal effectively with the Tibetan question.

Tibet has never been recognized by any country as an independent state. The Dalai Lama readily acknowledged in 1950 that there were times when Tibet sought, though rarely received, the protection of China. According to the Simla Convention of 1914, which was signed by Tibet and British India and was designed to reduce Chinese involvement in Tibet, China was to recognize Tibetan autonomy, to accept the so-called McMahon Line delimiting the border between Tibet and India, and to station no more than five hundred troops in Tibet. The Tibetan authorities, on the other hand, agreed to a statement to the effect that Tibet was under Chinese suzerainty and that it was part of Chinese territory. In the following years, British India, when dealing with questions involving Tibet, made clear to the Chinese that Tibet was under Chinese suzerainty, although in reality India almost invariably dealt with Tibet as an independent state. China, however, has never accepted the Simla treaty, and some Tibetans have argued that they themselves are not bound by it.

As far as the Chinese are concerned, Tibet is a Chinese territory. All the maps published in China in the past one hundred years showing China's ARs invariably include Tibet, as do all Chinese school textbooks published after the fall of the Qing dynasty. So when the CCP came to power and sent its armies to occupy Tibet in 1950, it saw this not as invading a foreign country but merely as establishing political control over an area traditionally considered Chinese territory. In 1954, India and China signed a treaty that accepted Chinese sovereignty in Tibet.

There were frequent reports in the late 1950s of Tibetans defying Chinese rule. The rapid and forceful introduction of a socialist system in Tibet was accompanied by the simultaneous abolition of traditional Tibetan socioeconomic institutions such as monasticism, serfdom, slavery, and forced labor. These institutions had been deeply rooted in Tibet for centuries. The nobles and monastic orders resisted the new measures by armed rebellion, culminating in the well-known 1959 uprising. In March 1959, Tibetan rebels seized the Tibetan capital of Lhasa for several days, only to be driven out by the Chinese troops stationed there. The Dalai Lama and many of his followers fled southward to India, where he still resides. The Chinese dissolved the Dalai Lama's government and handed over the administration of Tibet to the Preparatory Committee for the Tibetan Autonomous Region, which had been established in 1956. However, autonomy was not formally granted to Tibet until 1965.

Although very little is known about conditions in Tibet after the uprising, much time passed before Tibet became an AR. The granting of autonomy implies that the Chinese government has some degree of confidence in its control in a specific minority area, and the lengthy time that elapsed before the government granted autonomy to Tibet suggests that establishing Chinese rule in Tibet was not an easy matter. Not surprisingly, then, the administration of the Chinese in Tibet in the past three decades has produced mixed results. Although secular schools, health facilities, modern postal services, airlines, roads, industries, newspapers, radio, and so forth have been introduced into a region where none existed and a railroad linking Tibet with Qinghai has recently been completed, the standard of living of the Tibetan people has not improved much. In-migration of Han Chinese has led to considerable resentment among the local people who believe the Han capture the best jobs and control all the local administrative channels of power and authority.

China continues to have a serious problem with Tibet and the Tibetan people. Serious uprisings in 1988 and 1989 were brutally suppressed by public security forces and the army. Not since the Cultural Revolution have such repressive measures been used against Tibetans. These events demonstrate the failure in Tibet of China's minority policies and also suggest that China's long-term approach and solution to minority dissent may be one of simply eliminating minority peoples through forced acculturation and sinicization, the ultimate solution and outcome of Great Han chauvinism. Today, the Tibetan AR represents only about half the territory that traditionally was peopled by Tibetans and traditionally considered Tibetan territory. The remainder has been transferred administratively and incorporated into the territory of the neighboring provinces of Sichuan and Qinghai.

Minorities in South and Southwest China

Of the fifty-five officially recognized minority nationalities in China, twenty-five are found in the provinces of Guangxi, Guangdong, Yunnan, and Guizhou. Two large nationalities, the Zhuang and the Yi, with a combined 2000 population of 20 million, are concentrated in Guangxi and its neighboring provinces, while a substantial number of Miao, Bouyei, and Shui live in Guizhou. At least fifteen small minorities with a total population of more than 3 million are scattered in the hilly regions of Yunnan.

Because of the lack of historical data, it is extremely difficult to clarify the pattern of sequential occupation of China's southern and southwestern border regions by diverse minority groups. Before the establishment of the first Chinese Empire of the Qin dynasty in 221 BC, much of South China, including the middle Chang Jiang valley, was inhabited by various groups of Tai. Subsequently, the Tai moved southward as a result of Chinese expansion into the Chang Jiang valley. Gradually over the centuries, many Tai peoples migrated, primarily in small groups, into Yunnan and various parts of Southeast Asia.

China's largest minority nationality, the Zhuang, are believed to be ethnically a mixture of the Yue people, who were indigenous to the southeast coast of China, and a Tai group. By the time of the Song dynasty at least, the Zhuang had already settled in today's Guangdong and Guangxi areas, and the Tai had also established communities in southern and western Yunnan. The Mongol conquest of South China in the mid-thirteenth century prompted a massive migration of the Tai farther south into the territory of the Khmer Empire of Angkor in Southeast Asia. In the fourteenth century, the Tai founded the kingdoms of Siam and Laos in the valleys of the Chao Phraya and Upper Mekong rivers. Today there are more than 500,000 Tai in China who live mostly in the Xishuangbanna Tai Autonomous Zhou in southern Yunnan and in an autonomous zhou established for the Tai and the Jingpo peoples in western Yunnan.

Large-scale settlement by the Han in Yunnan and Guizhou did not begin until the Ming dynasty (AD 1368–1644) when the Chinese government encouraged the people in the Chang Jiang valley to migrate southward. Displacement of indigenous ethnic populations by the Han took place during the course of this southward migration. Many ethnic groups were forced into remote hilly regions that were much less attractive to the Han, who preferred a natural environment suitable for wet-rice paddy farming. In the lowlands where rice cultivation was possible, the indigenous peoples, such as the Zhuang and the Yi, were largely assimilated rather than displaced by the Han.

The consolidation of the southern border provinces by the CCP in the early 1950s was largely a peaceful act. Since then, some industrial enterprises have been introduced, and the region's traditional slash-and-burn agriculture has become a special target for reform. Increasingly, the Chinese way of farming—characterized by intensive cultivation, multiple cropping, vegetable gardening, and systematic application of organic manures—is replacing traditional shifting agriculture. As a result, the productivity of the land in the Zhuang areas in Guangxi has greatly increased, contributing directly to a higher standard of living.

Efforts at National Integration

China's minority nationalities have enjoyed considerable cultural freedom and experienced significant changes in their traditional ways of life. The ultimate goal of the

Chinese ruling authorities appears to be the integration of the minorities into the national socioeconomic and political systems. Several methods of integration have been used by the government. One is the settlement of millions of Han Chinese in minority areas where they then help develop the local resources and bring about a firmer Chinese control. Efforts at integration also include the recruitment of young minority people who are sent to centers of higher learning for education. After graduation, they are sent back to their homelands to assume leading positions. This policy has had very little success, though, with many of the minority nationalities, such as the Tibetans.

The development of minorities' written languages based on the pinyin system may be taken as another method of integration. Since all languages used in China employ the same alphabet for spelling, communication among ethnic groups and with the Chinese has become much easier. The introduction of written languages to minority groups and the publication of books and other materials, of course, greatly facilitate the spread of official doctrines.

Our lack of information does not permit us to estimate with confidence the extent to which minority nationalities have been successfully integrated into the Chinese socialist system and to say whether the standard of living of the Chinese minorities today is better or worse than it was prior to 1949. Few objective and in-depth studies, historical or contemporary, exist on these groups. And the Chinese press invariably reports only on the positive accomplishments made in the minority areas. Western journalists who have recently visited such areas tend to be extremely critical of Chinese policies and performance there, especially when reporting on Tibet and to a lesser extent Xinjiang. One Western journalist has reported that there is much mutual distrust between the Chinese and the Tibetans, that the Tibetans resent Chinese incursions into their religious life, that they long for the return of the Dalai Lama, and that there is poverty, starvation, and political imprisonment in Tibet. The 1988–1989 uprising as reported by Avedon (1989) and others offers very strong evidence that the Chinese goal of integrating the Tibetans into the Chinese system in the past forty years has failed.

Although it appears that the Tibetans, historically a very independent people with a strong and unique religious tradition, have passively and actively resisted CCP rule since the 1959 uprising, we should not automatically infer that the Chinese are having similar difficulties in other minority areas. Poverty is not a problem unique only to minority nationalities; it is, by Western standards, common in many parts of China. There are also indications that many local problems are not always known to the central government in Beijing and that no large-scale funds have been specially infused into the minority economies.

Today, it remains apparent that the Chinese government is unwilling to grant full political autonomy to its minorities; most of the top political positions in the ARs, for example, are held by the Chinese. Religious worship is now again permitted after it was violently denounced and severely prohibited during the Cultural Revolution (1966–1969) and its aftermath in the early 1970s; nevertheless, active and structured religious propagation is still not acceptable anywhere in China except under careful state control. Such policies and restrictions have made China's minorities unhappy. But with only about 8 percent of the nation's population and inhabiting largely environmentally marginal areas that require more financial input than elsewhere to obtain the same developmental results, the minorities are in a poor position to dispute and contest the policies of the central state that they oppose. However, with the possible exception of the Tibetans and Uygurs, all of China's minority nationalities, including the Zhuang, the Hui, and the Yi—which together make up the largest minority groups in China—appear to have at least passively accepted communist rule. Whether their full integration with the Chinese can be achieved in the future remains to be seen.

Questions for Discussion

1. China traditionally for most of its history has been focused on its internal development. How do you view its contemporary evolving self-image of its position as an emerging global power and member of the community of nations?
2. Do you see China today as an empire or as a nation-state or perhaps as some kind of political hybrid model that has merged the two and is evolving something new?
3. Is China's growing military power likely to lead to more confrontations with its neighbors and the United States, or do you foresee a future of increasing cooperation and mutual benefit for all sides?
4. What in your view are among the most significant accomplishments that the CCP has provided for the citizens of China during the past half century?
5. China has historically used a process of assimilation and sinification of frontier peoples for expanding the central state and its territory. Do you think this is a sustainable and workable model today for dealing with ethnic minority peoples such as Tibetans and Uygurs in western China?
6. Who are the Han Chinese, and what is the derivation of this term that the Chinese people use to describe themselves? What is "Great Han chauvinism," and do you consider it a problem in how the Chinese central state deals with minority nationalities?

Note

1. The Hui (Muslim) and Manchu nationalities are Mandarin speaking and are closely associated with the Han people. Together, these two groups accounted for about 11.5 million in 1982, about 17 percent of China's minority population. By 2000, the Hui population had grown to 9.8 million, and the Manchu had increased to 10.7 million (National Bureau of Statistics 2002). Such remarkable growth is accounted for in part by reclassification of ethnicity done through individual self-identification that was permitted in the 1990 census count.

References Cited

Amako, Satoshi. 1997. Asia since the cold war and the new international order: The historical perspective and future prospects. In *China in the Twentieth Century: Politics, Economy, and Society*, ed. Fumio Itoh. Tokyo: United Nations University Press, 159–67.

Avedon, John. 1989. Tibet today. *Utne Reader*, March/April, 34–41.

Basic Law of the Hong Kong Special Administrative Region of the People's Republic of China. 1991. Hong Kong: Joint Publishing.

Cavalli-Sforza, L. Luca, Paolo Menozzi, and Alberto Piazza. 1994. *The History and Geography of Human Genes.* Princeton, NJ: Princeton University Press.

Chan, Kam Wing. 2010. Fundamentals of China's urbanization and policy. *The China Review* 10: 63–94.

China through a Lens. 2004. Foreign policy. August 5. http://www1.china.org.cn/English/features/38193.htm.

Dreyer, June T. 1976. *China's Forty Million.* Cambridge, MA: Harvard University Press.

Du, Peng. 2000. The ethnic minority population in China. In *The Changing Population of China*, ed. Xizhe Peng and Zhigang Guo. Oxford: Blackwell, 207–15.

Fairbank, John K. 1992. *China: A New History.* Cambridge, MA: Belknap.

Gaubatz, Piper R. 1996. *Beyond the Great Wall: Urban Form and Transformation on the Chinese Frontiers.* Stanford, CA: Stanford University Press.

Ginsburg, Norton. 1968. On the Chinese perception of a world order. In *China in Crisis*, vol. 2, ed. Tang Tsou. Chicago: University of Chicago Press, 73–91.

Gladney, Dru C. 1990. Ethnogenesis of the Uighur. *Central Asian Survey* 9, no. 1: 1–28.

Goldstein, Avery. 2001. The diplomatic face of China's grand strategy: A rising power's emerging choice. *China Quarterly* 168: 835–64.

Harding, Harry. 1993. The concept of "Greater China": Themes, variations, and reservations. *China Quarterly* 136: 660–85.

Hsiao, Kung-chuan. 1960. *Rural China: Imperial Control in the Nineteenth Century.* Seattle: University of Washington Press.

Jacques, Martin. 2009. *When China Rules the World: The End of the Western World and the Birth of a New World Order.* New York: Penguin.

Lampton, David M. 2008. *The Three Faces of Chinese Power: Might, Money and Minds.* Berkeley: University of California Press.

Liang, Zhang, comp. 2001. *The Tiananmen Papers*, ed. Andrew J. Nathan and Perry Link. New York: Public Affairs Press.

Ma, Laurence J. C., and Gonghao Cui. 1987. Administrative changes and urban population in China. *Annals of the Association of American Geographers* 77, no. 3: 373–95.

Moseley, George V. H. 1973. *The Consolidation of the South China Frontier.* Berkeley: University of California Press.

National Bureau of Statistics. 2002. *Zhongguo tongji nianjian 2002* [China statistical yearbook 2002]. Beijing: China Statistics Press.

Office of the U.S. Secretary of Defense. 2010. Military and security developments involving the People's Republic of China. http://www.defense.gov/pub/pdfs/2010_CMPR_Final.pdf.

Pannell, C. W., and L. J. C. Ma. 1997. Urban transition and interstate relations in a dynamic post-Soviet borderland: The Xinjiang Uygur Autonomous Region of China. *Post Soviet Geography and Economics* 38, no. 4: 206–29.

Samuels, Marwyn S. 1982. *Contest for the South China Sea.* New York: Methuen.

Schwarz, Henry G. 1971. *Chinese Policies towards Minorities.* Occasional paper no. 2. Western Washington State University, Program in East Asian Studies.

Shambaugh, David, ed. 2000. *The Modern Chinese State.* Cambridge: Cambridge University Press.

Starr, S. Frederick, ed. 2004. *Xinjiang: China's Muslim Borderland.* Armonk, NY: Sharpe.

Toops, Stanley. 2003. Xinjiang (eastern Turkistan): Names, regions, landscapes, and futures. In *Changing China: A Geographic Appraisal*, ed. Chiao-min Hsieh and Max Lu. Boulder, CO: Westview, 411–23.

Vogel, Ezra, ed. 1997. *Living with China: U.S.-China Relations in the Twenty-First Century.* New York: Norton.

Wank, David. 1999. *Commodifying Communism: Business, Trust, and Politics in a Chinese City.* Cambridge: Cambridge University Press.

Wheatley, Paul. 1971. *The Pivot of the Four Quarters: A Preliminary Enquiry into the Origins of the Ancient Chinese City.* Chicago: Aldine.

Whitney, Joseph B. R. 1970. *China: Area Administration and Nation Building.* Research paper no. 123. Chicago: University of Chicago, Department of Geography.

Wiens, Herold J. 1967. *Han Chinese Expansion in South China.* Hamden, CT: Shoe String Press.

Wu, Cangping, ed. 1997. *General Report on China's Changing Population and Its Development.* Beijing: Higher Education Press.

Xinhua News Agency. 2002. Sixteenth National Congress of the Communist Party of China, 2002. November 18. http://www.china.org.cn/English/features.49109.htm.

CHAPTER 5

Population and Human Resources

People and human resources are the centerpiece and enduring reality that shape the images and perceptions most of us have of both contemporary and historic China. No theme is more compelling or profound of the mental portrait we hold of China than its people, their remarkable number, and their variation and distribution across its vast and changing landscape. The superlative of the world's largest population in a single country kindles a bright image of a boundless and industrious people who have created an enormous and growing consumer market as well as a powerful economic engine in manufactures and exports. At the same time, the faces of impoverished peasants in remote rural areas and scenes of millions of disenfranchised migrant laborers in the eastern cities remind us that China's huge population remains one of its most daunting challenges.

Throughout this chapter, we continue to examine the notion introduced at the beginning of the book that China's population and people are its greatest resource and treasure although representing perhaps its greatest question and problem as well. The latter issue focuses on labor absorption, decline in per capita arable land, future food grain production, growing stress on environmental resources such as water, insufficient educational investment and resources, and the need for improving the general educational level. The rapidly changing demographic profile of a population that is now growing older—and with a declining share of women—raises very significant social and economic issues that must be addressed in the future. Such changes continue to indicate that China's huge and still-growing population remains one of the most fundamental issues that confront and challenge its drive to develop and modernize.

About 21 percent of the human population lives within the borders of the People's Republic of China (PRC). The State Statistical Bureau (now the National Bureau of Statistics) of the Chinese government has officially reported, based on 2000 census results, that China's population in 2000 was 1,265,830,000 (excluding Hong Kong, Macao, and Taiwan; see table 5.1), an increase of approximately 137 million people, or 12 percent more than the 1990 population. The rate of net natural increase (the annual death rate subtracted from the annual birthrate) for 2000 was 7.58 per thousand, or 0.76 percent, and this dropped to 6.01 per thousand in 2003 (National Bureau of Statistics 2004). This 2000 census count continued the accurate enumeration of the Chinese

Table 5.1.   Demographic and Urban Trends in China, 1952–2008

Year	Total Population (millions)	Birthrate (per 1,000)	Growth Rate (net natural increase per 1,000)	Urban Population (millions)	Urban Share (%)
1953	587.96	37.00	23.00	78.26	13.31
1958	659.94	29.22	17.24	107.21	16.25
1964	704.99	39.14	27.64	129.50	18.37
1969	806.71	34.11	26.08	141.17	17.50
1975	924.20	23.01	15.69	160.30	17.34
1978	962.59	18.25	12.00	172.45	17.92
1980	987.05	18.21	11.87	191.40	19.39
1982	1,016.54	22.28	15.68	214.80	21.13
1985	1,058.51	21.04	14.26	250.94	23.71
1990	1,143.33	21.06	14.39	301.95	26.41
1996	1,123.89	16.98	10.42	373.04	30.48
2000	1,265.83	14.03	7.58	459.06	36.22
2005	1,307.56	12.40	5.89	562.12	42.99
2008	1,328.02	12.14	5.08	606.67	45.68

Sources: National Bureau of Statistics (2001, 2009).

population established in the 1982 midyear census. Together with the census of 1990, these three census counts are probably the most reliable estimates and counts ever taken of the Chinese people. In 2008, the population was reported as 1,328,020,000 (National Bureau of Statistics 2009). A new census was undertaken in late 2010. The new census results will likely be published in late 2011.

Counting China's People and Household Registration

For many centuries, China has employed various mechanisms to count its households and guarantee the security and behavior of its people. The socialist period brought with it an enhanced administrative capacity and witnessed an intensification of this population checking and control with the introduction of household registration books shortly after the Chinese Communist Party (CCP) took control of China in 1949. In 1958, following a decade of substantial rural-to-urban migration, a household registration (*hukou*) system was instituted through new state regulations that clearly and sharply delineated the population into permanent urban and rural components based on location, employment, and birth. Thereafter, being born a peasant farmer, one was almost surely destined to remain a peasant, and individual mobility and the freedom to move were sharply curtailed until the 1990s. The *hukou* system of household registration thus until recently provided a means of checking and controlling China's population as well as counting it. Inasmuch as there was close agreement between the 1982 census figures and the figures

Table 5.2a.   China: Estimated Population, Western Han Dynasty–People's Republic of China, AD 2 to 1949

Period	Year	Population
Western Han dynasty	2	59,594,978
Eastern Han dynasty	156	50,066,856
Three Kingdoms dynasty	220–280	7,672,881
Western Jin dynasty	280	16,163,863
Sui dynasty	606	46,019,956
Tang dynasty	742	48,909,809
Song dynasty	1110	46,734,784
Yuan dynasty	1290	58,834,711
Ming dynasty	1393	60,545,812
Qing dynasty	1757	190,348,328
Qing dynasty	1901	426,447,325
Republic of China	1928	474,787,386
People's Republic of China	1949	548,770,000

Source: Zhou (1979).

derived from the official household registration system, the *hukou* system, whatever its alleged shortcomings of over- or undercounting particular groups, appears to have been an accurate and effective means of keeping up with the size of China's population.

Whether such a system continues to be accurate in the face of the far-reaching economic reforms that resulted in relaxing the rules on rural migration in China and allowed for a much increased mobility of the rural population remains to be seen. While the 2000 census continues to be built on the earlier model of a close conjunction between enumeration and household registration, its validity is subject to increasing question. The location of people counted through the *hukou* system, as well as recent births and deaths among transients, makes the system less than fully reliable as a population enumeration method, given the increased mobility of the population during the past fifteen years and especially during the 1990s. It has been estimated that there is a so-called floating population of as many as 150 million people living in China's urban centers (Chan 2009, 2010).

China's official population data, both historical and contemporary, should be used with caution, however, because of incomplete coverage. Historically, population data were never collected for the sake of objective knowledge or national economic planning. Population registrations were carried out mainly for the purposes of taxation or conscription of able-bodied males for military corvée duties. To evade registration to avoid paying taxes or the draft was a common practice. Females and children, considered socially inferior and physically unfit for labor services, were largely unreported in dynastic times (Ho 1959). In the early twentieth century, census counts were not sophisticated enough for meaningful analysis of population changes. And until midcentury, data on such demographic variables as age-specific birth and death rates, migration rates, regional demographic characteristics, rural and urban differentials, and sex and age compositions were largely lacking. The figures cited in this book, especially historical figures on China's population, are meant to suggest rough orders of magnitude rather than precise demographic realities.

Historical Growth Patterns

Historically, China's population varied greatly (see tables 5.2a and 5.2b), and large discrepancies may have existed between official population data recorded in historical documents and the demographic truths. In addition to the reasons for the unreliability of data mentioned above, another important factor contributing to the varied historical figures was China's changing national boundaries in different dynastic periods following the rise and fall of the nation's political and military strength.

The earliest national population figures date back to the ancient Han period. In AD 2, the Han dynasty reportedly had a population of about 60 million. For the next sixteen centuries, there were drastic population fluctuations both temporally and spatially. Population tended to increase steadily during the more peaceful years and to decline in times of war or national disaster, in part also related to the expansion or contraction of the state area and the population under its control.

Chao (1986) explained the growth of China's population based on two main factors. Ever since the Zhou dynasty (1122–221 BC), China had maintained a traditional culture based on the Confucian values of a male-dominated society. Ideally, virtually

Table 5.2b. Official Data and Estimates of China's Population, Selected Years, AD 2 to 1848 (Millions)

Year	High	Low
2	59.6	
57		(31.0)
105	53.2	
156	56.4	
280		16.2
606	46.0	
705		37.1
755	52.9	
961		(32.0)
1109	(121.0)	
1193	(120.0)	
1381		59.8
1391		60.5
1592	(200.0)	
1657		70.2
1776	268.2	
1800	295.2	
1848	426.7	

Source: Chao (1986, 41).

all males married and were obligated to produce families with several sons to carry on the family name and to perform rituals associated with the veneration of ancestors. The families in turn were obligated to assist each male in forming his own family, even when resources were marginal, and to help ensure the economic viability or survival of the additional family unit.

Next, population growth and decline tended to parallel the cyclical pattern of dynastic rise and fall although not necessarily in the same sequence. Chao (1986, 30–31) estimated that under stable conditions and with an average life expectancy of thirty, the historical population growth rates prior to the Ming dynasty (AD 1368–1644) were probably between 0.5 and 1.0 percent per annum. He suggested that periods of stable growth were typically offset by extended periods of warfare and instability that saw substantial fluctuations in the population and probably at least until the fifteenth century kept the population more or less near or below the number it reached during the Han period (approximately 60 million; see table 5.2b).

Mallory (1926), drawing on a Chinese study, noted that between 108 BC and AD 1911 "there occurred 1,828 famines somewhere in China." This averages out to almost one famine a year in China since Han times. Mallory described and analyzed a number of factors, both environmental and human related, to account for these famines. One may conclude from Mallory's analysis that famine was historically, to a greater or lesser degree, a normal part of the mobility and demographic process in China.

The major exception to this may have been achieved during the Song dynasty (AD 960–1279), a period of substantial economic change, extensive trade expansion, and city growth. Although Chinese national territory was split between the Southern and Northern Song along the Huai River line after AD 1120, the combined population of the two empires may have been as great as 120 million, according to Chao (1986, 36). Thereafter, there was a period of decline during the Mongol invasions and conquest until the establishment of the Ming in 1368. By the end of the fourteenth century, peace and order had been restored, and an unprecedented four centuries of population growth began. During some periods, the annual rate of population increase apparently exceeded 1 percent.

The most rapid growth of population took place during the period from 1657 to 1848 when the population size jumped from perhaps as low as 70 million to 427 million (see table 5.2b). This growth was particularly noticeable during the two centuries from the mid-1600s to the mid-1800s under the benevolent despotism of the early Qing (Manchu) rulers when stable political and favorable economic conditions generally prevailed (Ho 1959). The national economy reached a new stage of development. The dissemination and widespread use of such newly introduced food crops as early ripening rice, "Irish" potatoes, corn, and sweet potatoes greatly enlarged the nation's capability to produce food. More and more new areas previously unsuitable for agriculture were now brought under cultivation. Perkins (1969) has described the sixteenth- and seventeenth-century arrival of potatoes, corn, and peanuts from the New World. While these were important because they could be grown on upland dry areas previously marginal or little used, Perkins noted the Chinese disinclination to eat potatoes based on taste preference and the modest use of corn prior to the twentieth century. Corn thereafter became more popular, especially as an animal feed, due to its widespread growth in the Northeast, a region little settled by Han Chinese until the twentieth century. Therefore, most improvements in food production likely resulted from an intensified use of improved varieties of rice and better cultivation practices within the traditional horticultural framework. Finally, the trend of population growth was again checked during the last century of the Manchu rule when a series of rebellions, the White Lotus (1795–1804), the Taiping (1850–1864), and the Nian (1853–1868), coupled with widespread floods, droughts, and famines, resulted in millions of deaths. Thus, when the Manchu (Qing) Empire ended in 1911, the population was probably around 400 million.

As Perkins (1969), Elvin (1973), and Chao (1986) have discussed, the relationship between food production and population growth after the Ming dynasty was complex. Environmental, cultural, economic, and political factors all came into play in a functionally and spatially variegated web of interrelated strands whose individual causal determinants are virtually impossible to disentangle from one another. Population and growth in food-grain output thus grew together historically but in an erratic, complex, and multifaceted manner, some aspects of which are discussed further in chapter 8.

Prior to the establishment of the PRC in 1949, the Nationalist government made several attempts to come up with a realistic population count. Many government institutions, including the Post Office Department, the Ministry of Internal Affairs, the Bureau of the Budget, and the Maritime Customs Office, carried out enumerations. But political instability, caused by internal warlord politics and the Japanese invasion, hindered them from satisfactorily implementing a registration system. There were conflicting reports regarding the actual size of the population. It is generally believed, however, that in the 1930s and the 1940s, the population of China was in the neighborhood of 475 million.

Population Distribution

The present pattern of population distribution in China is a consequence of expansion and migration by the Chinese in historical and recent times. Today, the vast majority of the Chinese people are concentrated in the eastern half of the country in several areas where the environment offers the necessary conditions for development of intensive agriculture, relative ease of transportation, and the growth of cities. Within the traditionally densely settled East, four major regions have especially dense populations: the North China Plain (primarily Beijing and Tianjin municipalities and Hebei, Shandong, Henan, and the southern Liaoning provinces), the middle and lower Chang Jiang (Yangtze River) Basin and Delta (Hubei, Hunan, Jiangxi, Anhui, Jiangsu, Zhejiang, and Shanghai), the Sichuan Basin, and the Xi Jiang (West River) and Zhu Jiang (Pearl River) basins and Delta (Guangdong and Guangxi; see figure 5.1 and table 5.3). In addition to relatively abundant precipitation and level topography, these floodplain areas have fertile alluvial soils produced by the river systems. In the South, the Chang,

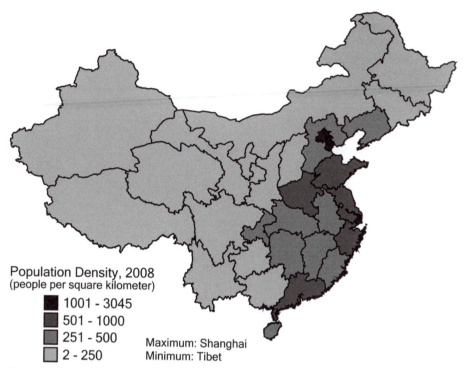

Figure 5.1.   China's population density, 2008 (calculated from National Bureau of Statistics 2009)

Zhu, and Xi and their numerous tributaries are navigable. These rivers have greatly facilitated movement and the development of trade. Most of China's major cities are found in these four fertile and productive regions.

Secondary population concentrations of somewhat lower density are found throughout the rugged uplands of much of South and Southwest China; in the remainder of the Songliao Plain of the Northeast provinces of Liaoning, Jilin, and Heilongjiang, which were settled later and where colder climate and extensive marshlands discouraged higher population concentrations; along the Hexi corridor in Gansu province; and in the oases in the Xinjiang Uygur Autonomous Region (AR; see figure 5.1), Xizang (Tibet) AR, Inner Mongolia AR, Qinghai, and other areas of Xinjiang and Heilongjiang, where minority nationalities reside. Insufficient rainfall, extremes of temperature, and poor soils have limited agricultural development and production in much of western China. In such areas, pastoral nomadism dominates.

Population growth during the period of Chinese socialism was highest in percentage terms in some of the sparsely settled interior provinces and regions such as Xinjiang. During the reform period, however, migration has been toward the coastal provinces, as rural people have left the farms in search of employment and improved opportunities in the cities and economically dynamic areas of the coastal regions with their expanded and growing ties to the global economy (Fan 2005; see fig 5.2).

Table 5.3. Total Population and Birthrate, Death Rate, and Natural Growth Rate in China by Region, 2009

Region	Year-End Population (10,000 persons)	Birthrate (%)	Death Rate (%)	Natural Growth Rate (%)
Beijing	1,755	8.06	4.56	3.5
Tianjin	1,228	8.3	5.7	2.6
Hebei	7,034	12.93	6.43	6.5
Shanxi	3,427	10.87	5.98	4.89
Inner Mongolia	2,422	9.57	5.61	3.96
Liaoning	4,319	6.06	5.09	0.97
Jilin	2,740	6.69	4.74	1.95
Heilongjiang	3,826	7.48	5.42	2.06
Shanghai	1,921	8.64	5.94	2.7
Jiangsu	7,725	9.55	6.99	2.56
Zhejiang	5,180	10.22	5.59	4.63
Anhui	6,131	13.07	6.6	6.47
Fujian	3,627	12.2	6	6.2
Jiangxi	4,432	13.87	5.98	7.89
Shandong	9,470	11.7	6.08	5.62
Henan	9,487	11.45	6.46	4.99
Hubei	5,720	9.48	6	3.48
Hunan	6,406	13.05	6.94	6.11
Guangdong	9,638	11.78	4.52	7.26
Guangxi	4,856	14.17	5.64	8.53
Hainan	864	14.66	5.7	8.96
Chongqing	2,859	9.9	6.2	3.7
Sichuan	8,185	9.15	6.43	2.72
Guizhou	3,798	13.65	6.69	6.96
Yunnan	4,571	12.53	6.45	6.08
Tibet	290	15.31	5.07	10.24
Shaanxi	3,772	10.24	6.24	4
Gansu	2,635	13.32	6.71	6.61
Qinghai	557	14.51	6.19	8.32
Ningxia	625	14.38	4.7	9.68
Xinjiang	2,159	15.99	5.43	10.56
National Total	133,474	12.13	7.08	5.05

Source: National Bureau of Statistics (2010).

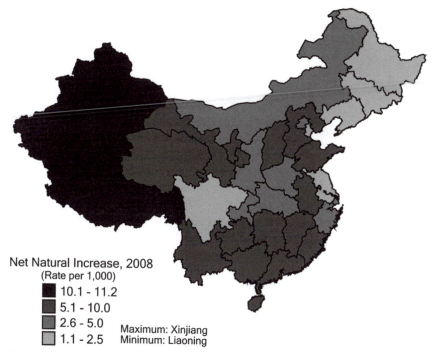

Net Natural Increase, 2008
(Rate per 1,000)
■ 10.1 - 11.2
■ 5.1 - 10.0
□ 2.6 - 5.0   Maximum: Xinjiang
□ 1.1 - 2.5   Minimum: Liaoning

Figure 5.2. China's natural population growth rate, 2008 (derived from National Bureau of Statistics 2009)

Growth in the interior may reflect patterns of internal migration of Han Chinese in an effort to populate remote border areas as well as higher birthrates in recent years among minority nationalities (see table 5.3). It is unlikely, however, to result in any fundamental change in the pattern of a densely settled eastern China and a sparsely populated interior.

Hukou Policy and Migration

Since 1983 and 1984, new laws that allow rural people to migrate to cities to work as service, factory, and construction laborers, although without the subsidies normally afforded state-supported urban workers, have further added to the densities of people in the eastern provinces where most of China's large cities are located. Numerous and variable estimates of the number and location of these temporary migrants have been proposed, with estimates as great as 150 million in total by mid-2009 with large cities such as Beijing and Shanghai having as many as 3 million each (Chan 2010). As temporary migrants, these workers are not counted as permanent urban residents, and most retain their rural *hukou* status. Recent policy changes are now altering the rigid *hukou* system of the past to allow rural residents to establish residency status in towns and small cities.

Clearly, the rules are changing to reflect the new realities of migrants who perform valuable and needed labor services to allow them to become legal residents where they work. It is becoming abundantly clear that the urban economies in many of China's larger cities and metropolitan regions could not function without the labor services of these migrants. Moreover, urban subsidies are diminishing in importance owing to efforts to improve the efficiency and profitability of state-owned enterprises (SOEs) or by simply allowing those that are inefficient to fail and close. Increasingly, these transient workers are becoming a part of the urban scene, and their temporary status is now evolving over time into a more long-term arrangement as befits their contribution to these burgeoning Chinese urban economies. More discussion on migration to cities and urban areas follows in chapter 10.

Population Size and Vital Rates

When the CCP came to power, it decided to conduct a national census to obtain information indispensable for the preparation of its first Five-Year Plan (1953–1957). The information sought was quite simple: the address of each Chinese household and the name, age, sex, ethnic status, and relationship to the head of the household of each household member. According to the census results, China in mid-1953 had a total of 582.6 million people, plus 11.7 million on Taiwan. Since this census was the first official population count China has made in the socialist period, it has been used widely as a basis for population estimates and projections and was a basis for more recent census counts taken in 1964, 1982, 1990, and 2000 (Banister 1987).

There are indications that even the Chinese leadership was unclear about the actual size of the population. In the 1970s, top-level Chinese leaders and other informed sources used national figures that differed by as much as 100 million. For example, officials at the Supply and Grain departments believed that China's population was 800 million, while the Ministry of Commerce stated that it was 830 million. In 1972, the head of the Chinese delegation to a UN environmental conference in Stockholm said that China had as many as 900 million people, a figure that was then widely used in the Chinese press until mid-1979, when the figure of 975 million (1978 population) was released by the State Statistical Bureau. Recent official information from China indicates that the rate of population growth in the 1950s and the 1960s averaged 2 percent per year.

Since the early 1970s, stringent policies of family planning have been enforced with substantial results. Vice-premier Chen Muhua disclosed in late 1979 that the rate of China's natural population growth had dropped from 23.4 per thousand (2.34 percent) in 1971 to 12.05 per thousand (1.2 percent) in 1978 (Banister 1984). Such a drastic reduction of the net natural increase in a few years is extremely rare in the history of the world's population evolution, outside of catastrophes such as massive famines. As seen in table 5.1 and figure 5.3, the rate had increased by 1990 to 14.26 per thousand, as some relaxation of the one-child policy occurred in rural areas. The Chinese government has since called for a further reduction to zero growth by the

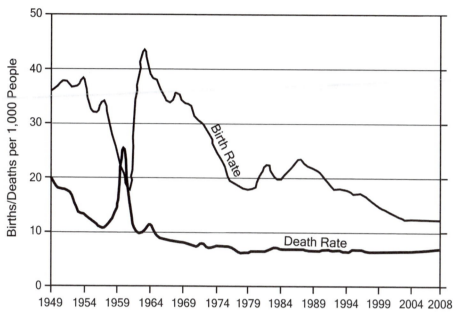

Figure 5.3. Birthrates and death rates in China, 1949–2008 (derived from National Bureau of Statistics 2009)

early twenty-first century, and a decline in fertility rates and birthrates has continued to reduce the natural growth rate. By 2000, the birthrate had declined to 1.3 percent with a rate of annual net natural growth of 0.69 percent, a figure that indicates continuing decline toward zero population growth. As seen in table 5.3 and figure 5.4, the birthrates continue to vary in different regions of China, with substantially higher rates in the western and interior provinces than in the east coast provinces. In 1999, the average fertility rate, based on a less than 1 percent sample, was 1.33 for the nation, well below the replacement level. A 2001 survey indicated a total fertility rate of 1.22 children per woman in urban areas, while the total fertility rate on average among rural women was 1.98. Thus, there is a significant gap between urban and rural families. The total fertility rate in 2008 was estimated at 1.8 children per woman, slightly below the replacement level for the whole population, although it has risen slightly over the past five years.

This evidence indicates China has gone through much of its demographic transition in a very short period, although whether such progress is sustainable in the absence of firm and coercive central policy remains problematic. A developing country experiencing the process of demographic transition or shift can often be seen to go through three stages of change, beginning with relatively high birthrates and death rates (say, 30 to 50 per thousand people per year) and ending with relatively low vital rates (5 to 15 per thousand per year; see fig. 5.3). The intermediate stage is a period of rapid population growth resulting from declining death rates and continued high birthrates. The rapid population growth in China resulted from good and sufficient

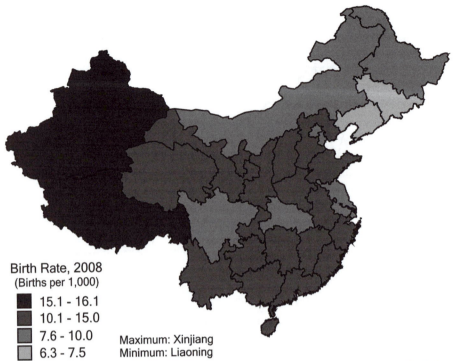

Birth Rate, 2008
(Births per 1,000)

- ■ 15.1 - 16.1
- ■ 10.1 - 15.0
- ■ 7.6 - 10.0
- □ 6.3 - 7.5

Maximum: Xinjiang
Minimum: Liaoning

Figure 5.4.   China's birthrate, 2008 (derived from National Bureau of Statistics 2009)

nutrition, improved health conditions, and better environmental sanitation, all of which help to reduce mortality. Meanwhile, the birth and fertility rates declined more slowly because of the persistence of traditional cultural values associated with Confucian ideals in favor of male children and often resulting in larger families, especially in rural areas, where the vast majority of China's people have always lived. Fertility reduction is necessarily a slow process, for it involves not only attitudinal changes toward the role of the family in the social fabric of the nation but also an awareness and adoption of new ideas and techniques of family planning. This parallels improvement in living standards, income, and the educational level of women. Such a reduction proceeds more rapidly and effectively in cities than in rural areas. China has been able to enforce compliance with family planning policies more easily in cities where people depend more on state support for housing, jobs, and other benefits.

Both birthrates and death rates in China have decreased appreciably since the 1949 revolution. In 1953, the census reported an impressively reduced crude death rate of 17 per thousand and a high crude birthrate of 37 per thousand, leaving a crude natural rate of increase of 20 per thousand. According to official data, between 1954 and 1957, death rates also dropped more rapidly than birthrates, resulting in large natural rates of growth. (Selected data for annual birthrates, death rates, and rates of net annual growth are presented in tables 5.1 and 5.3.)

Aside from the issue of food supply and nutrition, China's mortality and fertility declines are the result of two national programs focused on public health and family planning. The accomplishments of the early years of the PRC in public health are well known. In a relatively short time, China eliminated or greatly reduced such dreaded diseases as typhoid, smallpox, cholera, scarlet fever, tuberculosis, trachoma, and venereal disease. Parasitic diseases such as schistosomiasis, hookworm, and malaria were brought under control and today pose no serious threat to the population. These accomplishments were achieved through massive inoculation drives, pest-eradication programs, environmental cleanup campaigns, and preventive medicine. Public health programs today are carried out in every province, city, and county in China.

One of the saddest and most regrettable aspects of China's recent demographic history was the catastrophic human suffering that resulted from the late Chairman Mao Zedong's Great Leap Forward policy and program. In 1958, Mao announced a program to push China ahead forcefully through full communization of agricultural lands and property and other radical production and social policies. The peasants reacted unfavorably to these new policies that took private plots and animals away from them, and farm production, including that of food grains, plummeted in 1959 and 1960. Bad weather and inaccurate reporting of grain production further exacerbated the production declines. Food-grain shortages followed, and a massive famine swept the country (it was especially serious in 1960) (Ashton et al. 1984). So severe was this famine, coupled with the government's unwillingness to publicize the problem and seek help internationally, that as many as 35 million people may have died as a direct result of the government's failed policies and programs (Dikotter 2010). The tremendous demographic impact and suffering of this famine may be seen in decreased birthrates and increased death rates (see table 5.3 and figs. 5.3 and 5.5), with their devastating effect on the age cohorts born in 1960 and 1961. Only after the radical policies were rescinded and small private plots and animals were restored to the peasants did farm production increase and return some normality to both the rural economy and the way of life of China's rural people.

At an approximate average rate of natural growth of 20 per thousand between 1949 and 1982, China in those thirty-three years had a net gain in population of roughly 480 million (see table 5.1). Another 126 million were added between 1982 and 1990 and still another 137 million between 1990 and 2000 (Pannell 2003; Tien 1983). China continues to add about 7 million each year to its population, despite a reduction in birthrates owing to its enormous population. Despite recently declining birthrates, the large number of women in the childbearing years of fifteen to forty-five will ensure substantial population momentum and growth, but the size of the cohort is diminishing rapidly as relatively low birthrates have been in place now for a generation (see fig. 5.5). So rapid is the reported drop in birthrates and fertility rates that some observers now foresee a new potential problem in another generation of a rapidly aging population (Kinsella and Phillips 2005).

Family Planning

China expects to reach zero population growth at some point early in the twenty-first century, and 1.6 billion people as a maximum population reached around 2030 was a

China: Population 2005

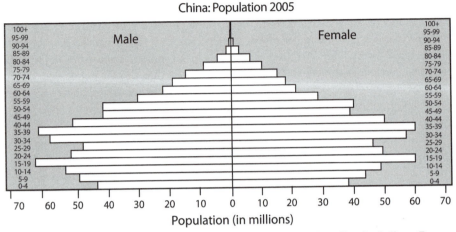

Figure 5.5. China, a population pyramid, 2005 (Bordon Dent, Jeffrey Torguson, and Thomas Hodler, 2009. Cartography: Thematic Map Design (6th ed.) McGraw-Hill: Dubuque, Iowa. Used with permission of McGraw-Hill)

possibility recently discussed (Wu 1997, 6). This goal, of course, remains ambitious, especially considering the fact that the nation has experienced only partial industrialization and urbanization, processes that are just now moving into full acceleration and that are yet theoretically associated with slowdowns in population growth in modernizing economies and societies. In a country where tradition continues to place a high value on male children, the target will be even more difficult to reach. The Chinese government, however, is now firmly committed to limiting the growth of its population, and it has provided effective if sometimes controversial leadership. The new social structure of the nation is such that it greatly facilitates family planning programs, and China continues to promote aggressively a reduction in fertility and births to achieve the goal of zero population growth (Wang 1999).

Failure to take effective action before the 1970s to stabilize the rapidly growing population was a shortcoming of Chairman Mao Zedong, who appeared to believe literally in the Marxist notion of surplus labor value and therefore that a large population would bring more producers. The Chinese leadership did not initially recognize the fact that an excessively large population could deter the growth and development of the nation's economy. Serious efforts were not made during the 1950s to reduce the birthrate, although limited proposals for family planning were made. Four separate family planning campaigns operated in China during the socialist period. Each successive one intensified the rigor with which the plans were implemented (Chen 1976).

Early Efforts and Policies

An initially cautious and limited approach to promoting birth control and to producing contraceptive devices was launched in 1956. The Great Leap Forward interrupted this program in 1958 because of its emphasis on late marriages. This was to some extent interrupted by the Cultural Revolution, although it is difficult to establish the Chinese government's official position on family planning during this time.

In 1971, China began to place much greater emphasis on family planning with its third family planning campaign. The Chinese press in the late 1970s carried extensive information on family planning programs and their effects on population growth. All newly married couples were expected and were legally obligated to practice birth control. The drive to lower the nation's fertility rate was implemented through a vast medical and public health system. It involved extensive face-to-face contact with the people by both trained medical personnel and paramedics, readily available contraceptives, and a variety of incentive programs (Chen 1976). More than a million "barefoot doctors," who normally received short-term medical training of between three and six months and who engaged in farm production while not performing their medical services, worked in China's countryside.

Dedicated medical personnel were then ubiquitous at all levels of China's rural administrative units. In the cities, the urban residents' committees were the basic administrative units where family planning programs were formally structured. At these basic levels of society, nurses, barefoot doctors, and midwives provided not only ordinary health care to the people but also modern contraceptives, sterilization, and abortion services that were readily available and free or at low cost. Contraceptives available to the Chinese are similar to those found elsewhere, and new methods have been frequently reported in the Chinese press. Xie (2000, 58) has noted that according to recent surveys in China, over 80 percent of couples were using contraception methods by the early 1990s. Most common methods included female sterilization through tubal ligation (41 percent), use of an intrauterine device (40 percent), and male sterilization (12 percent). Abortion had declined substantially from 20 million per year in the 1980s to 10 million in 1992.

Family planning, as noted above, has since the late 1950s been openly publicized through the administrative and educational systems. It is organized at the central government level by the Office of Family Planning of the State Council. Since 1981, the State Family Planning Commission is the principal agency responsible for carrying out China's population program (Xie 2000, 52). Leading cadres at the provincial, regional, and county levels transmit the state's policies downward to the rural communities and urban districts for enforcement. Each township has a Family Planning Committee that coordinates and supervises family planning work. Family planning subcommittees are responsible for such details of birth control as disseminating family planning information and contraceptives, organizing and leading study groups, persuading nonparticipants to adopt family planning, and checking the results of these programs. In the cities, similar activities are found in offices and factories as well as in residents' committees.

At these lowest levels of family planning, detailed records are kept on the number of the local population, the number of local women of childbearing age and their monthly cycles, the methods of male and female contraception used, and the projected number of births each year. Women who fail to practice birth control, who have too many children, or who get pregnant or give birth at the wrong time are likely to encounter unpleasant official pressure from cadres and sometimes social pressure from their colleagues and neighbors to conform to the rules.

The One-Child Policy and Program

In January 1979, a new family planning campaign emerged, and a one-child policy was formally put into effect for most of China's population. Since then, families have been encouraged to have only one child and to sign a pledge that they will have only one child. Many places have issued honor certificates to families that have decided to have no more than one child. The certificate provides economic benefits to those who qualify. For example, in many urban places, one-child families receive a "monthly health maintenance fee" (it originally amounted to a significant share of an average worker's monthly income) until the child reaches a certain age. Priorities for admitting children to nurseries, kindergartens, and schools are granted to these families, often free of charge. In some cities, such families enjoy priority in housing allocation and receive the same amount of living space as families with two or more children. In Shanghai, a retiree with only one child will get an additional pension payment to be calculated on the basis of 5 percent of his wage at retirement. These incentives, however, are canceled if a second child is born. In most places, families must return these acquired benefits to the state in installments if their one-child pledge is broken.

Education has been assigned a high priority in China's family planning. The masses are taught that there is no need to adhere to the old saying "Grain is stored against famine, and sons are brought up to ensure security for one's old age." It is also argued that improved medical care and public health services have greatly increased the probability of infant survival, and there is no need to produce many sons to ensure the survival of a few or even one. The masses are also told that successful family planning will help improve the health of both the mother and the child and will bring about a higher standard of living for the family. Since 1979, population planning has been justified on the basis that the nation's large population hindered the development of the national economy; created serious problems in availability of food, housing, education, and employment; and reduced per capita income growth and an already low per capita arable land supply. Yet the policies of economic reform, especially in rural areas with privatization of farms, have worked at cross-purposes with family planning policies. Peasants who work the land clearly see more children as a direct economic benefit.

Through the mass media, study classes, and small-group discussions, reasons for family planning are persistently explained to the people. Furthermore, health workers make frequent home visits during which pressure to adopt family planning measures is privately exerted. There are indications that the younger generation has been more receptive to family planning than the older generation was and that the urban population is more actively involved in family planning than the rural people. Local cadres must work closely with women of childbearing age to determine who shall have a child and when. Such intensive family planning programs, of course, affect personal freedom of action and frequently generate dissatisfaction. In the long run, however, a smaller population is seen to benefit the entire nation.

Compliance and Coercion

The policy of a single child per family and rigorous family planning introduced in the late 1970s has been controversial and unpopular in China. Perhaps the most controversial aspect of this are the coercive measures used to ensure compliance of all Chinese. In addition to the economic incentives and official pressures used to promote compliance, cadres and health workers, especially in the early 1980s, carried out forced abortions and sterilizations when family planning policies were violated. This created much resentment and criticism of the policies and their implementation, criticism that sometimes came from foreign sources (see, e.g., Aird 1990; Mosher 1983, 1993). Another unfortunate aspect of the policies was significant increases in the incidence of female infanticide, especially in rural areas, where peasants wanted to ensure the birth of a male heir.

While there may have been some overdramatization of the significance of this, vital statistics on the sex ratios of China's population in many counties offer direct evidence of inordinately large male populations, suggesting that the practice of female infanticide may have been more widespread than admitted during the 1980s (Li 1987). Other pernicious practices, such as forced abortion during the second or even third trimester of pregnancy and forced sterilization of women who have already had three or more children, have offended many foreigners who are concerned by the severity and rigor with which China has gone about its family planning and birth control program. Undoubtedly, freedom of choice has been denied to many if not most Chinese families, and some pain and suffering have been inflicted. Whether these severe and draconian policies and actions are justified by the magnitude of China's population problem and its aspirations for future development are complex and difficult questions for policy makers and officials. Thoughtful appraisals of this contentious issue and the related matter of the impact and significance of rapid fertility decline may be reviewed in Tien (1987) and Riley (2004).

As China enters the second decade of the twenty-first century, its demographic transition has been compressed to the point where its population will soon reach a threshold with a rapidly growing older population. Owing to the economic and social consequences of this age structure, policy makers have been giving serious consideration to ending the one-child policy in cities. At what point this occurs is unknown, but it does seem likely that China will soon relax its more draconian family planning rules and allow greater freedom of choice for families in choosing the number of children. It is unlikely that all rules will be terminated, but some greater flexibility in choice, especially regarding the limits of one child, appears likely. The *China Daily* (Chinadaily.com 2010) reported that a leading population official, Li Bin, indicated that China would continue its family planning policy for the immediate future and that the policy has succeeded in preventing 400 million births over the past thirty years.

Advances in Education and Literacy

China is a developing nation undergoing the throes of rapid but sometimes unequal economic growth and change. The nation's large and growing population poses a serious challenge to its hope of lifting itself out of the vicious cycle of underdevelopment. One cause as well as symptom of underdevelopment is that advanced educational opportunities are available to only a small share of the population. This affects directly the quality of the population. China's ambitious plans for the development of agriculture, industry, defense, and science and technology can hardly be achieved without the necessary technical manpower.

The financial resources of the Chinese government have improved, yet the various priorities for infrastructure projects, defense expenditures, and agricultural development, among others, create competition for state resources for various social and educational programs. Investment in education in recent years has improved, and China appears strongly committed to expanding educational opportunities for its people. In 2008, over 99 percent of Chinese children were enrolled in primary schools, and of those graduating, almost 100 percent advanced to junior middle school. Approximately 83 percent of those graduating from junior middle went on to senior middle or high school. A smaller yet significant fraction of these will advance to college or university, however, and students in higher education constituted a growing share of the total enrolled students in 2008. Further, in 2008, there were 2,363 institutions of higher education in China, and these institutions can accommodate a growing share of the nation's college-age students, compared to 2,600 institutions of higher education in the United States that can take 40 percent of U.S. college-age students. In 2008 in China, there were 20,210,249 students enrolled in all institutions of higher education (National Bureau of Statistics 2009). In the same year, 180,000 Chinese students were studying abroad, a quantity that has increased steadily over the past thirty years.

Literacy

Adult literacy has increased dramatically during the socialist period from about 25 percent of the population to more than 94 percent today. Although there are varying definitions of literacy, literacy in China today may be defined, for peasants and workers, as the ability to read and write 1,500 to 2,000 ideographs or characters (World Bank 2011). The Chinese Statistics Bureau defines literacy as the ability to read. A substantial effort to make older citizens literate was launched in the 1950s, and it had some success. Yet illiteracy, especially among women and rural people, remains a problem. In 1982, 19 percent of China's male population was classified as illiterate, whereas 45 percent of the female population was so classified, a sex-based difference greater than that found in some Muslim countries. By 2008, this had been reduced to a rate of 4.02 percent for adult males and 11.52 percent for females (National Bureau of Statistics 2009).

The disparity in educational enrollments for females has diminished. According to a sample survey, approximately half of all enrolled students in 2003 were females, and 41 percent of those enrolled in institutions of higher education were women (National Bureau of Statistics 2004). Thus, the earlier disparities in educational levels have changed remarkably, and women have made steady and significant progress at all levels of education in China during the past twenty years. Parity with males seems likely in the near future.

Women's Role in Modern China

The data on literacy and primary education presented above, in parallel with vital statistics on the sex structure of the population, indicate a somewhat lower status for women in China's socialist state. Despite substantial improvements in life expectancy, literacy, well-being, and nutrition under socialism, women lag behind men and continue to have an inferior status, especially in rural areas. While this may be expected in the countryside, given the patriarchal nature of the society and its culture and the established male-dominant position, females must still survive the perils of possible infanticide at birth, and they receive poorer nutrition, family care, and medical services and fewer educational opportunities during childhood. After their own marriages, rural women join new families where they are expected to work hard and contribute to the family economic production system as well as to produce at least one son. While not always relishing this role, Chinese women nevertheless have endured and persevered remarkably well (Lewis 2003).

China for more than two thousand years has been a patriarchal society in which males have been privileged and have enjoyed a dominant position. Women were subordinate and had few rights. The Confucian canon, on which the Chinese social order and value system have been based since the fourth century BC, institutionalized this idea of women's inferior role in society. Only in the mid-nineteenth century did things begin to change as new ideas about the role of women entered China and began to effect change. This was seen first in increased efforts to educate women and in early proposals to eliminate such archaic practices as foot binding and female infanticide. By the turn of the twentieth century, women's associations and new schools for women were widespread. In 1907, the Ministry of Education issued a regulation to govern the operation of women's schools (He 2003).

Following the founding of the Republic of China in 1911, a temporary constitution gave all Chinese rights of citizens, and this included women, yet the force of China's long history of social inequality impeded the move to full equality for women. In 1929, a new civil law went into effect that prohibited sexual discrimination and gave women rights to property, a significant advance. The communists came to power in 1949, and the 1954 constitution guaranteed equal rights for women. Supporting legislation (a marriage law) and related policies were aimed at strengthening and ensuring these rights.

Mao Zedong is credited with the saying, "Women hold up half the sky," a slogan that was popularized during the Cultural Revolution (1966–1969) and was designed to show the party chairman's official approval of the idea of full equality for women. During the Maoist years, women entered the labor force in cities and in state enterprises in large numbers. Consequently, their status as members of the brave, new society was greatly enhanced. Many also became CCP members and assumed leadership positions at various levels of government. Despite such progress, it remains clear that women have not yet achieved the same status in Chinese society as men, and this is especially true in rural areas. As noted, literacy rates for rural women are lower than male rates, and the daily lives of farm women remain arduous and challenging.

There is also evidence that as economic reforms led to closure on inefficient state factories, women were laid off or forced to retire early, and many of the gains of the Maoist years were lost once the force of economic reforms took hold and layoffs began. There have also been recent calls in the media for women to return to domestic roles given the reduced need of a large labor force (He 2003). Does this mean that the position of women in China's contemporary society may be retrogressing? Although there continue to be many challenges and difficulties for both urban and rural women in China, it seems unlikely that there will be any serious reversal of the steady advances of the past 150 years (Zheng 2000). One scholar has argued on the contrary that the role of women has in fact been much enhanced by the one-child policy, as it has resulted in considerably greater educational opportunities for single female children whose families now concentrate on ensuring the success of their only child, a daughter (Fong 2002).

One of the most serious and vexing issues related to gender in China is the matter of sex selection of fetuses, which is leading in some places to skewed gender distribution with greatly enlarged young male populations. If this is not ameliorated, it will lead to an unhealthy imbalance and an excess of "missing girls" in the future, with attendant serious social unrest. And, if because of the continuing treatment women receive as a result of the one-child policy their numbers diminish to the point where there is a real scarcity of available brides, women may come to be much better appreciated and treated in the future. According to the 1990 census, for every 100 women in China, there were 106.6 males, and this number had increased from 106.3 in 1982. Sample survey data from 1999 indicated that sex ratios were improving and adjusting in favor of more women as a share of the total population. The 2000 census report then indicated that the ratio had fallen back to 106.3; however, in the age cohort of 0 to 4 years of age, the male ratio had increased to 120.17, an indication that new technologies that allow for gender identification of a fetus may be encouraging more couples to abort female fetuses so that they may plan a male birth as their first—and perhaps only—offspring. If this is in fact the case, it is a disquieting prospect for China's future and suggests that there may soon be a large number of unhappy and frustrated males who are unable to find mates.

The fact that the national male-to-female ratio in China is not much different from that in other developing countries seems to contradict dire reports of female infanticide and mistreatment. Yet it does indicate that couples are taking advantage of new technologies in planning their families. The national statistics, moreover, mask regional variations that could clearly be seen in county-level data even as early as the 1982 census, where clusters of counties in Northwest China had male-to-female ratios of 112 to 100 or higher (Ronald Skeldon, personal communication, January 2001; see also Li 1987).

Human Resources and Employment

One serious consequence of overpopulation and underdevelopment is unemployment. Firm statistics are difficult to obtain, but unemployment among young people in China is known to be widespread and may be spreading among older, established workers as more and more SOEs are closed or restructured to improve operating efficiencies and profitability. This is especially true in the cities, where factory jobs are insufficient to meet the demand for employment and the service sector has not been sufficiently developed to absorb more unemployed.

China's economy has shifted from a system based heavily on agriculture to one more reliant on industrial and service-oriented activities. A parallel shift in employment is also occurring, and the agricultural sector's employment share has declined from 84 percent in 1952 to 40 percent in 2008 (see table 7.2 in chapter 7). Yet the problem here is the sheer size of a labor force that had grown to approximately 774 million in 2008 by keeping in step with population growth. Of this total labor force, 307 million (40 percent of the total) were in agriculture. This has resulted in serious under- or unemployment in rural areas as the Chinese economy converts to a greater focus on sideline, industrial, and service activities. Varying reports indicate there may be as many as 100 million to 150 million marginally employed or unemployed people in China's farm economy, and the number may increase as the population grows and the economy continues to modernize. This is truly one of the most serious problems posed by China's huge and growing population and one that may prove to be the biggest challenge to the future social and political stability of China (see Premier Zhu 2003, 9).

Migration

Migration, both domestic and international, has been largely restricted for most people during China's socialist period. It is true that during the first eight years of socialism, there was virtually an open policy on personal movement, and many rural people moved into towns and cities. This came to an end in 1957, and, as we previously described, a system of household registration (*hukou*) was implemented that effectively ended internal migration. Although there were subsequent periods of substantial but erratic migrations, such as were seen during the Great Leap Forward and the Cultural Revolution, internal population movements have continued to be controlled, with significant modifications, since 1994 (Chan 2004, 2009; Goldstein and Goldstein 1985).

In recent years and with increasing economic growth and demand for labor in the cities, coupled with far-reaching reforms and the evolution of China's market economy, movement of rural people to the cities has increased greatly, and estimates indicate approximately 150 million such temporary migrants (Chan 2010). The general policy of the state has been two pronged: rural folk should remain on their farms or in their villages and townships, and city dwellers should be accorded special privileges by virtue of their household registration in an urban location. Urban dwellers, also by virtue of their employment in SOEs or *danwei* (units), thereby have received many special subsidies, such as low rents in state-owned housing, education for their children, health services, low energy costs, and pensions. It is no wonder that many people wished to go to the cities and thus had to be controlled to keep them on the farms. As noted, these policies began to change as the central state increasingly recognized the contributions of rural migrants to the successful economic development of China's growing cities and coastal regions (Premier Zhu 2003, 9).

Yet to understand the real forces driving migration in China, it is necessary to look to the countryside and to consider the demographic and economic processes at work there. China's rural economy has been in rapid transition since 1978, and new methods of earning a living are everywhere, as are more efficient methods of cultivation on the farms. The rural areas have far too many people for the available jobs, and there is a serious problem of redundant employment. Approximately 50 to 55 percent of China's people are farm householders (depending on how these householders are defined), and, as we noted above, as the farm economy modernizes and is commercialized, an estimated 100 million to 150 million redundant farmworkers will need other employment in order to increase their marginal productivity (Chinadaily.com 2003). Many indeed have already moved to the towns and cities in search of better employment prospects, the major reason why people anywhere elect to move.

The 1980s and 1990s brought rapid economic growth and the fast changes typically associated with economic advancement, both in and around the cities and especially in the coastal provinces, as China's economic links with the global economy took hold (Fan 2002). Construction and factory workers were needed, as were workers to perform a variety of service activities. As these demands accelerated, farmers near and far were allowed to move to take these jobs as transient workers.

In 1984, China's policy makers modified the rules on migration to allow limited temporary migration of those who could provide their own grain rations. This resulted in a remarkable movement, largely from rural to urban areas, but it involved other shifts as well, and these introduced a new pattern of migration in socialist China. Over time, more and more people moved as new and better opportunities for jobs became available. Gradually, the changing economy and its demand for new workers in the growing cities and towns created a massive influx of human beings with an attendant relaxing and changing of the policies that determined, through enforcement of the *hukou* system, where people might live and work (Fan 2008).

Migrants moved from rural areas to towns and cities and in some cases even to other rural areas that offered new and better-paid opportunities. Rural-to-rural migrants are typically farmers from poor regions who migrate to other, wealthier regions and hire themselves out to farm the land of those villagers who have taken nonfarm employment in burgeoning local industries or other commercial and trading activities. Rural-to-town movement involves those laborers recruited to nearby towns to work in new factories or enterprises or those who go to these towns to open small businesses. Rural-to-city or town-to-city flows involve those who go to the cities to provide labor services as factory, construction, or transport workers as well as those who go to work as domestics or simply to seek whatever work is available. In addition, there is a flow of permanent migrants from rural areas to the cities who have had their household registration changed by virtue of an official job shift, such as results from entering a university or being assigned to an SOE.

Migrants are typically young and comparatively well educated. Both sexes migrate, but males dominate, and most are single. There are several variations of temporary migration. These variations are typically based on the migrant's length of stay, although many migrants become permanent residents by virtue of increasing liberalization of government policies and rules on transfer of residence. Others enter the cities without permits and remain there illegally. Migrants in some cities have reached significant thresholds and have created problems, although there is growing recognition of the need to improve the living conditions of these migrants who are providing valuable labor services. Inadequate housing, sanitation, employment, and health and educational services and a growing crime problem are among the serious issues associated with these temporary migrants in China's cities. As noted earlier, it has been estimated that 150 million temporary migrants are now in China's cities. In many cities, they may account for 10 to 15 percent of the total population. At the same time, the flow and presence of temporary migrants is crucial to China's urban economies, which offer many new employment opportunities in construction, factory, transportation, and service jobs (Chan 2010).

Early on in the economic reforms, government policy was content with the idea of temporary migration because if the economy soured and urban jobs dried up, the migrants could be pushed out and easily returned to their rural areas. Moreover, the state had no obligation to provide expensive subsidies to these temporary residents. Yet the scale of this migration, coupled with an enormous rural surplus labor force, has created the potential for serious social problems and possible political instability in China's cities if no better arrangements are provided to incorporate future migrants in a fairer and more equitable manner. Solinger (1993, 1995) argued poignantly on the fundamental inequity of the urban–rural dichotomy in the household registration system and lamented the second-class or marginal status to which the transient population was so clearly relegated. Yet given a very large surplus rural population, the economic advantages to the state, and the disincentives to allowing large-scale migration with its so many related social and environmental problems, it is easy to see why China would not allow a completely unregulated rural-to-urban migration. An increased recognition of the economic benefits of free mobility of labor and of more flexible policy and rules regarding such migration led some analysts to argue that the rules no longer mattered (Oi 1999; Solinger 1999). In January 2003, China's State Council issued new regulations and policies that gave rural migrants the legal right to work in cities and to begin to set up official residence in the urban places where they worked (Hutzler and Lawrence 2003, A12). Finally, the government had come to recognize the importance of the contributions of these migrants and their value to the urban economy. This indicates significant progress in the Chinese government's willingness to allow market mechanisms to work, and it is a crucial step forward in social policy and in improving the lives of millions of China's rural citizens. Yet China's policy makers continue to resist giving full benefits of urban citizenship to millions of these essential migrant workers. Chan (2010) has analyzed the provocative issues of how long this inequitable policy can continue and what its social and political implications are.

Population Growth and China's Future

As noted at the beginning of this chapter, China's people and its population are both its greatest resource and its greatest challenge, offering the nation's leaders a set of problems as the country seeks to improve its economic condition and boost its standard of living. For more than three decades, China's leaders and policy makers have recognized the significance and scale of this population question and problem, but the demographic momentum built up under the government's variable policies in the periods of political instability during the first two decades of socialist rule have led to such rapid growth that it will require another generation to slow. Tien et al. (1992) have argued that given China's level of economic development, the government has probably achieved about all it can in reducing fertility and thereby advancing the trajectory of its demographic transition. Yet the 1990s proved to be a period of remarkable economic advance for China, and this permitted the central state to proceed to enforce vigorously its one-child policy to the point that China's fertility rates and birthrates have continued to decline, leading to a rate of net natural increase as low as countries such as the United States.

Further reductions in fertility likely will proceed apace with continued economic growth and accompanying social change, as has been seen in Japan, Taiwan, and South Korea. Yet the draconian measures embodied in the policy of the one-child family cannot go much further without the full-scale national industrialization and urbanization that typically herald modernization and attitudinal changes about family size. These in turn will lead to long-term reductions in fertility rates and birthrates. If its trajectory of economic growth and related urban and industrial development continues in the pattern of the 1990s, China will likely approach and achieve zero population growth in the next two or three decades. Japan, South Korea, Hong Kong, and Taiwan offer good East Asian and Chinese examples of how this process works. They may, therefore, provide useful portents for the patterns likely to unfold in China should its current process of economic reform and political stability continue through the twenty-first century.

Should China's demographic profile of below-replacement fertility continue, this will raise serious questions about the future of Chinese society and the structure of the Chinese family. A generation of Chinese children will grow up with no siblings and few cousins. The next generation will have no aunts or uncles. Some parents may have little support or connection to their only child once that child marries out to another family, and the dynamics of family life will surely change profoundly as the number of family members contracts sharply.

Of perhaps even greater significance are the economic questions related to a rapid slowdown in population growth with the future possibility of population decline. While the effects of such a transition are just now beginning to appear, it is clear that even in one decade and certainly in two or three, if present demographic trends continue, China's annual added increment to the national labor force will be much smaller. At the same time, the proportion of its population sixty-five or older will be considerably larger, with all the related social and economic consequences. For example, in 2000, approximately 7 percent of China's population was sixty-five or older compared to 12 percent in the United States. In China, this 7 percent share is expected to double by 2026, a doubling that will require only twenty-six years, whereas in the United States the same doubling of the senior population will have occurred in 2013 but will have required sixty-nine years to do so. This kind of demographic compression of an older population is similar to what has occurred in Japan. As some have described it, however, there is likely to be a marked difference between Japan and China, for China is projected to be the first country "to get old before it gets rich," a portent of challenging social, economic, and political consequences to come (Kinsella and Phillips 2005).

This possibility might witness a reversal in the proportion of China's working population, with a sharp rise in its dependent population as a proportion of the total. The retirement age of state employees could lengthen, which would probably be a good thing. China conceivably could even enter a period when labor becomes scarce. Both the economic and the social ramifications suggest that China's official policy on limiting families to one child will change in the future, but other development indicators, such as improved incomes, higher levels of education and urbanization, and higher costs of living, will likely lead to a reduced preference among Chinese couples for more than one or two children. China will then have completed its demographic transition, and it will have entered a long-term phase of no growth or population decline.

Questions for Discussion

1. What has been the historical trajectory of China's population growth, and what are the key factors that have promoted decline or growth in the population over time?
2. Describe the demographic transition in China over the past sixty years and comment on key events and policies that have played a role in determining the pattern of population growth.

3. What is *hukou* policy, and why is it so important in explaining the processes of migration and economic and social development in China's urban centers?
4. Is China's large population an asset or a liability for the country's current and future development? Offer an explanation of relevant factors to support your response.
5. What does the "quality" of China's population have to do with its economic growth and development, and why is this important? What are the key factors that determine the quality of a population?

References Cited

Aird, John. 1990. *Slaughter of the Innocents: Coercive Birth Control in China.* Washington, DC: AEI Press.

Ashton, Basil, et al. 1984. Famine in China, 1958–61. *Population and Development Review* 10, no. 4: 613–45.

Banister, Judith. 1984. An analysis of recent data on the population of China. *Population and Development Review* 10: 3241–71.

———. 1987. *China's Changing Population.* Stanford, CA: Stanford University Press.

Chan, Kam Wing. 2004. Internal migration. In *Changing China: A Geographic Appraisal,* ed. Chiao-min Hsieh and Max Lu. Boulder, CO: Westview, 229–42.

———. 2009. The Chinese *Hukou* system at 50. *Eurasian Geography and Economics* 50, no. 2: 197–221.

———. 2010. Fundamentals of China's urbanization and policy. *The China Review* 10: 63–94.

Chao, Kang. 1986. *Man and Land in Chinese History: An Economic Analysis.* Stanford, CA: Stanford University Press.

Chen, Pi-chao. 1976. *Population and Health Policy in the People's Republic of China.* Washington, DC: Smithsonian Institution, Interdisciplinary Communications Program.

Chinadaily.com. 2003. Premier preoccupied with rural areas, unemployment, poverty. June 18. http://www.chinadaily.com.cn/highlights/nbc/news/318rural.html.

Chinadaily.com. 2010. Family-planning policy stays put. September 27. http://www.chinadaily.com.cn/china/2010-09/27/content_11350778.htm.

Dikotter, Frank. 2010. *Mao's Great Famine: The History of China's Most Devastating Catastrophe, 1958–1962.* New York: Walker Publishing.

Elvin, Mark. 1973. *The Pattern of the Chinese Past.* Stanford, CA: Stanford University Press.

Fan, C. Cindy. 2002. Population change and regional development in China: Insights based on the 2000 census. *Eurasian Geography and Economics* 43, no. 6: 425–42.

———. 2005. Modeling interprovincial migration in China, 1985–2000. *Eurasian Geography and Economics* 46, no. 3: 165–84.

———. 2008. *China on the Move: Migration, the State, and the Household.* New York: Routledge.

Fong, Vanessa. 2002. China's one-child policy and the empowerment of urban daughters. *American Anthropologist* 104, no. 4: 1098–1109.

Goldstein, Sydney, and Alice Goldstein. 1985. Population mobility in the People's Republic of China. Papers of the East-West Population Institute, no. 95. Honolulu: East-West Center.

He, Pequin. 2003. Women's rights and protection policy in China: Achievements and problems. In *Social Policy Reform in China,* ed. Catherine Jones Finer. Aldershot: Ashgate, 203–14.

Ho, Ping-ti. 1959. *Studies on the Population of China, 1368–1953.* Cambridge, MA: Harvard University Press.

Hutzler, Charles, and Susan Lawrence. 2003. China acts to lower obstacles to urban migration. *Wall Street Journal,* January 22, A-3.

Kinsella, Kevin, and David R. Phillips. 2005. Global ageing: The challenge of success. *Population Bulletin* 60, no. 1. Washington, DC: Population Reference Bureau.

Lewis, Jane. 2003. Women's rights and gender issues. In *Social Policy Reform in China,* ed. Catherine Jones Finer. Aldershot: Ashgate, 215–24.

Li, Chengrui, ed. 1987. *The Population Atlas of China.* Hong Kong: Oxford University Press.

Mallory, Walter H. 1926. *China: Land of Famine.* New York: American Geographical Society.

Mosher, Steven. 1983. *Broken Earth: The Rural Chinese.* New York: Free Press.

———. 1993. *A Mother's Ordeal: One Woman's Fight against China's One-Child Policy.* New York: Harcourt Brace Jovanovich.

National Bureau of Statistics. 2001. *Zhongguo tongji nianjian 2001* [China statistical yearbook 2001]. Beijing: China Statistics Press.

———. 2004. *Zhongguo tongji nianjian 2004* [China statistical yearbook 2004]. Beijing: China Statistics Press.

———. 2009. *Zhongguo tongji nianjian 2009* [China statistical yearbook 2009]. Beijing: China Statistics Press.

Oi, Jean C. 1999. Two decades of rural reform in China: An overview and assessment. *China Quarterly* 159: 616–28.

Pannell, Clifton W. 2003. China's demographic and urban trends for the 21st century. *Eurasian Geography and Economics* 44, no. 7: 479–96.

Perkins, Dwight H. 1969. *Agricultural Development in China, 1368–1968.* Chicago: Aldine.

Premier Zhu. 2003. Government work report delivered at the first session of the Tenth National People's Congress. March 5. http://www.chinadaily.com.cn/highlights/nbc/news/319zhufull.htm.

Riley, Nancy E. 2004. China's population: New trends and challenges. *Population Bulletin* 59, no. 2: 1–36.

Solinger, Dorothy J. 1993. China's transients and the state—A form of civil society. *Political Sociology* 21, no. 1: 91–122.

———. 1995. The floating population in the cities: Chances for assimilation. In *Urban Spaces in Contemporary China,* ed. Deborah Davis et al. New York: Woodrow Wilson Center Press, 113–48.

———. 1999. Demolishing partitions: Back to beginnings in the cities? *China Quarterly* 159: 1–34.

Taylor, Jeffrey, and Judith Banister. 1991. Surplus rural labor in the People's Republic of China. In *The Uneven Landscape: Geographical Studies in Post-reform China,* vol. 30, ed. Gregory Veeck. Baton Rouge: Geoscience Publications, Louisiana State University, 87–120.

Tien, H. Yuan. 1983. China—Demographic billionaire. *Population Bulletin* 38, no. 2: 2–44.

———. 1987. Abortion in China—Incidence and implications. *Modern China* 13, no. 4: 441–68.

Tien, H. Yuan, et al. 1992. China's demographic dilemmas. *Population Bulletin* 47, no.1: 1–42.

Wang, Gabe T. 1999. *China's Population: Problems, Thoughts, and Policies.* Aldershot: Ashgate.

World Bank. 2011. Literacy rate, adult total (% of population age 15 and above). Data.wb.org/indicator/SE.ADT.LITR.Z5.

Wu, Canping, ed. 1997. *General Report of China's Changing Population and Its Development.* Beijing: Higher Education Press.

Xie, Zhenming. 2000. Population policy and the family planning program. In *The Changing Population of China*, ed. Peng Xizhe and Zhigang Guo. Oxford: Blackwell, 51–63.

Zheng, Wang. 2000. Gender, employment, and women's resistance. In *Chinese Society: Change, Conflict, and Resistance*, ed. Elizabeth J. Perry and Mark Selden. London: Routledge, 62–82.

Zhou, Jinghua. 1979. Interview with a specialist on population. *Beijing Review* 22, no. 46: 20–22.

CHAPTER 6

The Production and Consumption of Culture in Postsocialist China

Christopher J. Smith

In his attempt to account for the diverse and apparently disconnected phenomena that seem to characterize the field of culture in contemporary China, Jason McGrath (2008) argues that all forms of production—in both the economy and the field of culture—are now being driven by the logic of capitalist marketization. McGrath suggests that in the first three decades after the communist revolution, all spheres of society and culture were driven from/by the party-state, but in the new millennium, it is clear that the driving power stems directly from flows of money, capital, and commodities. The irony of this is that where once state heteronomy determined all cultural production in China, in the current situation culture is still dominated and largely shaped by external forces, but now they are primarily market rather than state forces.[1] This may help to explain the current sense of disintegration and apparent loss of all cultural anchors that many China watchers have observed during the postsocialist era. As McGrath observes,

> The Maoist revolution . . . [itself] an alternative utopian vision of moder-
> nity, advocated the systematic and sometimes violent replacement of the
> old with the new, of "feudal" culture with revolutionary culture. . . . When
> postsocialist capitalist culture arrived, with its own need to continually
> revolutionize both production and consumption, it confronted a popula-
> tion that had already been cut off from much traditional culture while
> being immersed in the totalizing culture of revolution. . . . The loss of both
> traditional and revolutionary ideological reference points contributes to a
> persistent feeling of disintegration that accompanies the breakneck building
> of a new economy. (21)

As early as 1994, some scholars could already see a dangerous level of fragmenta-
tion, which Araf Dirlik (1994) described as the "total disintegration of ideological unity" in China, in which "every idea from the most contemporary to the most Reactionary . . . seems to find a constituency" (56). Geremie Barmé (1999) agreed, suggesting that in China by the end of the twentieth century, "fewer people in any group spoke the same language . . . and despite greater latitude, many were increasingly speaking at cross-purposes" (362). This would certainly have disappointed Mao

Zedong, who was hoping that intellectuals, workers, and peasants in the new society would one day all be speaking the same language. It was already becoming clear by the end of Mao's lifetime and certainly by the time Deng Xiaoping died, that this was not going to happen, and one decade into the new millennium, it was little more than a distant memory.

Some China experts have suggested that what they see happening in China today can be characterized as a mutation into an entirely new epoch, powered by China's entry into the global economy. The effects of this mutation are manifested in part as nostalgia for a lost communist past and in part as ambivalence in the sense that the Chinese people scarcely know where to turn for cultural and political guidance. This could explain why so many of them have become avid consumers of everything in the new era, including popular culture in its many manifestations, reminding us of Dai Jinhua's (2002) observation that "the image of happy consumers . . . has replaced that of angry citizens" in Tiananmen Square (221).

These sentiments are shared by many Chinese scholars and artists who feel that their country is now experiencing a serious "cultural deficit." Beijing artist Ai Weiwei, for example, is repulsed by having to watch other artists "celebrate their craven pragmatism and opportunism . . . [which reflects their] degraded standards and . . . lack of heartfelt values" (quoted in Barmé 1999, 363). Others are much more optimistic than this. Jianying Zha (1995), a Chinese cultural critic now living in the West, thought that what she saw in the early 1990s was the birth of an entirely new cultural form. Not all China scholars—and certainly not all Chinese people—found these new trends alarming: in fact, many felt that they were a major step forward in what is still, by many measures, a closed society. From this perspective, it is clearly possible to interpret what is happening in contemporary China, in cultural terms, as a positive and wholesome force: a virtual explosion of pent-up ambitions and desires among the long-suffering Chinese people. To bring some order and logic to an analysis of the new world of Chinese culture, it is useful to focus here on what McGrath (2008) interprets as two "countervailing movements" at work in the production of culture in China since the early 1990s: one involves what McGrath calls a "deterritorializing" trend from heteronomy to autonomy in the relationship between culture and the ideology upheld by the party/state, and the other is a simultaneous "reterritorialization" process, occurring as culture is commodified and subjected to market mechanisms and profit margins.

This chapter attempts to make some sense out of the confusing mix of elements that has come to distinguish contemporary culture in China. In simple terms, two trends are obvious. The first trend consists of the habits and practices of everyday life, where there has been a significant shift away from the heroic and utopian values of the revolutionary era toward the more quotidian pursuit of wealth and the pleasures of consumption (Tang 2000). In the 1990s and continuing through the first decade of the new millennium, China has been evolving through successive stages of what might best be called an "incomplete modernization" (McGrath, 2008, 205). During this period, there has been a rhetoric of progress throughout, but unlike the situation in the communist period, it is no longer either possible or desirable to define an end product. The other trend in China, operating within a narrower definition of culture as artistic expression, is a movement away from the "highbrow" to the "lowbrow" end of the culture spectrum, from what are traditionally considered to be elite cultural practices to the mass reproduction of popular culture that (some would argue) has no apparent depth or significance (Kraus 1995). Significant efforts have been made to maintain some of the cultural practices within such domains as oil painting, classical music, literary novels, ballet, spoken drama, and avant-garde poetry. The problem in China today, however, is that in spite of these continuing efforts, people attempting to generate a cosmopolitan—and uniquely Chinese—form of art in such domains are finding that they are almost as disadvantaged as those trying to maintain a public interest in China's traditional or elite art forms. The major reason for this—and the principal dynamic in the production of cultural and art forms in contemporary China—is the pervasive impact of market forces. To quote Kraus (1995) again, "High-minded novelists now have more difficulty finding publishers; ballet and spoken drama troupes and symphony orchestras are forced to cut their payrolls, as well as to find ways of earning additional funds" (183). The most immediate impact of the reform era on the elite forms of Chinese art (in both their contemporary and their traditional manifestations) has been the reduction—and in many cases the complete withdrawal—of state subsidies for the art forms considered most important by would-be cosmopolitans in China.

Cultural Trends in Postsocialist China

One way to begin looking at cultural change in China is to focus on the way Chinese people have been spending their time and money in the understanding that this might be able to tell us something about what they consider important in their lives. It is important to point out here, of course, that this endeavor restricts us almost entirely to looking at trends in urban China, in the booming east coast cities, where almost all the benefits to be experienced in the new era have been situated. During the mid- to late 1990s, time budgets and daily activity diaries first offered researchers a chance to explore different dimensions of everyday life. In a review of such research in contemporary China, Wang Shaoguang (1995) referred to the notion of "private time" to describe what people do to amuse or entertain themselves when they are not working. In Wang's discussion of the trends emerging in the 1990s, he reported some dramatic changes since the end of the Cultural Revolution not only in the quantity but also in the quality of private time. Beginning in the 1980s and continuing to the present day, the evidence suggests that the party/state no longer feels quite the same need to define what are and are not suitable leisure-time pursuits. Rather than requiring citizens to toe the line defined by official state ideology, people are now allowed to spend their spare time however they want provided that their behavior does not directly challenge the state or threaten the social order.[2] What this means is an effective uncoupling of public and private time: in essence, the people have been liberated on the understanding that they will exercise restraint and presumably in the clear knowledge that the state can at any time redefine where the line is to be drawn between what is acceptable behavior and what is not.

Wang's data, collected from a variety of official sources and from research reports compiled by Chinese scholars, suggest that the average Chinese city dweller had significantly more free time in the 1990s than was the case in the previous three decades. The reasons, Wang suggests, include the widespread introduction of household consumer durables that saved time in and around the home; a reduction in the length of the working day, associated mainly with the abolition of political study sessions and voluntary political work; and an increasing efficiency in the reform-era workplace. The average Chinese family in 1980 had two hours and twenty-one minutes of "leisure time" per day, and this had increased to four hours and forty-eight minutes by 1991 (Wang 1995, 158).[3]

The reestablishment of the private sector in the Chinese city has allowed the expansion of a huge variety of commercial and service industries, including fast-food restaurants, home-appliance repair shops, and "decoration" companies that are able to perform a wide range of previously ignored or do-it-yourself activities in the home (Ikels 1996). These new ventures have reduced the burden of shopping, food preparation, repair work, and housing maintenance for many urban dwellers, some of whom have been lucky enough to use their newfound affluence to search out new and varied leisure pursuits.

Two other aspects associated with the postsocialist era have had positive effects on the supply and diversity of leisure activities in China. One has been the presence of foreign investors who have been willing to pour capital into a variety of new activities or enterprises, including filmmaking, music productions, nightclubs, restaurants, and bars, all of which offer new diversions for the people. At the same time, this has broadened the range of what is acceptable as cultural production and expanded the demand for freelance artists, who no longer have to depend on the state for their livelihood. The other important change is in the fiscal situation of China's culture "industries" (Wang 2001). All state units and enterprises involved in cultural production, including those in the communications business (newspapers, magazines, radio, and television), now have to be economically self-sufficient, which translates to being responsive to market forces and to appeal to new consumer tastes (Smith 2002a). Programs and events that smack of the heavy-handed ideological content required in the past are quickly dropped and replaced (Ikels 1996, 261).

Spending Money

To get a feel for the essence of new cultural developments, it is also useful to investigate how the Chinese—and, more specifically, those who are doing well in the new era—spend their wealth. It is clear that many things that were hard to come by, even taboo, in the old days are now common, as a walk around any of the new supermarkets in the glitzy neighborhoods of Shanghai and Shenzhen will demonstrate.[4] The nature of some of these changes is illustrated in table 6.1, which shows the increase in the average level of income and retail sales from 1978 to 1997. The magnitude of the increases is startling: in less than two decades, urban income more than tripled, the amount of living space almost doubled, and the percentage of households using natural gas increased from 14 to 76 percent. Also helping to make life at home much easier

Table 6.1. Consumption Categories and Media Exposure in Urban China, 1978–1997

Category	1978	1985	1997	2008
Per capita living space (square meters)	3.6	5.2	8.8	20.3
Natural gas (% of homes)	14	22	76	89.6
Consumer durables per 100 households:				
Washing machines	6	48	89	94.65
Refrigerators	0	7	76	93.63
Sofas	89	132	205	—
Televisions (color)	—	17	100	132.89
Media and communication (national)				
Television broadcast coverage (5)	—	—	88	97.0
Radio broadcast coverage (%)	—	—	86	96.0
Publications				
Newspapers	186	1,445	2,149	1,943
Books	14,987	45,603	120,106	275,668
Magazines	930	4,705	7,918	9,545
Urban telephones (100,000s)	28	48	554	2,316.0
Long-distance phone lines (1,000s)	19	38	1,146	n/a
Mobile phones (1,000s)	0	0	1,323	641,245

Source: Adapted from Tang and Parish (2000, table 2.1, 39) and State Statistical Bureau 2009, chapters 9 and 21.

and more comfortable were huge increases in the ownership of consumer durables: automatic washing machines, refrigerators, sofas, and color televisions. These upward trends are likely to continue as urban households try desperately to "keep up with the Zhangs," and in real terms, as Tang and Parish (2000) point out, "Improved objective conditions meant that family life was far more comfortable in the 1990s than at the end of the 1970s" (42).[5]

Data released by the Chinese government in 2003 show that the living standards of both the urban and the rural populations continued to improve into the new millennium. The annual per capita disposable income of urban households, for example, was 7,703 yuan in 2002, an increase of more than 40 percent in a five-year period (see table 6.2). The Engel coefficient, which measures the proportion of expenditures on food to the total consumption expenditures, was 37.7 percent for urban households, down from 44.5 in 1998—again this suggests that more of total household income was available for the consumption of "luxuries" or for saving.

Tang and Parish (2000) also call attention to what amounts to a very significant shift, from what might be defined as an era of scarcity, in which people in China could only fantasize about the things they did not have, to an era of plenty, in which many now have access to almost everything they want. In the not-too-distant past, the hero or heroine of Chinese society was the person with the most effective *guanxi*, the one able to get the best deals, find the real bargains, and generally make the most of his or her connections in an otherwise luxury-scarce environment. Such finely tuned networking skills

Table 6.2.   Increases in Urban and Rural Incomes in China, 1998–2002

	Unit	1998	1999	2000	2001	2002	2008
Per capita disposable income of urban population	Yuan	5,425	5,854	6,280	6,860	7,703	15,780.8
Per capita net income of rural population	Yuan	2,162	2,210	2,253	2,366	2,476	4760.6
Engel coefficient of urban households	%	44.5	41.9	39.2	37.9	37.7	37.9
Engel coefficient of rural households	%	53.4	52.6	49.1	47.7	46.2	43.7

Source: China Statistics Press 2009, chapter 9.

are no longer as prized as they once were; in fact, the only requirement today for living the good life in China is having plenty of money. More important, it is now becoming clear that in contemporary urban China, shortages and underproduction are no longer the problem; the anxiety these days—for the state's economic planners at least—is that demand will not be great enough to soak up all the products being made. This may help to explain why the state seems to be so fixated on increasing the rate of urbanization in China, which—it is assumed—will result in the establishment of new household units that have to be filled and furnished with domestically made consumer durables. All this is consistent with the state's push during the past decade in the direction of privatized housing, as is discussed in chapter 10 of this book.[6]

As all this suggests, the exemplary economic figure in today's China is much more likely to be an information technology executive, a salesperson, or a marketing agent than a peasant or a soldier. The contemporary fantasy—what Elizabeth Croll (1994) described as the people's "dreams of heaven"—is for every household to have an array of consumer durables appropriate to its status (Tang and Parish 2000, 43). To make sure they move in the direction of achieving this goal, urban consumers are now surrounded by almost nonstop and totally pervasive advertising: on the streets, online, on television, and in publications of all types. While the Chinese exercise their long-overdue right to spend lavishly, the rest of the world hopes that they will continue to spend, helping to pull the rest of the world out of its economic slump (at the time of writing, early October 2010). In a recent *Time* magazine story coming from Xian, for example, we see pictures of customers streaming into the showroom to look at the new "Wuling" minivans manufactured by a joint venture between General Motors and two Chinese local carmakers. Local dealers could barely keep the new vans in stock, which should come as good news for the rest of the world:

> As debt-laden consumers in the U.S. retrench, increasingly wealthy Chinese consumers could become one of the most important sources of growth for the global economy. Shoppers in China are opening their newly stuffed wallets wider than ever. Passenger car sales surged 76% in October from a year earlier . . . while overall retail sales jumped 16.2%. Such spending has contributed to China's robust recovery from the global economic crisis. Gross domestic product grew a hefty 8.9% in the third quarter from a year

earlier . . . [and] there is reason to believe such eye-popping spending can continue. As more regions of China's vast hinterland join in its amazing economic boom, more and more of the country's 1.3 billion people can afford cars, refrigerators and flat-panel TVs items not too long ago considered luxuries for a fortunate few. (Schuman 2009)

In the same story, Chen Baogen, the mayor of Xian, reports that his city of 8 million had been lagging behind the thriving cities on the coast but that now incomes are growing quickly and consumption is taking off: in the first nine months of 2009, retail sales increased by 19 percent, well above the 14.8 percent growth posted in China's cities nationally. It is worthwhile here to point out two national trends that illustrate the directions in which the new Chinese economy is heading. The first is the growing evidence of urban/rural inequality: with each passing year in the new millennium, the gap between the average income in the cities and the countryside has widened (see figure 6.1). The other national trend, which shows no sign of diminishing, is the evidence that the Chinese, in spite of all the new consumption that is taking place, are still saving at a much greater rate than most other people in the world. Cornell University economist Eswar Prasad calculates that China's average urban household saving rate reached 28 percent of disposable income in 2008, which was actually 11 percent higher than in 1995. At the present time, private consumption accounts for only 35 percent of China's gross domestic product (2008 data), which is down from 46 percent in 2000 (this figure is about half the rate for the United States and lower even than in India, where the figure is about 57 percent) ("The Spend Is Nigh" 2009). Wisely, perhaps, Chinese consumers are still uncertain about the future, and with social safety nets remaining relatively weak, many families still prefer to save to take care of their parents in old age and to pay rising medical bills and prepare for retirement.

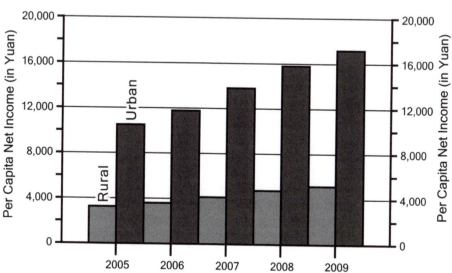

Figure 6.1.   The urban-rural income gap in China, 2005-2009 (National Bureau of Statistics of China; see *China Daily* (online): March 2, 2002, http://www.china daily.com.cn/china/2010-03/02/content_9521611.htm)

One trend that is of obvious importance in a discussion of culture and cultural change in China is the way the Chinese people eat, what they eat, and where they eat it. As incomes have risen in China, the structure of spending on food and eating habits has changed very significantly (Ma et al. 2002). The data available suggest that as incomes rose during the 1990s, households spent a smaller proportion of their total income on food, but there was a corresponding increase in the amount of meat and fish they ate and a sharp decline in their consumption of grains (mainly rice and wheat). Not surprisingly, the ratio of meat and fish to grains consumed increased as a function of rising incomes among urban households: in households with per capita incomes below 1,000 yuan, grains represented 28 percent of total purchases, with meat and fish accounting for 21 percent and 6 percent, respectively, but in households with per capita incomes above 4,000 yuan, grains made up only 14 percent, while meat and fish accounted for 38 percent and 13 percent, respectively.[7] Ma and his colleagues also point out the beginning of a very significant trend toward eating away from the home in urban China, a trend that more than doubled, from 6 percent of all food expenditures in 1991 to 13 percent in 1999. As the authors suggest, this trend, which looks likely to continue as average incomes rise in China, will have a significant impact on the urban landscape. They observe, "Powerful forces, such as rising income, are not only changing the pattern of people's diets, the new habits of consumers also have their own effect on the nation's urban environment . . . [which has] spawned a huge increase in the catering industry and created a highly visible venue in which people are able to meet, entertain themselves and do business" (7). In the spirit of the reform era, all these new "private time" activities have blossomed in response to the changing economic climate. The demand of the newly empowered Chinese consumer for a more varied lifestyle, one that includes eating in restaurants, is obviously an important dimension of this new trend.

Also apparent in table 6.1 is an increase in the consumption of media of all types in urban China and a sizeable increase in the amount of electronic interaction through regular and mobile telephones and the Internet. Increased media exposure is a sure sign of the coming of a modern age in any country, in part because it allows consumers a greater degree of flexibility in what they can do in their spare time but also because it offers people a much greater opportunity to imagine different possibilities for themselves and their family members (we are reminded again here of Croll's "dreams of heaven"). Television coverage in all urban areas of China is now almost universal, and the hours of programming (and watching) have increased noticeably. Much of the new programming (on television at least) consists of what many cultural critics refer to as "lowest-common-denominator programming," including soap operas and historical dramas (Dai 2002). Critical dramas, documentaries, and satirical comedies have also been aired over the past few years, with enough intellectual and political content to suggest at least a minor role for television in sharpening the critical faculties of its viewers.[8] The line beyond which such programs are not allowed to go, however, is still drawn by the state, which is not surprising if we bear in mind that all media and publication outlets are still publicly owned and operated.

The consumption data also show that readership in all contexts (including books, newspapers, and magazines) has increased significantly during the reform era, although it seems likely that in the late 1980s and early 1990s, television began to attract some of those who might have otherwise been readers in the future or who had been readers in the recent past. The proliferation of newspapers of all types in China during the past decade—including party propaganda outlets, respectable dailies, and a galaxy of tabloids—implies that far more topics and issues are now being covered, including even gossip about the personal lives of some of the country's top leaders. One is tempted to think that this absolute increase in the availability of publications has allowed the Chinese people to voice a greater range of opinions on and about public issues (Lynch 1999). In some spheres, the state has encouraged the press to be hypercritical, especially when such a stance supports official policies—for example, in the state's quest to reduce official corruption; in its position on some environmental issues; in its struggle to oppose certain organizations, such as Falun Gong; and in its attempts to wipe out crime—but in many of the more politically controversial spheres, coverage ranges from the perfunctory to the nonexistent (Smith 2002b).

A Brief Aside: China and the Internet

At this point, it is important to add a comment about the growing importance of the Internet in China, which is illustrated in figure 6.2a—showing the average rate of change in Internet usage throughout the first decade of the new millennium—and figure 6.2b—which provides some global comparisons in Internet access. Although this is a topic that deserves a chapter to itself, it is important at least to make some mention of it here. In a recent review of Guobin Yang's (2009) book *The Power of the*

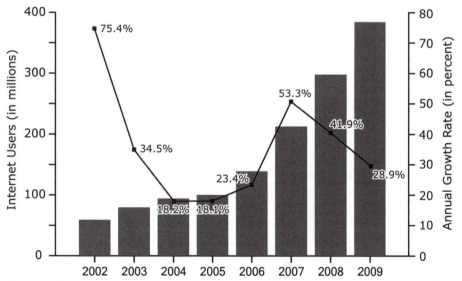

Figure 6.2a. Overall growth and annual rate of change in Internet use, China 2002–2009 (Statistical Survey Report on Internet Development in China (January 2009): China Internet Network Information Center, http://www.slideshare.net/ zhouzixi/the-23rd-statistical-survey-report-on-the-internet-development-in-china)

Internet in China, which focuses more on the role of the Internet in the realm of citizen activism, Rebecca MacKinnon begins with a story about a Chinese man who posted a message on Twitter at 5:00 a.m. on July 16, 2010, reading as follows: "Pls help me, I grasp the phone during police sleep. . . . I have been arrested by Mawei police, sos." The message was signed by "amoiist." Within hours, it was "retweeted" by hundreds of people in China and around the world, and although there were no other "tweets" from "amoiist," his short postings—sent from a mobile phone while the policeman watching him slept—were sufficient to get him bailed out of jail quickly. People in the caller's network immediately contacted his family and friends in the city of Mawei. Other news on Twitter made it apparent that the police had taken "amoiist" from his place of work the day before; he was arrested along with several others for posting information online about the alleged gang rape and death of a young woman, which had been "officially" listed as a death resulting from a hemorrhage after a pregnancy. Bloggers everywhere immediately rallied for the release of "amoiist": one of them organized a campaign in which hundreds of people mailed postcards to the Fuzhou detention center where he was being held with the simple message, "Guo Baofeng, your mother wants you home for dinner." Others started a fund-raising drive to pay for his defense, and after two weeks, Mr. Guo and two other bloggers arrested around the same time were released. "I used Twitter to save myself," he wrote on his blog. The massive online reaction, he believes, helped to free him.

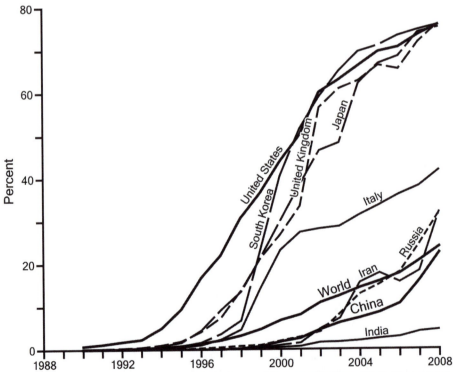

Figure 6.2b.   Internet users as a percentage of the population: Global comparisons, 1990–2008 (World Bank: World Development Indicators, May 2010)

MacKinnon points out that had this happened in 1990, before the Internet became widespread in China, Guo would never have got his message out in so short a time, with such a relatively rapid response: "Now people like Mr. Guo with Black-berries and Twitter accounts can instantly rally people inside and outside China to help them. Not only do the tools exist, but a community has developed around the tools, a community that spreads information and mobilizes people to act" (see http://rconversation.blogs.com/YangGB_MacKinnon.pdf). She tells this story simply to illuminate the changes that Yang (2009) describes as a social revolution in which China's citizens now have access to a technological revolution that he (Yang) thinks might eventually lead to political and institutional change in China. Yang reports that despite China's authoritarian political system, censorship, and surveillance, the Chinese Internet has become a highly contentious place, with fierce and passionate debate being common. Although the Internet still cannot be used for overt political organizing, Yang offers many examples of how citizens have used the Internet to expose and bring down corrupt officials throughout the land. As he indicates, this influences how government officials behave in the knowledge that their actions might easily produce a vast set of responses. New organizations of all stripes have made use of the Internet to raise awareness for their causes and expand their membership. Yang suggests that the contentious nature of Chinese Internet culture actually predates the Internet itself and reflects a long tradition of social protest throughout Chinese history.[9]

To conclude this section, it is useful to take a step backward to consider the magnitude (and the significance) of the cultural changes that have been recorded in China in the postsocialist era. In not much more than two and a half decades, many of China's citizens (and especially those living in the biggest cities) have witnessed a shift from Maoist asceticism to the middle-class enjoyment of capitalist luxuries. As Judith Farquhar (2002) has pointed out, this can be expressed in terms of the satisfaction of new "appetites" that not only were not realizable in the old days but also could not even be talked about in public. In the mid-1970s, it was totally unacceptable to introduce the topic of individual appetites, or their indulgence, into public discourse in China.[10] By comparison, beginning in what Farquhar called the "roaring nineties," the enjoyment of personal appetites was being extravagantly displayed in public places, often with an air of defiance.

The Politics of Cultural Production in Postsocialist China

Thus far, we have considered only the material (and measurable) aspects of cultural change in contemporary China. It is also apparent that artists and other cultural producers in China today can realistically dream of a more positive future after a long period in which they were held captive in a metaphorical space between the old controls of the CCP and the newer mechanisms of the market. As Richard Kraus (1995) explains this situation, "The changing relations between plan and market force[d] changes in the lives and visions of China's artists and in the pleasures offered . . . [to] urban audiences" (173). Although the present and the future look quite positive for cultural and artistic production—with enormous swaths of new territory opening up—Kraus reminds us that there is a downside to these new trends. For some artists, the new era is much less secure because of the withdrawal of state support. The "market," while offering more freedoms in terms of what work can now be done, means a lot more start-up work and excessive competition in searching out new audiences and patrons. There is also a drift toward the "coarsening" of public life (to use Kraus's term), in which there is a massive drift toward the "lowbrow" end of consumer preferences. A major problem facing Chinese artists today is in deciding what type of art is to be produced (and consumed), with many would-be artists being forced to decide whether they can live with what Kraus refers to as the "cheapening" process associated with this drift toward the lowest common denominator of mass taste. If they choose to shun the path of the market, artists face the prospect of banishing themselves to obscurity and poverty.

In the Maoist era, the arts were regarded primarily as a tool for mass mobilization of the population, and the major task for the CCP's propaganda officials was to encourage and provide financial support for artworks that adhered to the party line. Kraus (1995) points out, however, that propaganda work was never as centralized in China as it was in the old Soviet Union, and the job of censorship was often left to writers, editors, and local officials in the hope that they would effectively censor themselves. This worked most of the time to suppress "unorthodox" or dissenting ideas and productions, but it began to come apart at the seams when or if the sense of common cause among the people wavered. This is precisely what has happened in contemporary China, with many new films dealing with controversial topics (such as HIV/AIDS, homosexuality, political dissidence, and coal-mining accidents) being banned almost automatically. Surprisingly, this situation has not choked off all the art and culture operating at the margins of political acceptability, primarily because the huge expansion in all media outlets that has accompanied the economic reforms overloaded the CCP's control mechanisms (Lynch 1999). For this reason and also (according to Kraus at least) because artists' professional organizations began to meet with some measure of success in protecting their members from political criticism, cultural productions and artworks that were far from being politically correct began to appear in many parts of China throughout the 1990s and into the new millennium. Although such events were liable to be attacked sharply by the government's censors, particularly immediately after the Tiananmen Square massacre (Kraus 1995, 177), the net result has been the emergence of an impressive new body of artistic and cultural work (Wang 2001).[11]

From a historical perspective, it is important to note that a great deal of China's politics during the past century has been centered on the task of reforming culture, based on the belief that traditional Chinese culture is fundamentally feudal and must be radically restructured, often using Western norms as the model. Only this way, so the old argument goes, can China be transformed into a "modern" state. As early as the 1920s, the emerging CCP, believing that modernization could occur only by replacing feudalism with cosmopolitan culture, began to prepare for a broad purge of traditional Chinese culture. Once the party was entrenched at Yenan, however, and later when it finally came to power in 1949, this situation changed significantly. From that time on, party leaders realized that the only way to radicalize the peasants was to force urban intellectuals to honor them by developing national forms of culture that were heavily rural based (Kraus 1995).[12] Some of the individuals whom Kraus (1995) refers to as "radical cosmopolitans" within the CCP were hoping that, after gaining power and mobilizing the peasantry, the party would eventually be able to return the country to a May Fourth type of urban, international culture.[13] With hindsight, we now know that this did not happen or not until long after Mao's death. Ironically—and we can perhaps imagine Mao turning over in his grave—it was market reform that was finally to "internationalize" Chinese culture, and this certainly did not happen in the way CCP intellectuals had anticipated. The huge amount of new popular culture notwithstanding, an argument can still be made that the May Fourth call to bring cosmopolitan culture to China has been reinvigorated in new ways during the postsocialist era.[14]

In addition to a proliferation of domestic cultural productions, in the new millennium the Chinese public has been exposed to a wide range of foreign art forms—often arriving in a haphazard order, juxtaposing "high" art (like the ever-popular events featuring the Three Tenors) with more "lowbrow" television dramas and blockbuster movies. In his evaluation of these developments, Kraus (1995) concludes that "diversity coupled with diminished political control has freed space for a newly autonomous aesthetic domain in which artists explore the implications of art separated from politics" (178). As the CCP slowly relaxes its grip, artists and critics from all walks of life have been involved, with little obvious input from the state, in momentous debates about what signifies beauty and aesthetic standards in China (Wang 1996). Reduced artistic dependence on the state means that instead of focusing on works that are social and political in nature, artists and cultural producers have been increasingly likely to approach the once-dangerous territory of personal exploration into the domains of emotion, sexuality, and psychology. As a way of exploring some of these contemporary artistic and cultural developments more fully, the remainder of this chapter examines two dimensions of artistic and cultural production in China, focusing first on literary works and then on filmmaking.

Chinese Literature in the Postsocialist Era

After Mao's death in 1976, there was a gradual relaxation in the state's hold over literature (and the arts more generally) in China. Memories of earlier repression and even of the murder of artists during the antirightist movement of 1957 and throughout the Cultural Revolution helped produce what began as a trickle and ended up as a torrent of the so-called literature of the wounded (sometimes also referred to as "scar" literature). Some of this literature has been marketed outside China, with one or two authors becoming household names, but much of it is either too repetitive for Western tastes or, in the case of exiled Nobel Prize–winner Gao Xingjian, too obscure to sell well. The third plenum of the eleventh party congress in 1978 invited China's intellectuals and artists "to liberate their thoughts, to break into previously forbidden zones, and to not fear a return to repressive policies" (Link 1983, 20). In spite of these words, the state has seen fit to tighten up on writers at intervals, as was the case in the early 1980s and again in the period after the Tiananmen Square massacre in 1989, when there was a shift back to the "pre-thaw" climate that had formerly faced intellectuals and artists.

Also emerging at about the same time as the scar literature was a new genre sometimes referred to as "reportage": writing that was more confrontational in its efforts to expose government corruption and official instances of wrongdoing. Much of this work was intentionally direct, but sometimes writers tried to be more elliptical in their work, in part, perhaps, as a deliberate tactic to confuse the state's censors. In one section of a poem called "Reply," for example, Bei Dao refers obliquely to the dark days of China's recent past (referring to them as the "Ice Age"), and at the same time he asks why the aftereffects of such darkness have lasted for so long.[15] Bei Dao's work is often interpreted as a deliberate example of resistance literature. In a poem published on the cover of an issue of *Tintian* (Today), a literary journal published in exile by Tiananmen-era expatriates, Bei observes, "Power depends on yesterday," but "Literature always stands facing Today" (Schuman 2009). And in the opening stanza of what is now considered to be his most famous poem, "The Answer," which has become something of a rallying call for the Tiananmen Square protestors, Bei writes that "the gilded sky is swimming with undulant reflections of the dead" ("The Spend Is Nigh" 2009). We can assume that from the safety of exile, Bei Dao is free to say what he really means, and as Jing Wang (1996) observes, his words here tell us "worlds about the antagonistic and agonistic relationships between 'Power' and 'Literature,' and about how the Chinese literati are constantly mapping out, consciously or unconsciously, their own positionality against a rival who is both real and imaginary" (197).

Among the prose writers most closely associated with the reportage literature of the late 1970s and 1980s and writing in a much more journalistic style than Bei Dao is Liu Binyan, who was a former *People's Daily* correspondent and whose best-known work is *People or Monsters*, and the one-time minister of culture for China, Wang Meng. Wang—often writing under a pen name but well known to most of his readers in official circles—produced many stories dealing with different types of official corruption, arrogant officialdom, and party factionalism, and he continued writing into the 1990s, supporting humanism and criticizing the authoritarian nature of the Chinese state. The fact that he was allowed to continue writing throughout this period was indicative of the new era's tolerance and acceptance, but in the mid-1990s, ironically, critics began accusing him of abandoning his earlier principles and selling out to market forces.

The first decade of the reform era (the 1980s) was marked by an often seething conflict between those writers who relied on the new commercial culture and others who felt that Chinese literature had strayed too far from its own culture. The struggle between these two positions has been described as a culture war (Schoppa 2002, 431) and also as part of what Jing Wang (1996) calls China's era of "High Culture Fever," which she defines as a "post-revolutionary fever about knowledge and enlightenment . . . an era in which miracles and superstitions triumphed over rational forces" (38). Wang also observes that the early and mid-1980s ushered in what amounted to a "symphony of unmitigated optimism" in China (37), accompanying the state's modernization program, which was steering the country toward previously unimagined prosperity. She argues that during this period China's intellectuals "not only collaborated with the Party in its reconstruction of the socialist utopia, but . . . busily proliferated their own discourse on thought enlightenment" (37). Wang concludes her interpretation of this period by suggesting that there was movement toward the perfection of the "twin projects of modernization and enlightenment" in China, a movement that was to some extent halted by the more dystopian views of the future that dominated cultural productions at the end of the decade (views that were both manifested in and represented by the Tiananmen Square tragedy).

China's literary critics seem to agree that the period from the late 1980s to the mid-1990s was dominated by the works of Wang Shuo, who has been described as the epitome of the individualist in a nation caught up in the often sordid business of getting rich. Wang became an icon of popular culture in the new China and a continuous best-seller; many of his books were translated into television series and movies (often with himself as the scriptwriter). One view of Wang's literary output was that it "captured the crude vitality of the entrepreneur unbound, the loose world of the modern criminal, and the boredom and amorality that occasionally led good girls into the arms of bad men" (Barmé and Jaivin 1992, 217). His stories are usually populated by marginal characters who are street smart and who operate at the very border of respectability, which explains why they are referred to in Chinese as *liumang* characters, the word *liumang* translating roughly to "hooligan."

Barmé (1999) suggests that Wang Shuo's writing serves as "shorthand for certain urban attitudes that have been quintessentially expressed not in the tones of overt dissent but more often in the playful and ironic creations of popular culture" (63). In spite of his appeal to the masses, Wang clearly has a serious side; Barmé, in fact, refers to him as "a playful writer of serious intent," arguing that he has a perceptive knack for being able to see the funny side of the current situation. Wang Shuo's books and the films and television shows made from them offered popular entertainment for the masses throughout the 1990s, and he is still popular today although primarily as a scriptwriter or occasional film director. It is also apparent that Wang's work is a form of "coded" political writing, which generally remains acceptable to the censors in spite of its satirical edge. In this sense, his books represent a more compelling form of art than much of what is currently available in the mass market, primarily because they have acted as a "literature of escape" during the upheavals of the reform era (Barmé 1999, 63).

One way to interpret the popularity of the *liumang* culture is in the context of the constant turmoil the Chinese people have been living through during the past two decades. In spite of an expanding economy throughout the 1980s and the obvious fact that the power of the CCP had been undermined on several fronts, the evidence suggests that many Chinese people were at the time still confused by the seeming contradictions surrounding them. In many ways—and especially in the economic realm—people were much freer than before, but their words and actions were still closely supervised by the state, and in reality they still had no political freedom. The Chinese were living in two very different worlds: the openness of the capitalist marketplace had been superimposed on what was still a secret society dominated by an authoritarian elite. This resulted in what Barmé (1999) calls an "existential malaise" that was accompanied by significant psychological tensions (62). The cultural uncertainty this produced was, according to Barmé, "part and parcel of the existential sociopolitical crisis that the Communist Party variously attempted to ignore, escape from, avert, and even co-opt" (62). Unlike the situation in the recent past, however, today it is no longer an option for the state simply to squash all potential dissidence by military action or to counter it with traditional party propaganda, and it is this new space of uncertainty that is inhabited (and exploited) by Wang Shuo and his motley crew of characters.

The typical *liumang* character is a person who rides around the city on his bicycle (later such characters were more likely to be on motorbikes, as with the courier Mardar in Lou Ye's 2000 film *Suzhou River*, and today the *liumang* is no doubt driving a car!). He is constantly on the lookout, although languidly, for any action that might be coming his way, and all the time he looks even more sinister by his reflective sunglasses, ensuring that no one can see his eyes. As harmless as this character may seem, there is, perhaps just below the surface, the possibility of a much darker persona. John Minford (1985) even suggests that the *liumang* is more than just a shady character; he could be any one of a variety of urban ne'er-do-wells: a "rapist, whore, black-marketeer, unemployed youth, alienated intellectual, frustrated artist or poet—the spectrum has its dark satanic end, its long middle band of relentless grey, and, shining at the other end, a patch of visionary light. It is an embryonic alternative culture" (quoted in Barmé 1999, 63). Barmé (1999, 64) also points out that the *liumang* personality has a history in China's urban culture that dates back to the late Qing dynasty, when the term was first used to describe the petty criminals and minor hooligans who hung around the dockland areas in Shanghai as the city's underworld culture expanded. Also originating in this era was the connection between the *liumang* and a variety of sexual misdemeanors, ranging from premarital sex all the way to gang rape. Barmé even notes that the phrase "to play the *liumang*" is still sometimes used in everyday Chinese speech today to describe overt sexual harassment (64).

It is also possible to expand the notion of the *liumang* and to apply it in a more directly political way, either as a form of resistance or as a way of describing the malaise that is rampant in the contemporary Chinese city, as manifested in its growing rates of urban crime and corruption (Farrer 2002). Wang Shuo, however, for the most part seems to use the term more loosely than this to describe the often unscrupulous and unprincipled activities of petty entrepreneurs in the new China. *Liumang* characters seem to have no time for cultural pursuits, they operate at the fringes of and sometimes just beyond the law, and in most cases they appear to have no moral or social conscience. But Wang's usual hero is not a rapacious capitalist who reinvests profits to expand production; he is much more likely to be the type who takes all he can find and squanders it until it is all gone—as if there was no tomorrow. In some circumstances, he might behave like the small-time crook who skims off a percentage of everything that passes through his hands, takes every opportunity to eat and drink lavishly at someone else's expense, and exploits every bit of power that is available to him.[16]

In a typical Wang Shuo story, all these themes may be intertwined, and according to Barmé (1999), Wang also throws in a measure of subtle satire, often aimed at CCP officialdom, which occasionally results in his work being banned by the censor. In Wang's 1989 story "An Attitude," for example, the hero is arraigned in court for "practicing literature without a license" (an obvious parody). The judge asks him how he would behave if he someday became king of the cultural world in China, to which he answers that he would allow those who submit to him to prosper, but those who cross him would perish. Presumably with tongue in cheek, he also says he would be loyal to his partners whatever happened, and even if he were forced to purge them, he would be sure to let them down gently! We can assume here that Wang is referring to the situation that existed in China before and during the Cultural Revolution, when writers could be censored, excommunicated, and even jailed for what they wrote.

In other works, Wang pokes fun at China's past and even at important elements of Chinese culture and nationalism. In what might be seen as an allegory of China's obsession throughout the 1990s with hosting the Olympic Games, his novel *Please Don't Call Me Human* mixes historical fact and sheer fantasy to create a bizarre mockery of China's past and present. A group of profiteers in contemporary Beijing are searching for a new athletic hero to restore China's damaged pride after a humiliating defeat in an international wrestling competition. The person they eventually select—a character by the name of Tang Yuanbao—is by no means a warrior; he is, in fact, a simple Beijing pedicab driver who just happens to be related by birth to one of the leaders of the Boxer Rebellion (the pun is probably intended here). The group has to prepare Tang for his first competitive bout, and as the leader points out,

> My primary concern is with the image of Tang Yuanbao. He'll be representing the entire Chinese race, and just winning a match won't be enough. This must be a rout in every sense of the word. That way he takes glory with him wherever he goes, the subject of veneration every step of the way, a national hero. . . . Comrades, we must act prudently. Just beating up some foreigner won't do it. Our ultimate purpose is to establish a national model. Do not, under any circumstance, underestimate the importance of our work, for what we are engaged in now is an enterprise of historic proportions—the mere thought makes me shudder. (Wang Shuo 2000, 45)

Once he has been selected, Tang is introduced to the public in a parade, followed by a banquet at a special restaurant where the food is considered to have almost supernatural qualities. Wang here manages to poke fun at the Chinese obsession with banqueting and feasting. In preparing the diners for their meal, one of the group leaders notes that

> every dish on today's menu has profound ties to Chinese culture, and will cause us all to ponder our great nation. Why not call it "cultural cuisine," the partaking of which can be the equivalent of completing a spirited course in Chinese culture. In the history of the world, ours is the only civilization whose food has been passed down without change for generations, which is why for millennia China has stood tall among the nations of the world. We can cut off our queues, unbind our feet, even change into Western suits, but we cannot stop eating. That has formed a national characteristic, instilling pride in us as descendants of China's earliest rulers. Our ancestors took great pains to keep us from forgetting our roots. Now dig in. (Wang Shuo 2000, 73)[17]

Wang Shuo's work has been heavily criticized on a variety of counts, including charges that his views on the new sexual culture in China effectively act as a call for Chinese people to abandon all forms of sexual inhibition. But miraculously, he has more or less been able to stay out of real trouble and has even become something of an urban legend: quotations from his books have been mass-produced and printed on T-shirts. An essentially positive picture of Wang Shuo and his work is put forward by Chinese cultural critic Dai Jinhua (2002), who argues that Wang's heroes may actually represent a new type of Chinese person (*xinren*) at the end of the millennium, someone who was able to come to grips with living in a society in which ideologies were clearly in a state of flux. The benefit to Wang's readers, according to Dai, was that almost for the first time a modern writer was telling them stories that they could relate to and that allowed them some release and satisfaction.

By the end of the 1990s, as Wang Shuo began to fade out of the picture, young writers in China began to explore new territory, with some of the most popular and talked-about works steering clear of Wang's fondness for hoodlum life, choosing instead to explore the once-forbidden topics of sex and sexuality. Earlier in the decade, the publication of Jia Pingwa's (1993) *The Abandoned Capital* (1993) caused a sensation, selling more than 500,000 copies in just three months, and although the story is tame by Western standards, being cast in what amounts to soft pornography, it is a description of the sexual exploits of a middle-aged writer. In his interpretation of Jia's book, Barmé (1999) suggests that it offers a reasonably representative reflection of what any Chinese person could see happening during the 1990s, when life in the new cities had obviously left behind most of the former values of Chinese civilization and communism and was focusing instead on a new era of urban experimentation and alienation. Barmé quotes what one Chinese cultural critic said in describing this situation in the reform era: "All we dream of now and hope for in the future are money and sex. The unprincipled process of moneymaking and sexual gratification has gravely undermined the pillars of civilized society" (184).

In spite of this advance warning, most Chinese readers were probably unprepared for the next development in this direction, when new books written by young, sexually liberated women began to appear in the bookstores (Jiang 2003). The most popular example of this genre is Zhou Weihui's (2002) blockbuster novel *Shanghai Baby*, which was officially banned in China but was widely available through black market outlets (the ban on the book no doubt contributed significantly to its vast popularity). As one critic commented, more significant even than the heroine's detailed exploration of her sexual exploits was "the novel's examination of life, freedom, love, and death" in the new Shanghai. This same critic ultimately questioned whether China's women could in real life hope to find anything like the sexual and emotional liberation enjoyed by Zhou's heroine. Other critics panned Zhou's book, describing it as derivative and shallow. In a review appearing on the Amazon.com website, for example, a reader described the book as problematic, suggesting that it "wants very much to be . . . shocking . . . [but] it contains little more . . . than some random sex scenes." On the other hand, the same reviewer felt that the book did "accurately portray how beautiful, cold, and ultimately helpless it feels to be young."[18]

More sensational still was the reaction to Hong Ying's novel *K: The Art of Love*, which is set in China in 1935 and contains long and very explicit descriptions of the lovemaking between a westerner and a Chinese woman. The woman, Cheng Lin (*K*), is the wife of the head of the English department at a major Chinese university. It turns out that she is adept in the "Daoist art of love," but because her husband is impotent, she begins to coach a junior member of the faculty: a British visitor to the university. The most startling thing about the book, described on the cover as "the Chinese *Lady Chatterley's Lover*," is that it was based loosely on the lives of real characters, and in 2001, a Chinese woman living in London actually filed a lawsuit against Hong Ying, claiming that the book had caused her significant "spiritual damage" by libeling her late parents (the fictionalized head of the department and Cheng Lin). As was the case with Zhou Weihui's book, publicity (generated by the lawsuit) guaranteed that Hong Ying's book would be a best-seller. The main title of the book, *K*, is the same in the Chinese edition as in the English edition (even though the letter *k* is both unpronounceable and incomprehensible in Chinese). What all of this amounts to, in the words of one reviewer, is "a kind of avant-garde trifecta, combining experimental gestures, sex, and the glamour of fact" (Lanchester 2003, 25).[19]

To draw this section to a close, it is useful to look at one final piece of writing that viciously (though humorously) brought attention to the state of affairs in the great Chinese rural backwaters: Yan Lianke's award-winning novel *Shouhuo* (translated as *Enjoyment*, or, more realistically, as *Grin and Bear It*) was published in 2004 and won a few awards before being officially banned. The story is set in a village that is so cut off from society that it has had virtually no contact with the outside world and hardly appears on any maps. In addition, and perhaps as result of its dire poverty and isolation, almost the entire population suffers from one or another physical deformity or disability. To launch an effort to "stand up" and be counted in the new age of profiteering, a county official forces the disabled villagers to become part of a traveling freak show. Audiences can pay to let firecrackers off next to a deaf man or race a one-legged villager, all of which was intended to raise enough money to bring Lenin's embalmed body from Russia back to the village in an effort to cash in on communist nostalgia and attract tourists to the village. McGrath (2008) refers to the scheme as "a perfectly, darkly comic illustration of postsocialist marketization" (206), emphasizing the fact that in the new China, everything is up for sale. In the end, the plan fails, and the villagers drop out of the wider society. The county official, wanting to join them, maims himself so that he can become one with the disabled villagers. As McGrath notes,

> The greatest irony may be that this fantasized Chinese space that can . . . [hope to] escape Chinese modernity is populated entirely by deformed people, who still in some sense appear less sick than the society they seek to escape . . . [and] in seeking to step off the juggernaut of modernity [in fact having been left far behind by it] . . . the Shouhuo villagers in fact suggest the utopian potential of a state of abjection—in the original meaning of the word as being cast-off, rejected, or excluded. Their physical abnormalities are a metaphor for their desire to exist outside of modern history, a desire as aberrant as it is perhaps sane. (207)

The destitution of the villagers, although carried to extreme limits in the service of art, realistically portrays the actual conditions faced by many of the millions left behind in the Chinese countryside: the other side of the economic miracle, left far behind by neoliberalism, Chinese style, what David Harvey (2003) refers to as "accumulation by dispossession" and others, not quite so politely, call "gangster capitalism." All over China, this process has involved a direct, involuntary transfer of wealth from the many to the few, as all kinds of collective enterprise, in the cities and in the countryside, have been closed down or sold off to private interests with little compensation for the workers or previous owners.

Chinese Filmmaking in the Reform Era

China's film industry has experienced a major transformation since the end of the Cultural Revolution, both qualitatively and quantitatively, as part of the nation's push toward modernization. During the Maoist era, the CCP exercised significant control over all films that were made in China. For the party, film was considered more than just art: it was an integral part of what Mao Zedong referred to as "cultural work," which he considered to be "a forceful weapon to unite and educate the people as well as to fight against and annihilate the enemy" (Mao Zedong 1975, 111). Mao believed that literature and the arts should be evaluated from the perspective of the "two services" (serving politics on the one hand and the people on the other) and the "two hundreds" (letting "a hundred flowers blossom" and "a hundred schools of thought contend"). In both cases, the former of these considerations was considered the primary goal, with the latter representing the means of achieving those goals. With the establishment of class struggle as the party's major task before and during the Cultural Revolution, greater emphasis came to be placed on a film's political orientation than on its artistic merit or its aesthetic quality. Not surprisingly, what suffered during this era was both the quality and the quantity of films produced and the diversity in the type of films that could even be considered. Relatively few films were produced during the Cultural Revolution, and those that were released were stamped with a uniformity dominated by the dictates of socialist realism.

As the reforms took hold, China's leaders gradually came to accept that political correctness need not be the only or the major criterion for determining the merit of a film. There is still a threat of censorship for films that are directly critical of the government and its policies or films that deal with topics that the government feels are either potentially damaging to the social order or "unhealthy" in some way, which is usually the case for films made about homosexuality in China, like *Lan Yu* (directed by Stanley Kuan, a Hong Kong–based director) and *East Palace, West Palace* (directed by Zhang Yuan). Filmmakers are no longer required to sing the praises of the CCP to get their films approved, and some have actually been very skillful at using their work to send out a quasi-political message. Sometimes, however, even the most subtly portrayed political criticism is detected by the censor, as was the case with Chen Kaige's *Yellow Earth* (1984), which paints a striking picture of peasant life in northern Shaanxi province before 1949.

In addition to the controversy surrounding its political message and in comparison to the unambiguous moviemaking of the recent past in China, *Yellow Earth* represented a breath of fresh air in the filmmaking industry, in part because it broke dramatically with the propagandist paradigm that had dominated filmmaking in the prereform era. Silbergeld (1999) has suggested that the making of *Yellow Earth* represented a shift away from a focus on the Cultural Revolution to a focus on China's cultural roots, following a trend set in the literary field. *Yellow Earth* was also the first "independent-minded cultural critique of the entire socialist experiment in China," departing in almost every possible way from the formulaic films of the recent past. As McDougall (1991) has suggested, *Yellow Earth* represented an entirely new era in Chinese filmmaking, in part because of its sharp reversal from and rejection of the canon of socialist realism. In the past, anything in a film that looked or sounded like art for art's sake, ambiguity, or abstraction had to be eliminated if the film was to escape the censor's knife. Central to the Maoist aesthetic standard during the revolutionary era was the purging from film (and, in fact, from all forms of artistic production) of the so-called middle characters (Silbergeld 1999, 18), which meant characters that a Western audience might think of as "real" people or as people with complex and unresolved motives and attitudes. In *Yellow Earth*, this would describe *all* the characters: the peasants, for example, are portrayed not as the vanguards of the revolution but as near-silent, superstitious victims of their environment. In most films made in China before the early 1980s, the peasants were depicted as heroes; they were usually tall, uncomplicated people, strong and resolute, suntanned and healthy, always looking ahead, and never looking sideways. These visual characteristics were also evident in the socialist realism posters and artwork that were common in China before the coming of the reforms.[20]

In *Yellow Earth*, there are no such characters; there are, in fact, no heroes, no villains, and no theatrics—only ambiguity, long shots, and extended silences—and probably for that reason, public reaction to the film was initially underwhelming, to say the least.[21] Most Chinese filmgoers found the peasants portrayed in the film to be sullen and their silences too long to be interesting. *Yellow Earth*, in other words, looked too much like a foreign art film to please Chinese audiences, and ever since its release in China, cinema critics have argued with each other about the meanings of the film. Silbergeld (1999) raises the possibility that the film's ambiguity is deliberate: if Chen had wanted to launch a criticism, in other words, the only way he could do it was to try to fool the censor by making the whole thing hopelessly ambiguous, which he apparently succeeded in doing.

In some ways, the transition that occurred with the making of *Yellow Earth* reflected similar transition that was taking place in the world of art in China during the reform era. The mid-1980s marked the appearance of a wide range of avant-garde works produced by Chinese artists, some of which were funded and exhibited by foreigners (including Hong Kongers) who were willing to invest in up-and-coming artists from the mainland. After *Yellow Earth*, cinematographer Zhang Yimou branched out on his own and within a few years had become the most successful and internationally best known of China's Fifth Generation filmmakers. He exhibited a fondness for exotic scenes, peopled by either barbaric or tragic characters with exhilarating or suffocating stories behind them. Although much of his output met with mixed reviews in China, internationally his films have won him more awards than all other Chinese film directors combined. One of Zhang's most common themes is primitivism, which is evident in the films he made during the late 1980s and early 1990s. Primitivism as a literary mode expresses a form of nostalgia for a precivilized way of life (see Silbergeld 1999) and is analogous in some ways to the "roots-searching" genre of post–Cultural Revolution literature and poetry (Wang 1996). The assumption here is that there was a paradise-like era some time in China's past: a golden age when innate instincts and passions prevailed over the dictates of reason, somewhat akin to our own idea of the "good old days," when life was lived more spontaneously and instinctually with no trace of the anxieties and frustrations we associate with the modern world.

A primitivist plot generally features characters who have escaped from the complications and alienation of modern civilization into the simplicities of a long-lost and presumably idyllic life. Zhang's masterpiece, *Red Sorghum* (1987), is such a film, presenting the audience with a sparkling example of primitive, natural existence. The film is populated with simple, lusty characters who lead unrestrained, almost barbaric lives but who are portrayed as somehow heroic and admirable. The boy narrating the story in the film tells how his grandfather rescued his grandmother from bandits, how they later consummated their relationship in the sorghum fields, and then how they took over a winery, making liquor from the grain (sorghum). All the men in the film, from the sedan-chair bearers to the workers at the winery, are rustic and uninhibited. They drink to their heart's content, sing loudly, and behave badly. In their remote and primitive land, the characters revel in being free from social restraints, as if they have reverted to the innocence of wild animals, satisfying their no-longer-hidden urges without taking social values into consideration.

In Zhang's next films, he replaced the charm of the simplicity and masculinity characterizing *Red Sorghum* with a much more depressing human landscape. In *Judou* (1990), for example, instead of the wide-open spaces and carefree lives of his former protagonists, Zhang presented the audience with the claustrophobic atmosphere of a dye works, which had a suffocating effect on the characters who worked in the factory. The heroine is deprived of the freedom the grandmother in *Red Sorghum* enjoyed, though Zhang, by presenting modern life and progress so negatively, may again be showing his nostalgia for a life unfettered by restraints or traditions—a life that has long since disappeared in China. Inside the factory compound, that primitivism is still in evidence, but this time it takes on the dimensions of ignorance and decadence. Instead of the "noble savage" types populating *Red Sorghum*, the characters here were shown as frustrated, incomplete, and unfulfilled individuals. In *Judou*, Zhang moves to the other extreme: from the triumph of primitivism to a scathing criticism and denunciation of the evils that this primitivism is associated with. It is perhaps significant to note that this shift may have worked to make Zhang's later films both more controversial and less commercially successful at home; *Red Sorghum* was an immediate success both in China and overseas, but *Judou* was banned in China for some years.[22]

The Chinese (but world-renowned) cultural critic Dai Jinhua (2002) has been scornful of the movies and the motives of Zhang Yimou and the other Fifth Generation filmmakers in her discussions of postcolonialism in contemporary China. In their discussion of Dai's work, editors Jing Wang and Tina Barlow suggest that Dai is using the term *postcolonialism* in this instance to describe the peculiar cultural conditions that existed in China during the 1980s and 1990s and that she places the burden of Western imperialism on the directors themselves. Dai argues that Chen Kaige and Zhang Yimou have effectively been captured or "colonized" by Western cinematic aesthetics, which have trapped them in "an orientalism that internalizes the fantasy of the Other/West for an imaginary China that is premodern, a splendid spectacle of exoticism and an ancient land ruled by repressed desire" (Dai Jinhua 2002, 14).

In the postsocialist era of state-subsidized filmmaking, China's budding directors faced a serious obstacle: the realization that although filmmaking is an art form, it is also a part of everyday commerce and industry. Put simply, to make a film good enough to sell at home and abroad is an extremely expensive and financially risky business. Some of the Fifth Generation filmmakers rejected the avenue of pure artistic narcissism that was beginning to characterize Chinese art and literature in the 1980s and 1990s, choosing instead to focus their work on the Western (especially European) market for art films, much of which was channeled through a series of international film festivals. The assumption (and the reality in many cases) was that a good review—and preferably a prize at one of the many summer film festivals—would not only increase sales abroad but also attract future investors willing to encourage the making of new Chinese films.[23]

Dai suggests that this strategy worked; in fact, it may have worked too well because as filmmakers were fleeing through "the narrow gateway," they often fell into a different trap: "Since securing foreign investment . . . had become the . . . focus, the prerequisite for filmmaking became the representations of an Orient that was palatable and intelligible for Western viewers." By internalizing this Western cultural perspective, China's filmmakers were forced or forced themselves to reconstruct their narrative subjects according to Western expectations. As Dai reports, "Films . . . that managed to squeeze through these narrow gateways . . . simply fell under the yoke of one discursive power in their attempt to escape from another" (Dai Jinhua 2002, 50).

Although the state still determines the total number of films to be made domestically each year in China and also how many films will be imported, there have been some important structural changes in the industry in the postsocialist era. Film studios, for example, are now treated as enterprises, with the director rather than the party secretary overseeing production and being responsible for all financial affairs. Studios now have to raise all their own revenues rather than relying on the state for support, and although the film distribution companies are mostly still state run, they now pay the studios according to sales rather than a fixed allocation as in the past. As their contribution to the reform-era restructuring process, filmmakers are now required to form their own groups to contract with the studios if they wish to make a film, and from that point onward all executive decisions are their own. This also allows directors to seek out potential investors, including entering into joint ventures with overseas film interests, with other film studios, and with nonmedia enterprises. In 1993, films that had begun life with direct investment from the state-run studios made up less than 20 percent of the total number produced in that year. In the past, films were shown to the public mainly free of charge or for very low admission prices, meaning that producers and directors in the film industry did not have to work particularly hard to make a film popular; the only real requirement was that the state's censors be happy that the state's political message had been presented. Today by comparison, the box office is directly in control of a film studio's future, and as the costs of filmmaking have increased, so have ticket prices. Audiences have also become more discriminating, and with growing competition from television and other cultural or recreational outlets, film studios have had to pay special attention to their potential viewers.

A persistent theme in the story lines of the contemporary Chinese film industry has been the production of films that portray individuals trying to change their lives and social status, often with characters that are willing to "jump into the sea" (i.e., plunge into the world of private business, clearly reflecting the zeitgeist of the postsocialist era), as well as films that focus on marriage and sexual relationships or on other themes familiar to Western filmgoers, such as love triangles and extramarital affairs, themes that were strictly taboo in the old prereform days. Another consistent trend running through the reform era to the present day has been the Chinese film industry's focus on film scripts depicting social and interpersonal struggle. Earlier, such films would probably have originated or been inspired by issues of class conflict or questions of political belief, but in the contemporary era, filmmakers have been freer to focus on social or psychological dramas, individual quarrels between citizens, and conflicts between citizens and the state—the situation so poignantly portrayed in Zhang Yimou's much-acclaimed *The Story of Qiu Ju*. It is also important to note a shift away from the depiction of traditional families in Chinese films, with an increasing prevalence of single people as heroes and heroines; this would be consistent with the industry's shift toward investigations of the self and personal issues.

The huge potential demand (and ability to pay) of urban Chinese audiences has also resulted in a significant shift toward films that have an urban focus, with a corresponding shift away from films set entirely in rural areas. In addition, a new genre of filmmaking involves the issue of rural–urban interactions, focusing on some of the problems and frustrations encountered by rural people when they travel to cities (or when they do a significant amount of business in nearby cities). This is the situation in Zhao Xiaowen's film *Ermo*, which features a determined rural woman who sells her homemade noodles in a nearby city. Her dream is to earn enough money to buy the biggest television for sale in the city, and to this end she spends much of her free time gazing enviously through the window of the television shop in town. The story line can be interpreted as a critique of urban China's postsocialist infatuation with consumption and getting rich. Ermo's "unwitting desire . . . for freedom from poverty and rural isolation" drives her "pursuit of modernity in a box [the TV] . . . that has been earned with her feet and her blood" and which subsequently "enslaves her body and soul" (Silbergeld 1999, 91).

Another film featuring the rural–urban transition is Zhang Yimou's *Not One Less* (1999), which was his first film to deal with contemporary China. The film was notable at the time because it was set in a poor village in a remote and mountainous part of the countryside and because the actors in the film were mostly local people playing themselves. Wei Minzhi is a thirteen-year-old girl who is called on to teach the grade-school class when the village's only teacher is called away to visit his sick mother. The village is desperately poor, and Wei has only one piece of chalk for every day the teacher is away. Because the school is funded on a per capita basis, she is told that she must try to keep all the students at school (hence the film's title). In poor areas, it is not unusual for parents to take their children (especially the girls) out of school, either to help at home, to work in the fields, or to be sent to the city in search of a job, so Wei becomes convinced that keeping all her students in class is more important than anything she can teach them. Unlike many of Zhang's films over the years, *Not One Less* is not a melodrama or a tearjerker (like *To Live*), it does not have gorgeous scenery and evocative colors (like Zhang's primitivist films *Judou*, *Red Sorghum*, and *Yellow Earth*), and it does not feature glamorous lifestyles and elements of high fashion (as in *Shanghai Triad*). It is rather a matter-of-fact look at a desperately impoverished area, in the shadow of a city that becomes increasingly attractive to the local people. As such, the film looks at a way of life that is very far removed from the new wealth associated with the cities and the coastal regions of postsocialist China.[24]

The consistent popularity of crime stories as film plots in China suggests that viewers were interested in topics that were very different from their everyday lives (escapism) and that the public was well aware of the growing prevalence of crime in Chinese society, especially in the cities. The near obsession with issues of crime and corruption, with the result that some of the most popular characters are portrayed either as law enforcement officials or as criminals, represents a clear shift away from characters playing the part of peasants and military personnel, as was the case in many of the films made in the prereform era. Many films have focused on life within a criminal or semicriminal urban environment, including the film voted as outstanding Chinese language film at the 2000 Hong Kong International Film Festival, *Suzhou River*. This was the only entry from mainland China and the second film of a young director, Lou Ye, who is one of the so-called Sixth Generation of Chinese filmmakers, these being graduates of the Beijing Film Academy who began making films in the 1993–1995 period. The film resembles a contemporary film noir (Hitchcock style), set in the ugliest and most run-down neighborhoods of present-day Shanghai. Most of the action takes place along foul Suzhou Creek, which winds through decaying warehouses and decrepit factories on its way to Shanghai's glittering waterfront. Some film critics have recognized a connection between the characters of *Suzhou River* and those who normally inhabit Wang Shuo's stories—marginal, *liumang*-like characters.[25]

Although they make or have made very different films, Fifth and Sixth Generation filmmakers have at least one thing in common: their uneasy relationship with the censor, which has helped to create a truly independent film culture in China. For the most part, Sixth Generation filmmakers do not usually sit around waiting for the censor to do his or her work; they go out and raise their own funds to make whatever films they want to make. It is not surprising, therefore, to find that their major characters are often silent, sullen resisters who refuse to give in to the system. In Jia Zhang-ke's film *Pickpocket*, for example, the leading character is a thief with scruples who, unlike some of his old burglar friends—*liumang* characters again—is unable to parlay his talents into economic success in the new era. He is spurned by an old gangster friend, harassed by the police, and cursed by his father, who says he should have drowned him when he had the chance. When he is finally arrested for theft, he is handcuffed to the cable of a telephone pole in a public square and left there to be stared at like a creature behind bars in a zoo (Lu 2003). Sometimes the heroes of Sixth Generation films are loveable villains, as in He Jianjun's film *Postman*, which begins with the leading man reading the letters he should be delivering. When he learns that many of the letters are heartfelt pleas for human contact, he starts to write replies to those same letters; it is an illegal act but one with a prosocial message.

Dai Jinhua (2002) suggested that in 1994 there really never was such a category as Sixth Generation filmmaking in China. She argued this for a number of reasons, including her belief that the new school of directors had not (at that time) produced enough good work that was clearly categorizable as a new genre of filmmaking. At the time, she also hoped that when (or if) a new group of filmmakers emerged in China, they would be able to claim something more original than just being the next group of graduates from the same institution as their predecessors. As she put it, "I harbored the optimistic, perhaps chimeric, expectation that when social transformations shattered cultural heroism, a new generation of filmmakers might be able to appear in their own names, rather than in the name of any new 'generation'" (78).

Later on, Dai reluctantly admitted that a Sixth Generation of filmmakers had in fact emerged in China, but she suggested that their appearance was particularly badly timed. With the exception of Wang Shuo's work, the climate of opinion in China in the early 1990s was clearly not favorable to the type of films these new directors wanted to make. What the people wanted to see was dictated by the new norms of mass culture, which in the visual domain included only what Dai referred to scornfully as "quasi-film" phenomena (television and video). At that time, Dai claimed, art films were "box-office poison," but with some of their better works, the more famous Fifth Generation filmmakers had been able to sidestep this problem because they were often generously funded by international investors who had a specific idea of the Chinese films they wanted to see being made. In Dai's (2002) words, "Riding high on the crest of the wave of orientalism, the Fifth Generation undoubtedly produced a China fever . . . a hunger for Chinese film in European and American art film festivals" (80), but there was little left in the way of funding for the younger hopefuls. When these young directors learned that no one was willing to fund their film projects, they were forced into the margins of artistic space; in fact, many of them "ended up in various venues, joining the nomadic Beijing artist groups: making their living with TV programs, advertisements, and MTV, or doing temporary work on various film production teams, yet still committed to their dreams" (81).

The first Chinese filmmaker to break out of this predicament was Zhang Yuan, who managed to scrape up just enough cash to make what would become the first Sixth Generation film, *Mother*. The film was so poorly funded that Zhang ended up shooting most of it crudely in black and white. It was also produced entirely on location, and all the actors were amateurs. Zhang followed this with his better-known film *Beijing Bastards*, in which he collaborated with China's rock star Cui Jian, and then later with his tour-de-force *East Palace, West Palace*, which dealt with the topic of homosexuality in Beijing. These were all entirely self-financed productions, and they represented a new genre of films in China that would become known primarily as "independent productions." Dai Jinhua's (2002) version of the troubled history of the Sixth Generation leaves the impression that these were a small number of young film directors who wanted to remain independent of both state funding and global capital and who would eventually "break away from commercial culture's ambush of art films" (84). Although the nature and the content of the new films was a subversion of the official system of film production in China, the creative styles adopted by the new directors, Dai argues, were forced on them more by circumstances than by actual artistic choice, and in this sense the new independent filmmakers were operating very much in the style of many young filmmakers outside Hollywood, especially in the global "South."

In spite of this, a few of these new films actually made it to the important art film festivals around the world, and some of them did well. Dai (2002) suggests, however, that their success was in part a result of the thirst or demand for Chinese films that had been started by the Fifth Generation directors and in part a result of the fact that most of the films were labeled as "underground film" (90). She argues that the reviews of these films tended to bypass their actual artistic qualities (or lack thereof), concentrating only on their political significance. (This would be similar in a number of ways to the warm reception the West gave to films from Eastern Europe at around the time of the demise of the communist ruling powers.) What Dai concludes from all this is that the filmgoers and critics at art film festivals around the world were more interested in where the films had come from—and in the fact that they had been shot "underground" or independently—than in their actual artistic merit. According to Dai (2002), because the films of Zhang Yimou and his imitators had satisfied the West's "old orientalist" tastes, the West again "privileged the Sixth Generation as the Other" (90), reflecting Western liberal intellectuals' anticipations or expectations of Chinese cultural conditions in the postsocialist era. Created as a mirror image, this expectation again validated Western intellectuals' mapping of China's democracy, progress, resistance, civil society, and marginal world position. What she is suggesting is that Western intellectuals disregarded not only the cultural reality displayed directly in these films but also the filmmakers' cultural intentions.

Western critics, so Dai's argument stresses, used the work of these new directors to create an imaginary China, a China that they wanted and hoped to see emerging (but that was in fact not emerging). The result was that the films of the Sixth Generation became what Dai calls "scenes in the fog" in the sense that they were highly exposed in the outside world but little known and little seen inside China. Dai reports, for example, that even as one of China's best-known film critics, she heard about most of the films in question only from overseas publications and reports sent to her by friends from abroad. There were of course some benefits that would accrue from this situation in that a path had been paved for the new young directors of the future to follow. As Dai (2002) concludes, "If the 'Zhang Yimou style' used to be a narrow door through which Chinese directors could move on their 'march toward the world,' independent filmmaking now became a shortcut to the powerbase in Western cinema" (91).

This dilemma—between making film as art and making money from film—is clearly evident in the production of independent films in contemporary China. To illustrate this point briefly, it is useful to focus on the controversial film *Frozen* (1996), which was (and still is) banned in China and had to be smuggled out of the country for distribution in the West. *Frozen* provided an early window on the underground culture in China in the immediate post–Tiananmen Square era and an entrée into the world of avant-garde art in Beijing. Most of the cultural productions depicted in the film fall into the category of extreme performance art, which is clearly a fringe activity in terms of drawing crowds and generating income. Some of the performances depicted in the film—including one in which two men eat a bar of soap each with a knife and fork—are attended by a mere handful of other artists and a smattering of baffled passersby. The antihero of the film is a young performance artist who decides to make his own suicide the object of his last work of art. In the middle of summer, he melts a huge block of ice with his own body heat and appears to die of hypothermia. The film's ending is open to different interpretations, and some critics suggest that the artist may actually have faked his death. As an independently made film, *Frozen* is now only legally available outside China, and the filmmaker was forced to use a nom de plume, Wu Ming (anonymous, although everyone in the know acknowledges that he is Wang Xiaoshuai, who later directed a much more popular and more politically correct film called *Beijing Bicycle*).

It has been difficult for many filmmakers to steer a steady course between "selling out" by pandering to the mass tastes of the market and producing works that are too "artsy" to be commercial. In the realm of independent cinema in China, the new master of this is clearly Jia Zhangke, who has now made an entire series of films that focus a sharp eye onto the cultural and industrial wastelands of postsocialist China. Many people in China find his films too slow to watch and often difficult to interpret, but they have been enormously popular among foreigners, especially in the film festivals around the world. Jia first attracted attention globally with a series of films that have become known as *The Hometown Trilogy*, providing an inside look at a gritty industrial city and the lives of marginalized individuals as they struggle to navigate the radically transforming terrain of postsocialist China (*Xiao Wu*, 1997; *Platform*, 2000; and *Unknown Pleasures*, 2002). This notoriety has won over a small but influential audience worldwide, helping him to get support for new projects from sponsors in many different countries. At a retrospective of Jia's work at the Museum of Modern Art in New York in 2010, the publicity flyer describes Jia as the "spiritual grandson" of the Italian neorealist filmmakers of the immediate post–World War II era. Like the Italian auteurs, "Jia's primary interest is the relationship between character and environment—in particular the industry-scarred landscape" (Lauer 2010). Speaking about two of the trilogy films, a press release has this to say about Jia's early films:

> *Platform* [2000] . . . remains his most lauded work . . . [but] Jia's best film is its follow up, *Unknown Pleasures*. . . . More than just another "disaffected youth picture," *Unknown Pleasures* carries the confusion and contradictions facing China's "birth control generation" on its shoulders, with a final scene—a masterpiece of irony—that perfectly summarizes China's generational sea change. The cheesy Mandopop songs that float in and out of the film and throughout all of the director's works provide an ironic counterpoint to the proceedings: the days of rigid Maoist indoctrination may have passed but pop's insipid, sentimental lyrics play like a new kind of proto-capitalistic mass opiate. It's the contradiction that lies at the heart of both Jia's films and China's transitioning state. (Lauer 2010)

One of Jia's later films, *Still Life* (2006), pursues a similar theme but in a different setting. Both main characters are searching for absent spouses: coal miner Sanming's wife left him sixteen years ago, but he has only recently come from his native Shanxi province (where the trilogy films are shot) to look for her. The setting for *Still Life* is Fengjie, on the Yangtze River in Sichuan province, just upstream from the Three Gorges Dam project. He learns that the address his wife gave him—and most of Fengjie in fact—has now been flooded by the reservoir project. He decides to wait for her and gets a job with a demolition crew working on dismantling what remains of the city. The other character, Shen Hong, is looking for her husband: he disappeared two years ago to work in a factory near Fengjie. When she eventually locates him, she lets him know that she now has a lover and that she wants a divorce, but Sanming has hope of reuniting his family still. As film critic Shelly Kraicer (2010) tells us, *Still Life*, like Jia's earlier films, offers a remarkable picture of devastated lives and landscapes in the new China:

> Landscapes are treated . . . [with] long, slow, 180-degree pans that turn vast fields of rubble, waste, and half-decayed, soon-to-be-demolished buildings into epic tableaux. In style, these images seem partially derived from traditional Chinese scroll painting, but have nothing to do with them in content. It is precisely the spectacular ugliness of the physical devastation of the urban environment around the Three Gorges that captures the camera's gaze: an anti-still life that monumentalizes destruction, giving it an awful, sublime grandeur normally reserved for scenes of natural beauty.

Conclusion

A simple way to interpret cultural change in postsocialist China is to think of it as a series of transitions or transformations, working somehow in unison. The first of these was the shift from the "plan" (socialism) to the "market" (capitalism). As artistic and cultural productions are liberated from one source of domination (the state), they are captured by another equally powerful source (the market). In this chapter, by looking at changes in the way people are now spending their time and money, we have seen at least some aspects of the way this shift has been manifested in the patterns of everyday life in postsocialist China. The chapter also showed, making use of a number of brief case studies, how some of the developments occurring in society at large were parlayed into the world of literature and film in the new China. A second transformation involved the downward movement from "high" to "low" forms of culture, as market share and profits become the primary standards of appraisal for new cultural forms. The third transformation would involve the notion of convergence, as art and culture in China are assumed to become increasingly westernized through the mechanisms of transnationalism and globalization.

As attractive as such simple notions of transformation may be, as Wang (2001) notes, in truth they do not have much explanatory value, immersed as they are in Cold War ideology, permeated with elite notions of cultural aesthetics, and insistent on simple binaries such as plan/market, high/low, and East/West. Wang argues that simple dichotomies of this type are based on clichés, and she is convinced that further exploration will reveal many divergent trends within Chinese culture during the past few decades. Her major criticism, however, is that all three interpretations of what has happened in postsocialist China fail to take into consideration the vastly important role the state still plays in shaping culture and cultural trends. The CCP has, at least from the early 1930s on, been concerned about developments in popular culture, has clearly realized the importance of using culture for propaganda purposes, and has seen the need to control the form and expression of culture (Goodman 2001). In the contemporary era, the CCP has certainly adjusted some of its perspectives, but it is still fundamentally concerned with the management and articulation of popular culture, although, as Wang suggests, this has involved a subtle shift away from the coercive control of culture (of the pre- and post-1949 era) in the direction of greater legal regulation during the reform era.

In making this argument and as an alternative to the simple dichotomies mentioned above, Wang (2001) identifies two significant trajectories along which we can see the notion of "culture" moving within the past two decades in China. One of these she refers to as the "popularization of the discursive construction of *xiuxian wenhua*," or "leisure culture" (71). This is a situation in which the state has effectively been able to create a new nation of consumer-citizens as a result of very well calculated policies. She suggests that, freed from the need to mobilize the people through its own political ideology and refusing to allow any significant democratization that would capture the imagination of the people, the CCP has invited and enabled the Chinese people to become members of an egalitarian consumer (re)public. This has been part of what Wang describes as a determined state campaign to "democratize society's access to cultural goods" (71), which has been achieved, she implies, in a number of ways, using the system of socialist legality to accommodate a modern culture of consumerism—for example, by implementing a Customers' Civil Statute, a Tourists' Civil Statute, and a Law of Consumer Rights Protection. The state also passed into law a forty-hour workweek in 1994, effectively creating the reality of a two-day weekend for the first time in China: the so-called double leisure day (*shuangxiu ri*). What the people could or should do with their newfound leisure time became the hottest topic in newspaper and magazine columns, and the state, at all levels, was not hesitant about making its own recommendations. On the streets, in factories and offices, and even in government departments, "a leisure culture fever swept over all major cities" (75).

To help provide the wherewithal for leisure consumption, interest rates were lowered several times between 1996 and 1998, encouraging the people to spend more and save less. This was seen as a part of the state's larger goal of boosting China's "spiritual civilization" to keep it in line with obvious advances that had already been made in China's "material civilization" (Bakken 2000). The state, in other words, was busy trying to get the people to be more "modern" and "civilized" by teaching them how to spend their leisure time profitably (with an emphasis on "spend"). These policies had multiple goals: in addition to helping build the Chinese character and develop a modern state, they were also part of a deliberate attempt to increase consumer demand for commodities as varied as automobiles, computers, and sporting gear of all types.

Wang's notion here is intuitively attractive and almost convincing, but it suffers from an exposure to the realities of China in the late 1990s and the new millennium. The most obvious contradiction in her argument is that in spite of the state's attempts to "democratize" consumption, its own economic reform policies have worked to increase massively the extent of inequality in China. This has been the case not only between the cities and the countryside, as foreign investment and state policies continue to favor the coastal provinces, but also, more disturbingly, within the cities. Behind the facade of appearances, a much more significant transition is taking place in China: socialist space, both in the cities and in the countryside, which was characterized by full employment, secure jobs (sometimes with fringe benefits), and minimal income and lifestyle differences (Solinger 2000, 2002), has been transformed into postsocialist space. This new space has for some time now been experiencing the negative "externalities" of the reform era, including—in the cities anyway—loss of state-employment security, dangerously high unemployment rates, rising crime rates, and new and previously unimaginably high levels of income polarization. Millions of workers have been laid off from their jobs in state-owned enterprises, and at the same time millions more poor peasants are entering the cities in search of jobs and are willing to accept those jobs at lower-than-market rates without any benefits. It is not clear exactly how or when such people will be able to enjoy the benefits of what Jing Wang calls the new "egalitarianism" of consumption. This portrayal of China's cities, dire as it is, is matched and perhaps exceeded by what is happening—and not happening—in the countryside, as rural/urban inequality reaches previously unimaginable levels.[26]

The second trajectory that Wang (2001) points to—and one that seems to offer a more robust account of cultural change in postsocialist China—is the emergence of a new cultural economy (*wenhua jingji*), which has resulted in the "collapse and convertibility of cultural capital into economic capital" (71). In addition to its focus on increasing leisure pursuits among the Chinese people, during the past two decades the state has concerned itself with turning culture into capital. A distinct pattern for this was established in Guangdong province after 1994. Guangdong had been traditionally known as a place to get rich, but as most of the guidebooks indicated, there was very little in the province to attract tourists (especially domestic tourists). In just a few years, however, the province succeeded in reinventing itself as a cultural vanguard by launching a range of shows and festivals—featuring both "high" and "low" cultural form—under the auspices of the Guangdong Provincial Institute of Cultural Development Strategies. The link between culture and business, the latter being what Guangdong was traditionally famous for, was made effectively, and the new catchphrase became "utilizing culture to promote business." A levy was imposed on major entertainment venues in the province, as well as on television stations, newspapers, and magazines, a levy that was earmarked as a "construction fee" for the building of cultural enterprises, and philanthropic donations to opera houses, symphonies, and ballet troupes were made tax deductible.

Another key innovator in this regard was the city of Beijing, where an explicit link was forged between cultural history and tourism, with the preservation (and often the entire rebuilding) of key historic sites. Huge new investment projects were announced, with plans to construct large-scale cultural establishments such as museums, libraries, and culture-focused shopping malls. The Department of Culture in Beijing publicized plans to rebuild or recover lost cultural venues such as run-down theaters and cinemas, and a plan was announced to build a special cultural zone composed of five locations of major historical significance across the city. As bold as these plans were, it is worth noting that Beijing—and other Chinese cities, especially Shanghai—was simply being more entrepreneurial and image conscious in its search for new investment funds and tourists, a strategy that it learned from many other cities around the world. In the process, however, the redevelopment of historical sites and scenery has helped Beijing buck the trend toward cultural homogenization and globalization.

There is a danger, perhaps, when discussing changes in Chinese culture of assuming too great a role for the state in the control of culture during the prereform era and too little a role during the postsocialist era. As Goodman (2001) points out, "Even at the height of the Mao-dominated era . . . creators of culture did not have to be previously state-sanctioned in order to have access to public outlets such as museums, journals, or publishing houses" (247–48). This was clearly a different situation from that in the Soviet Union; writers in China were not required to be members of an official writers' association, and the initiative for writing and translation was never fully monopolized by the state. In the postsocialist era, by contrast, it is unusual to find agents of popular culture that have had *no* contact at all with the state and its infrastructure. Many of them have previously worked for the state and have taken their *guanxi* (connections), as well as their training and expertise, with them into their cultural exploits. In countless instances, new cultural activities have close structural relationships with the state, and this is especially true in the new era of decentralizing decision making down to the local level. As Goodman notes, "Local government provides access to capital (funds, equipment, and buildings), labor, and political protection. . . . The booming collective sector of the economy is largely at the level of local government and ensures financing and a network of influence to support . . . [cultural] activities that want to grow beyond the small scale of private enterprise" (249–50). Goodman uses such evidence to suggest that rather than being created from below, much of the development of new cultural trends in China today is better described as having been subverted "from above" (250). In this sense, he is agreeing with Jing Wang's claim that the state has developed a new ruling technology that is not content to let popular culture go its own way. As Wang (2001) argues and as has been emphasized throughout this chapter, "leisure and pleasure" in China, even in these days of market triumphalism, "are not easily disentangled from politics and state sponsorship" (99).

Questions for Discussion

1. What are some of the most obvious cultural changes that mainlanders have experienced in the postsocialist era?
2. How have Chinese people reacted to their new wealth in terms of the way they spend both their money and their time?
3. One of the major concerns about countries like China that have developed rapidly in the era of globalization is that many people (especially those in the countryside

and the peripheral regions) are being "left behind." Is this the case in China, and what evidence is there to suggest that the Chinese government is aware of and is willing and able to respond effectively to this concern?

4. What have been some of the most significant trends in the world of literature and film in the new China?

5. In the realm of cultural production, is there any evidence of significant resistance to the economic and political forces currently dominating everyday life in China?

6. How do the films of Jia Zhangke (especially his "Hometown Trilogy") represent postsocialist China? What do you think is the reaction of the Chinese government to the way Jia (and other contemporary filmmakers) are showing the new China to the rest of the world?

Notes

1. This may explain why McGrath (2008) refers here to the "relative autonomy" of culture in contemporary China (9).

2. Another way to put this is to say that today the state no longer has an "official" ideology that governs people's everyday lives. As we have seen throughout this book, in the Dengist era (post-1978), most of the principles associated with Maoist-style communism were discarded: ideas about a collective utopian future were replaced with a quotidian concern with everyday life, while at the same time economic reforms increasingly stressed the importance of consumption, getting rich, and the endless expansion of markets. The rigid class identities of Mao's time disintegrated, both informally as a result of new economic and geographic freedoms, and formally as an edict of the regime. To justify its existence in such circumstances, the CCP started to redefine its hegemonic project as a determined attack on China's poverty and backwardness, which was to be achieved primarily by stimulating rapid economic growth (Blecher 1997).

3. Charlotte Ikels (1996) reinforces this observation in her book about Guangzhou, *The Return of the God of Wealth*, where she reports that the official working week was cut in 1994 from forty-eight to forty-four hours and again in 1995 from forty-four to forty hours (222). As she observes, "The amount of discretionary time . . . has increased as people have been relieved of tedious tasks that occupied their non-work time in the past. . . . They no longer have to stand in long lines for rationed goods, struggle to get things repaired, or scout state stores for clothing. . . . The elimination of rationing, the return of the small repair business . . . and the expansion of commerce have made it easier to meet basic needs and still have time left over" (222).

4. Again it is important to stress that nothing like the same level of consumer affluence and availability is available in most parts of rural and interior China. After spending two years in a medium-size town in Sichuan province on the Yangtze River, Peter Hessler (2001), in his book *River Town*, offers some interesting observations on this issue. Hessler and another young American were employed to teach English grammar and literature to a group of students in a small teacher-training college. As a way of getting their students to write in English, they asked them about their lives and their aspirations, and at one point in the book, Hessler describes what happened when his colleague gave an in-class assignment in which the instructions were for the students to "write about anything you want." As Hessler recalls, "The students . . . had written about anything they wanted, and what he had was forty-five shopping lists. I want a new TV, a new dress, a new radio. I want more grammar books. I want my own room. I want a beeper and a cell phone and a car. I want a good job. Some of the students had lists a full page long, every entry numbered and prioritized" (26). It became apparent that Chinese people

living in the interior regions are only just starting to define some of their dreams about being consumers, whereas in the not-too-distant past, such desires would have been muffled either because there was no chance of becoming wealthy enough to make such purchases or because the public expression of such desires would have been politically incorrect.

5. A survey conducted among Chinese business people indicated some very expensive tastes, at least when it comes to cars. The magazine *Zhongguo Qiyejia* (Chinese Entrepreneurs) surveyed two hundred entrepreneurs: 28 percent of those surveyed said that Mercedes-Benz tops all brands and is their first choice when it comes to purchasing their own cars, with BMW close behind, followed by Audi and Lexus (see http://ce.cei.gov.cn/enew/new_f1/fk00fb40.htm, accessed December 2000).

6. The state's anxiety about consumer spending may also explain in part why it has pushed so hard to promote the domestic automobile industry and increase the rate of private car ownership as a way of forcing Chinese consumers to spend rather than save their earnings, even in the face of the potentially disastrous environmental consequences of such a policy (Johnson 1997; see also chapter 3).

7. A survey conducted in 1998 by the State Statistical Bureau (now the National Bureau of Statistics) covering twelve cities and five thousand people provided a snapshot of urban Chinese consumer habits at that time. Somewhat unexpectedly, spending on food was above the national average, with 56 percent of the households surveyed spending US$36 to US$84 a month on food. The major break with the past, however, was in some of the food items that people were spending money on. Ready-to-eat frozen Chinese dishes were consumed by 54 percent of households (refrigerator ownership was 88 percent). Instant noodles were another favorite, with a consumption rate of 73 percent (one-fifth of those surveyed ate at least ten packs of instant noodles a month). Coffee drinking was also on the increase, with just over one-third of those sampled drinking two to six cups a week, while 8 percent were daily coffee drinkers. Another dramatic break with the past was in the new popularity of bottled water, which was drunk by 65 percent of the respondents (for a further discussion of these trends, see http://www.apfood online.com/magazines/2000/mar/art02.html, accessed November 2000). The data reported on here are part of what can be described as a more general shift—among all Chinese people but especially those living in the biggest cities—toward the "commercialization" of food consumption in China (Gale et al. 2005). Rural households in China have traditionally consumed food mostly grown on their own farms, and while they still rely on self-produced grains, vegetables, meats, and eggs, rural households are now purchasing more of their food as they enter the mainstream of the Chinese cash economy. Gale and his colleagues reported that purchases of food by rural Chinese households increased 7.4 percent per year between 1994 and 2003.

8. Among the most significant of these was a 1988 multiseries documentary called *Heshang* (River Elegy), which was probably the most watched and debated series in the history of Chinese television (according to Barmé 1999, 23). The main impetus behind the *Heshang* series was a conclusion reached by its producers that China's traditional culture and its long history (symbolized by the Huang He [Yellow River]) was acting as a drag on the future and that what China really needed was to abandon the traditional, inland, earthbound worldview, which it had had for so long, and to substitute for it a new orientation toward the sea (the "deep blue") and the outside world. For details about the series, see Su and Wang (1991).

9. All this change has been welcomed by almost everyone in China, but none more so than by old China hands—foreign China scholars who came first to China in the early 1980s—who often amused themselves by telling stories about the difficulty of living in China. One of the most frustrating experiences would be trying to make phone calls, which led to the near impossibility of keeping in touch with the outside world. Just fifteen years ago, several apartment compounds might have shared a single pay phone. All this has changed today, with an increase

in telephone availability, coupled with a rapidly rising use of cell phones and the Internet. Again we might suspect that in the long run, this new openness will have an impact on the quantity and the quality of public consciousness about all sorts of issues, including the environment, the abuse of women and children, the problem of urban unemployment—and not least of all, the ability to follow through in terms of activism on any number of issues. It is evident, in other words, that people in China have not only much more to do in their spare time today and much more spare time in which to do it but also much more information about the world and whatever interests them. An optimist might suggest that the logical outcome in this situation will be an increase in Chinese awareness about social and economic problems and a greater concern that their government should be doing more to solve such problems.

10. In the Maoist state, as Farquhar (2002) observed, "It was much more proper to speak of past suffering (in the old society), future utopia (when communism is achieved), and, in the present, work, production, and service. For at least two decades . . . wishes and discomforts could not be spoken of casually and privately, [and] the existence and indulgence of non-collective appetites were almost an embarrassment" (3). We should remember here that an excessive concern with individualism and consumerism is diametrically opposed to the principles of orthodox socialism. The CCP continues to urge the people to consider doing public service, to maintain high standards of morality, and to pursue selfless behaviors, highlighting the huge disparity that has now emerged between what was acceptable behavior in the past and what is acceptable today (Bakken 2000). During the past two decades, commercialization and entrepreneurialism have resulted in major changes in the traditionally solid ideological facade of the socialist regime and in the realm of centralized cultural production. As one Chinese scholar suggests, this new development has been met by "an almost visible sigh of relief, a not-so-quiet celebration of the demise of overpoliticization and the end of Ideology" (Tang 2000, 273).

11. The primary trend throughout the reform era has been for diminished state control, and as artwork and cultural production have become increasingly commodified, artists of all types have come to rely less on securing financial support from the state and more on the sale of their paintings, books, shows, and presentations. A review of what sells and what does not sell in contemporary China indicates that the market has been kinder to some art forms than others: poetry of all types and classical Chinese artworks for the most part have not sold well. But with these exceptions, the demand for art and cultural productions is more robust today than ever before in China, and a great deal of experimental art and increasingly risqué literature has emerged, especially in the big cities of the coastal provinces. It is also quite revealing to note that Zhang Yimou was selected to produce what for most Chinese has become known as the "the greatest show on earth": the televised coverage of the opening and closing nights of the 2008 Olympic Games in Beijing. For a highly critical account of this event, see Tarantino and Carini (2010).

12. A visually stunning example of this trend was portrayed in Chen Kaige's mid-1980s film *Yellow Earth*, which is set in northern Shaanxi province in the late 1930s. Northern Shaanxi, which is part of the dry and barren Loess Plateau, is considered the cradle of Chinese (Han) civilization, but it was also—at the time the film is set—close to the heart of the emerging communist movement, some two hundred miles to the south in Yenan. The action takes place ten years before the revolution, and the communists are attempting to win over the peasants to their cause, which involves expelling the Japanese, defeating the nationalists, and bringing about the modernization of China, roughly in that order (McDougall 1991). The premise of the film is that operatives of the Eighth Route (Red) Army are being dispatched into the countryside to collect peasant folk songs that will be transformed into battle songs for the soldiers as they march to war.

13. The May Fourth movement of 1919, which was the culmination of more than two decades of intellectual ferment and debate over China's perennial weakness internationally, is generally considered to have been the event that gave birth to modern Chinese nationalism. Some scholars refer to it as China's equivalent to the European Enlightenment (see, e.g., Hutchings 2001, 306–7).

14. Examples of this include the immensely popular but controversial *Heshang* (River Elegy) television series (see above); the introduction of abstract, Western styles of painting and music; and the new vitality of a literary modernism that, according to Kraus (1995), "rejects the . . . [socialist] realism imposed over several decades" (180).

15. This poem is quoted in Wang (1996, 197–98).

16. In a more nuanced *liumang* performance, such as Mardar's role in the film *Suzhou River*, the character in question is dragged rather reluctantly into illicit activity, in this case kidnapping, by the people he hangs out with. Mardar's particular transgressions have a high price, however, when his girlfriend apparently drowns herself and he is sent to prison for his role in the crime.

17. On the subject of food obsession in China, see also Mo Yan's (2001) often hilarious but sometimes quite shocking book *The Republic of Wine: A Novel*. For a review of the book, see Goldblatt (2000).

18. Other China scholars have attributed more significance to Zhou's journey of self-discovery, arguing that it can be interpreted as a metaphor for the ongoing struggle of young people in China and their need to reconcile their own personal desires with the collectivist orientations of Chinese culture. Weber (2002), for example, suggests that *Shanghai Baby* illustrates some of the important dimensions of what amounts to a process of identity construction in the new postsocialist era, identifying seven dimensions of this new identity that are clearly exemplified in Zhou's book: the desire for self-expression, the drive to be economically successful, the generation gap between youth and their parents (and China's leaders), the attraction of hedonism and consumerism and the apparent need for immediate gratification, the call for female sexuality and empowerment, an admiration of all things foreign, and a desire to express criticism of the party/state in public. Weber concludes that Zhou's book offers a "powerful metaphor" for the dilemmas facing China's young people, who are trying to deal with the multiple demands associated with rapid social change (366).

19. Lanchester (2003) suggests that such a lawsuit could have been filed only in China, which is one of the few places where the law allows dead people to be libeled. This is an unfortunate trend because Chinese writers certainly do not need a new way for books to be banned. Hong Ying lost the lawsuit and was ordered to pay damages and make a public apology in the press. Lanchester concludes on the absurdity of this whole situation: "A lawsuit heard in Manchuria, between two people who live in London, over a novel published in Taiwan, giving a fictionalized version of events which happened three quarters of a century ago between people all of whom are dead—welcome to the world of the twenty-first-century Chinese literary novel" (25).

20. As Silbergeld (1999) notes, "If an actor wanted to praise, it was done with an arm outstretched to the sky. If he despised something, he pointed to the earth strongly. If he got excited, he put his hand to his heart" (19).

21. *Yellow Earth* shifted the emphasis from the determined and heroic human characters of socialist realism filmmaking to the vaguer, ambiguous uncertainty of the modern era. The film won only one award in China, for photography (the cinematographer was Zhang Yimou), and it failed miserably in its first domestic showings in part because it was considered by Chinese audiences to be impenetrable. After the film created a sensation at the Hong Kong Film Festival in 1985, however, it was rereleased in China and became one of the most popular films of the year.

22. In *Raise the Red Lantern* (1991), a later film, instead of showing the glories of primitive ways of life, Zhang (1997) points out the wretched existence that women endured under the oppression of Chinese feudalism and patriarchalism. Again, he is presenting a longing for a simple, more primitive way of life, but this time he is demonstrating the hopelessness of those who seek such freedom in a world of structural and permanent inequality.

23. In Dai's (2002) words, "Initially an indicator of the success of Chinese art films, winning awards soon also became a means of survival, providing a chance to secure foreign investment, co-production, or other forms of assistance. Ironically, this narrow gateway became the sole opening for directors who wanted to keep a cultural foothold in art, evade the commercial tide . . . and thus avoid the mainstream model" (50). If we accept Dai's interpretation here, it may have been as a gesture to his audiences or perhaps as just a way to become more commercially successful at home that Zhang Yimou's more recent films (e.g., *The Story of Qiu Ju*, *Not One Less*, *The Road Home*, and *Happy Times*) have contemporary themes, with stories that are less obviously obsessed with the past and with primitivism. The first three of these films are set in remote parts of rural China, with images of the starkness and remote beauty of the land and of settlements that are clearly premodern. In these barren landscapes, the people are often shown as being close to nature, hardworking, uncomplaining, and basically content with a life that offers little more than getting enough to eat. Life in such villages is depicted as congenial; the people are for the most part kind. Zhang then switches his locale to urban China in *Happy Times*, but he remains consistent in sticking with the themes of poverty and a concern for China's underclasses, in this case middle-aged workers made redundant from their once-lucrative factory jobs.

24. As the plot develops, one of Wei's students, Zhang Huike (played by himself), runs away to look for work, so she sets off for the city, determined to get him back in the classroom. Again, unlike almost all of Zhang Yimou's earlier films, this one has a Hollywood-style happy ending, but even so, it is also able to insert some sharp social commentary. When the girl finally arrives in the city, after walking most of the way, she searches for several days unsuccessfully but by sheer perseverance manages to get herself on a local television news show, which helps her find the boy and bring him back home safely. The village is then showered with goodwill and gifts from city people and the television station, and the film ends with a statement about the importance of keeping schools open in poor parts of the countryside—a direct reference to the growing disparity between the city and the country in postsocialist China.

25. *Suzhou River* has also been described as a brooding tale of love and loss. It is set in Shanghai but certainly does not portray any of the glamour usually associated with that city's rapid economic modernization. In sharp contrast to such glamour, the film shows Shanghai as a dark and dangerous city whose inhabitants strive endlessly for fulfillment in the liminal world of the riverside. The film begins as the narrative of a lonely videographer who makes a careful study of the human traffic passing along the river in front of his balcony. He falls in love with a young woman who performs as a mermaid swimming in a giant tank in a seedy bar down by the riverside. Despite his initial euphoria, the narrator is deeply troubled by his lover's unexplained silences and periods of absence. She slips in and out of his life, leaving suddenly to lose herself in the human traffic of the city streets. At this point, another story is blended in, and a young man appears, claiming that the mermaid character is actually his former lover (who had jumped into the Suzhou River from a bridge and drowned). To make matters more confusing, the two women are actually played by the same actress, so there is some real basis for the confusion: the audience is left at the end of the film asking whether there were in fact two women or just one. Although *Suzhou River* is a unique film in a number of ways, it shares some characteristics with other films made by Sixth Generation directors. As one reviewer has commented, "In general,

the Fifth Generation made pretty films set in the rural past; the Sixth Generation makes gritty films set in the urban present. Emperors and concubines have been replaced by the grungy malcontents of Zhang Yuan's *Beijing Bastards*. . . . Its anomic punksters spit out obscenities in sync sound and groove to hard rock. A night at the Peking Opera gives way to an all-nighter in the Beijing mosh pit" (Corliss 2003, 1).

26. For a number of different perspectives on this situation, see Whyte (2010).

References Cited

Bakken. Borge. 2000. *The Exemplary Society: Human Improvement, Social Control, and the Dangers of Modernity.* London: Oxford University Press.

Barmé, Geremie. 1999. *In the Red: On Contemporary Chinese Culture.* New York: Columbia University Press.

Barmé, Geremie, and L. Jaivin, eds. 1992. *New Ghosts, Old Dreams.* New York: Times Books.

Blecher, Marc. 1997. *China against the Tides: Restructuring through Revolution, Radicalism, and Reform.* London: Pinter.

China Statistical Data. 2003. Statistical communique of the People's Republic of China on the 2002 national economic and social development. http://www.16congress.org.cn/e-company/03-03-20/page030102.htm.

Corliss, R. 2003. Bright lights. *Time Asia,* January 31.

Croll, Elizabeth. 1994. *From Heaven to Earth: Images and Experience of Development in China.* London: Routledge.

Dai Jinhua. 2002. *Cinema and Desire: Feminist Marxism and Cultural Politics in the Work of Dai Jinhua,* ed. J. Wang and T. E. Barlow. London: Verso.

Dirlik, Araf. 1994. *After the Revolution: Waking to Global Capitalism.* Hanover, NH: Wesleyan University Press.

Farquhar, Judith. 2002. *Appetites: Food and Sex in Post-Socialist China.* Durham, NC: Duke University Press.

Farrer, James. 2002. *Opening Up: Youth Sex Culture and Market Reform in Shanghai.* Chicago: University of Chicago Press.

Gale, Fred, Ping Tang, Xianhong Bai, and Huijun Xu. 2005. *Commercialization of Food Consumption in Rural China.* Washington, DC: U.S. Department Agriculture, Economic Research Service, ERR-8, July.

Goldblatt, H. 2000. Border crossings: Chinese writing, in their world and ours. In *China beyond the Headlines,* ed. T. E. Weston and L. M. Jensen. Boulder, CO: Rowman & Littlefield, 327–46.

Goodman, D. S. G. 2001. Contending the popular: Party-state and culture. *Positions* 9, no. 1 (Spring): 245–52.

Harvey, David. 2003. *The New Imperialism.* Oxford: Oxford University Press.

Hessler, Peter. 2001. *River Town: Two Years on the Yangtze.* New York: Perennial.

Hutchings, G. 2001. *Modern China: A Guide to a Century of Change.* Cambridge, MA: Harvard University Press.

Ikels, Charlotte. 1996. *The Return of the God of Wealth: The Transition to a Market Economy in Urban China.* Stanford, CA: Stanford University Press.

Jia Pingwa. 1993. *The Abandoned Capital.* Beijing: Beijing Publishing House.

Jiang, H. 2003. The personalization of literature: Chinese women's writing in the 1990s. *China Review* 3, no. 1 (Spring): 5–27.

Johnson, Todd. 1997. *Clear Water, Blue Skies: China's Environment in the New Century.* Washington, DC: World Bank.

Kraicer, Shelly. 2010. Chinese wasteland: Jia Zhangke's *Still Life.* http://www.cinema-scope.com/cs29/feat_kraicer_still.html.

Kraus, Richard. 1995. China's artists between plan and market. In *Urban Spaces in Contemporary China: The Potential for Autonomy and Community in Post-Mao China,* ed. D. S. Davis et al. Washington, DC: Woodrow Wilson Center Press, 173–92.

Lanchester, J. 2003. Looking for trouble in China. *New York Review of Books* 50, no. 4 (March 13): 25–27.

Lauer, Andy. 2010. Tomorrow, today: The films of Jia Zhangke at MoMA." March 5. http://www.indiewire.com/article/tomorrow_today_the_films_of_jia_zhangke_at_moma.

Link, P., ed. 1983. *Stubborn Weeds.* Bloomington: Indiana University Press.

Lu, T. L. 2003. Music and noise: Independent film and globalization. *China Review* 3, no. 1 (Spring): 57–76.

Lynch, D. C. 1999. *After the Propaganda State: Media, Politics, and "Thought Work" in Reformed China.* Stanford, CA: Stanford University Press.

Ma, H. Y., J. K. Huang, S. Rozelle, and F. Fuller. 2002. Getting rich and eating out: Consumption of food away from home in urban China. Working paper 02-E11. Center for Chinese Agricultural Policy, Beijing.

Mao Zedong. 1975. *Selected Works of Mao Zedong.* Hong Kong: Modern Chinese History Archives.

McDougall, B. S. 1991. *The Yellow Earth: A Film by Chen Kaige.* With a complete translation of the filmscript. Hong Kong: Chinese University Press.

McGrath, Jason. 2008. *Postsocialist Modernity: Chinese Cinema, Literature, and Criticism in the Market Age.* Stanford, CA: Stanford University Press.

Minford, John. 1985. Picking up the pieces. *Far Eastern Economic Review,* August 8, 30.

Mo Yan. 2001. *The Republic of Wine: A Novel.* New York: Arcade.

National Bureau of Statistics. 2004. *Zhongguo tongji nianjian 2004* [China statistical yearbook 2004]. Beijing: China Statistics Press.

Schoppa, R. K. 2002. *Revolution and Its Past: Identities and Change in Modern Chinese History.* Upper Saddle River, NJ: Prentice Hall.

Schuman, Michael. 2009. Will China's consumers save the world economy? *Time.* November 15. http://www.time.com/time/world/article/0,8599,1938591,00.html (accessed October 12, 2010).

Silbergeld, J. 1999. *China into Film: Frames of Reference in Contemporary Chinese Cinema.* London: Reaktion Books.

Smith, C. J. 2000. *China in the Post-utopian Age.* Boulder, CO: Westview.

———. 2002a. From "leading the masses" to "serving the consumers"? Newspaper reporting in contemporary urban China. *Environment and Planning A* 34: 1635–60.

———. 2002b. Postmodernity in new millennium China? *Asian Geographer* 21, nos. 1–2: 9–32.

Solinger, D. J. 2000. The potential for urban unrest. In *Is China Unstable?* ed. David Shambaugh. Armonk, NY: Sharpe, 79–94.

———. 2002. Labour market reform and the plight of the laid-off proletariat. *China Quarterly,* 170: 304–26.

"The Spend Is Nigh," *The Economist,* July 30, 2009, http://prasad.aem.cornell.edu/doc/media/Economist.30July09.pdf.

Su, X. K., and L. X. Wang. 1991. *Deathsong of the River: A Reader's Guide to the Chinese TV Series "Heshang."* Translated by R. W. Bodman and P. P. Wang. Ithaca, NY: Cornell University Press.

Tang, W. F., and W. L. Parish. 2000. *Chinese Urban Life under Reform: The Changing Social Contract*. Cambridge: Cambridge University Press.

Tang, Xiaobing. 2000. *Chinese Modern: The Heroic and the Quotidian*. Durham, NC: Duke University Press.

Tarantino, Matteo, and Stefania Carini. 2010. The good, the fake and the cyborg: The broadcast and coverage of Beijing 2008 Olympics in Italy. *International Journal of the History of Sport* 27, nos. 9 and 10 (June): 1717–38.

Wang, Jing. 1996. *High Culture Fever: Politics, Aesthetics, and Ideology in Deng's China*. Berkeley: University of California Press.

———. 2001. Culture as leisure and culture as capital. *Positions* 9, no. 1 (Spring): 69–104.

Wang Shaoguang. 1995. The politics of private time: Changing leisure patterns in urban China. In *Urban Spaces in Contemporary China: The Potential for Autonomy and Community in Post-Mao China*, ed. D. S. Davis et al. Washington, DC: Woodrow Wilson Center Press, 149.

Wang Shuo. 2000. *Please Don't Call Me Human*. Translated by Howard Goldblatt. New York: Hyperion East.

Weber, I. 2002. Shanghai baby: Negotiating youth self-identity in urban China. *Social Identities* 8, no. 2: 347–68.

Whyte, Martin King, ed. 2010. *One Country, Two Societies: Rural-Urban Inequality in Contemporary China*. Cambridge, MA: Harvard University Press.

Yang, Guobin. 2009. *The Power of the Internet in China: Citizen Activism Online*. New York: Columbia University Press.

Zha, Jianying. 1995. *China Pop: How Soap Operas, Tabloids, and Bestsellers Are Transforming a Culture*. New York: New Press.

Zhang, X. D. 1997. *Chinese Modernism in the Era of Reforms: Cultural Fever, Avant-Garde Fiction, and the New Chinese Cinema*. Durham, NC: Duke University Press.

Zhou Weihui. 2002. *Shanghai Baby*. New York: Pocket Books.

CHAPTER 7

A Preface to China's Changing Economic Geography

Central Planning, State Policy, and the Transition to a Market and Global Economy

China, the world's most populous country, today has one of the world's largest economies as well. After several centuries of stagnant or slow economic growth, China in the twentieth century ushered in a cataclysmic period of revolution, strife, and far-reaching and radical political change. Finally, after a time of intense internal struggle and the death of party chairman Mao Zedong in 1976, more moderate leadership emerged and reform of the economy soon followed.

After twenty-seven years of erratic political change and sometimes chaotic economic performance during the Maoist period (1949–1976), China changed dramatically once its leaders decided at the Eleventh Communist Party Congress in 1977 to proceed with economic reforms and to allow market incentives to help stimulate economic growth. In 1978 at the beginning of these economic reforms, the size of China's economy as seen in its gross domestic product (GDP) was approximately US$52.8 billion (National Bureau of Statistics 2009). By 2007, this had increased approximately sixty-five-fold to US$3.44 trillion, and the per capita GDP in 2008 had increased roughly sixty times to approximately US$3,381 (see table 7.1).[1] Today, China is recognized as the world's second-largest economy, after its economic growth drove its annual GDP beyond that of Japan (Monahan 2011). This kind of remarkable economic growth has improved and transformed profoundly the lives of hundreds of millions of Chinese while also altering dramatically the landscape of city and countryside and reorienting the regional framework of production and distribution (Naughton 2007).

The remarkable renaissance and rise of China as a global as well as a regional economic engine must be considered and examined in the context of a rapidly changing world economic system based on new technologies of production and distribution that both lead to and reflect new spatialities in the framework of China's economic geography. These new spatialities, as Dicken (2007) describes them, reflect the distinctive locational and functional elements of production in an economic system that itself

179

Table 7.1. China: Growth in Gross Domestic Product (GDP), 1952–2008 (Current Prices)

Year	GDP 100M (yuan)	GDP Per Capita (yuan)
1952	679.0	119
1960	1,457.0	218
1970	2,252.7	275
1975	2,997.3	327
1978	3,645.2	381
1980	4,545.6	463
1985	9,016.0	858
1990	18,667.8	1,644
1995	60,793.7	5,046
2000	99,214.6	7,858
2005	183,217.4	14,053
2008	300,670.0	22,698

Note: U.S. dollars converted at the rate of 8.28 yuan/US$1 from 1998 to 2005. In mid-2005, China's government allowed a modest increase in the value of the yuan relative to other major currencies. In May 2006, the rate was approximately 8 yuan/US$1, and in October 2010, the rate was approximately 6.65 yuan/US$1.

Source: National Bureau of Statistics (2009).

is shifting in response to the interactions and relationships between the state and enterprises as seen on a variety of levels and scales, including a significant and growing global impact.

In the case of China, once the reforms of the late 1970s and 1980s began to take hold, a fundamental spatial redeployment of production and distribution was set in motion. New economic regions emerged that sought to link China's new production centers, which were geared toward the global exchange economy, as the country began to alter its economy from a command and direct system to one that was designed to employ market forces and incentives to advance the rate of economic growth and accelerate trade with the global trading system. These new policies and the resultant economic production coincided with extraordinary advances in technology both in production and in transportation and logistics. Some of these transport and shipping systems were already in place in Hong Kong, for example, in what was emerging as one of the world's largest container ports, and China moved quickly to take advantage of the existing modern transport infrastructure in the Pearl River Delta region. In retrospect, it is no surprise that the first new special economic zones (SEZs) were all located in that region and that these have served as a model for China's spatial reorientation from interior China to the coast as the country reorganizes its regional and distribution focus toward a rapidly growing global economic system.

Structural Shift and Spatial Outcomes

Structural shift or change of China's economy was put in motion during the early stages of communist rule in 1952. However, the pace of the change was muted and

Table 7.2. China's National and Agricultural Labor Force and Production Output by Sector, 1952–2008

Year	Population (millions)	National Labor Force[1] (millions)	Agricultural Labor Force[2] (millions)	(%)	Value of Output of Sector (% share of GDP)		
					Primary	Secondary	Tertiary
1952	574.82	207.29	173.17	83.5	50.5	20.9	28.6
1957	646.53	237.71	193.09	81.2	40.3	29.7	30.1
1975	924.20	381.68	294.56	77.2	32.4	45.7	21.9
1978	962.59	501.52	283.18	70.5	28.1	48.2	23.7
1980	987.05	423.61	291.22	68.7	30.1	48.5	21.4
1985	1,058.51	498.73	311.30	62.4	28.4	43.1	28.5
1990	1,143.33	647.49	389.14	60.1	27.1	41.6	31.3
1995	1,211.21	680.65	355.30	52.2	20.5	48.8	30.7
2000	1,265.83	720.85	360.43	50.0	16.4	50.2	33.4
2005	1,307.56	758.25	329.70	44.8	12.2	47.7	40.1
2008	1,328.02	774.80	306.54	39.6	11.3	48.6	40.1

[1]All individuals employed or self-employed in urban and rural areas.
[2]All individuals working in agriculture, forestry, animal husbandry, and fisheries.
Source: National Bureau of Statistics (2009).

Table 7.3a. China's Structure of Production, 1985–2008 (% Share of Output)

Economic Sector	1985	1990	2000	2005	2008
Primary	28.4	27.1	15.1	12.2	11.3
Secondary	42.9	41.3	45.9	47.7	48.6
Tertiary	28.7	31.6	39.0	40.1	40.1

Source: National Bureau of Statistics (2009).

Table 7.3b. China's Employment Structure, 1985–2008 (% Share of Employment)

Economic Sector	1985	1990	2000	2005	2008
Primary	62.4	60.1	50.0	44.8	39.6
Secondary	20.8	21.4	22.5	23.8	27.2
Tertiary	16.8	18.5	27.5	31.4	33.2

Source: National Bureau of Statistics (2009).

modest owing to erratic policies, political events, and related economic performance. This is evident in the slow pace of the shift in farmworkers to nonfarm activities in the period 1952–1975 (see table 7.2). The reforms of 1978 accelerated the structural shift. This shift may be tracked in two simple ways. One way is to consider the structure of production or output by comparing the percentage share of production attributed to different sectors of the economy. A glance at table 7.3a indicates a continuing decline in the share of the value of output accounted for in the primary sector, mainly agriculture, of the economy since 1985, to the point where this sector accounted for only 11.3 percent of the value of China's total production by 2008. The share of China's working population in agriculture has dropped sharply in the past decade, and by 2008, only about 40 percent of the workforce continues to work in this sector (see table 7.3b and figure 7.1). While the decline is striking, nonetheless there were in 2008 still over 306 million workers in the primary sector, a matter that is taken up later in this chapter as we discuss the importance of the primary sector of the economy in absorbing labor.

While the value of agricultural production as a share of total production has declined steadily, there has been significant growth in both the secondary and the tertiary sector shares in their contribution to the value of total output, and the secondary sector, led by industry (including mining and construction as well as manufacturing), now accounts for almost half the total value (48.6 percent). This will not come as a surprise to anyone who has been following China's economy in recent years or indeed anyone who shops in a Wal-Mart or Target store. Over the past decade, China has become the manufacturing center of the world for a host of products, initially producing low-value consumer goods, such as textiles and garments, shoes, sporting goods and equipment, small tools and hardware items, lamps, fans, and light fixtures.

Increasingly, however, the complexity, value, and sophistication of these products has increased as China attracts more investment from abroad to take advantage of the country's low-cost and relatively productive labor by setting up manufacturing plants. Thus, computers and peripherals and a large array of electronic products are

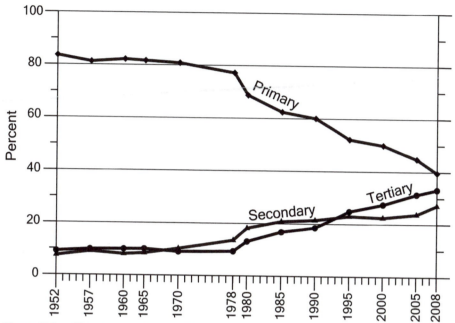

Figure 7.1.   China's employment by sector, 1952–2008 (National Bureau of Statistics 2009)

being increasingly produced in China, along with auto parts and equipment. Auto assembly plants of many global firms based in major industrial countries, such as the United States, Germany, and Japan, are rapidly expanding their production in what is today the world's most rapidly developing automobile market (Studwell 2002). This outsourcing of manufacturing came initially from China's nearby neighbors, such as Hong Kong and Taiwan, but more recently it has come from Korea, Thailand, Japan, Mexico, and the United States. China's tertiary or service sector has also grown and now accounts for roughly one-third of the value of its production, and the trend can be expected to continue as the economy grows and matures to the point where producer services increase in importance relative to other activities.

By any standard, this growth and structural shift is a remarkable accomplishment and has had the effect of transforming the lives and well-being of hundreds of millions of China's citizens. The number of citizens in China who have been pulled out of abject poverty is startling and may be in the tens if not hundreds of millions. Moreover, the Chinese worker's ability to seek new employment and to shift locations to where the jobs are has become an increasingly important part of the evolving economic and social scene in contemporary China. At the same time, it is certainly true that this new wealth and economic growth has not been shared equally among China's people or its regions.

The Chinese government's decision in the late 1970s to introduce far-reaching economic reforms had an especially dramatic effect on the coastal regions of China, as noted, where incomes have grown rapidly and the cityscapes have come to resemble those in Japan and Korea. In addition to these changes in the structure and growth of the economy, there have also been significant, related outcomes in the geography of this economic growth and change.

Is China's Economic Growth and Transition Different?

It is the goal of this chapter to examine and explain the character of China's recent remarkable economic growth, to describe and provide insight into the spatial changes and outcomes related to this economic growth—that is, China's changing economic geography—and to examine the various scales on which this has occurred. As with so many other aspects of change in China, the story of this economic change must be viewed in the context of political change and within an evolving framework of policy and the implementation of that policy at the level of the central state and of the larger regions (such as the provinces) and, perhaps most significant, at the local level. All these levels and scales matter, but to local people, it is perhaps what is happening in their city, town, or neighborhood that matters most. It is at this level that life is lived and business is done. In this chapter, we try to include discussion of economic change at all these levels and on all these scales.

One of the more challenging yet intriguing issues in pondering China's economic growth and geography in the twenty-first century is the lack of an adequate theoretical framework for examining and analyzing the processes of Asian economic growth and change. As Yeung and Lin (2003) have discussed, mainstream economic geographers have sadly ignored for the most part the remarkable economic advance that has occurred in Asia in the past quarter century, thus leaving a kind of conceptual vacuum in which those interested in understanding the processes of Asian economic advance, especially in spatial terms, have had to use conventional, theoretical "lenses" derived from Western experience. Yeung and Lin do note the recent emergence of what may be called China's economic transition theory, although they argue that it has not been extended into economic geography studies of Asia.

Recently, Lin (2009) has examined and analyzed the special role of land in China's economic history and development and has argued again for a distinctive and somewhat idiosyncratic function in the manner that land has been used in China's recent and current economic and urban transition. This is an essential aspect of the broader conceptual framework of a distinctive Asian and Chinese transition mechanism in economic growth and development and one that demands further scrutiny and study by those interested in developing countries generally as well as those in Asia (Oshima 1987).

In this chapter, we use Western conceptual approaches to economic change, such as the conventional notion of structural shift in an economy, as well as the idea of comparative advantage and regional specialization of production. Yet the reader should keep in mind distinctive economic conditions in Asia and other conceptual approaches while asking what is the best and most appropriate means to understand recent growth, change, and development in China.

China's economic advance continues in a manner that has had and is likely to continue to have an enormous impact on the lives not only of the Chinese but of all people in the world. China's immense size makes it a key player on the global stage, and as its economy grows, the impact of China's size will become ever clearer to people everywhere. One example is China's increasing participation in various international organizations such as the United Nations and its various subagencies. China's entry into the World Trade Organization in 2002 provides the most striking illustration of international recognition of China's growing economic power and its role as a key player in the world economy. The enormous growth in China's trade and its activity as an export powerhouse, especially its enormous export surplus with the United States, testifies to its might. This huge trade surplus has allowed China in 2010 to accumulate more than US$2.5 trillion in foreign currency reserves, a situation that makes China a key player now exceeding Japan in its impact on global capital markets and the world economy. At the same time, such a trade imbalance is becoming an increasingly contentious political problem between the United States and China.

As the twenty-first century unfolds, China's role as a powerful economic engine as well as a major political and military force will become increasingly evident. Its economic role will be seen in its commercial power as market, producer, and, more recently, investor; in its industrial role as manufacturer and innovator; and in its service role as collector, distributor, and processor of goods, services, and information. China's economic power is growing, and the next half century will likely witness an extraordinary increase in the commercial and industrial force exerted by this rising power.

Chinese Socialism and Central Planning

Following a successful communist revolution in 1949, China emerged as a Marxist state and brought with it a radical new vision for the planning and operating of its national economy. Its model at the time was the Soviet Union, which had evolved a highly centralized mode of economic operation. In this model, the central state, through a State Council and various ministries, created a centrally planned economic system in which bureaucrats in central offices drew up plans for allocating resources, thereby determining the structure of economic production as well as the location of that production.

Such planning and production could not be achieved overnight, so a phased approach was adopted, but clearly the ultimate goal was full ownership of the means of production by the state, by which the state would centrally plan and direct the economic system for the entire nation. As Eckstein (1977) explained in his study of the Chinese economy, the Chinese Communist Party (CCP) leadership in 1949 initially had three main options in operating China's economy following the successful 1949 revolution: 1) a free market option, in which prices would be set and resource allocations would be made based on market forces in equilibrium, more or less based on supply and demand for products; 2) a market socialism option, wherein the state would own and control the means of production and provide central planning but would allow some other pricing system based on market forces to determine the allocation of some resources or segments of the economy; and 3) a command economy option, in which the allocation of goods, services, and factors of production would all be determined by central planners (party bureaucrats) through an administrative bureaucratic apparatus rather than by market forces. As we shall see in a brief chronological survey of actual events, all three of the options have been in play at various times during the more than half century of communist rule in the People's Republic of China.

Chronology of Planning

Following a brief period of rebuilding and restoration of economic production in the early 1950s, the leadership of the Chinese central state began to move toward a centrally planned model, based largely on the Soviet approach to economic planning. In 1953, the first Five-Year Plan was introduced with the trappings of a command economy in which the central state, through its State Council and State Planning Commission and various ministries and bureaus, planned and organized production, allocated resources to support this production, and began to set prices.

Initially, things went reasonably well owing to good weather and a surge in farm production. The communist revolution appeared to have been a good thing, and conditions were improving for many people, most of whom lived in rural areas. This initial success emboldened the communist leadership and especially party chairman Mao Zedong to seek more radical approaches to advancing the cause of socialist egalitarianism in Marxist China. Beginning in late 1957 and proceeding rapidly thereafter, Mao promoted radical new policies in an attempt to move Chinese society to the left to fulfill his vision of a Marxist revolution and to move away from what he perceived to be the more conservative, technocratic approaches to economic growth that other senior leaders, such as Liu Shaoqi, were advancing. In this way, China's economic planning shifted to the Great Leap Forward, a mass social and political movement that sought to propel China to a stage of economic production that would rival that of Great Britain, and to do this very rapidly as a demonstration of the power of a people's revolution.

The Great Leap Forward involved a sharp advance to full socialism in which all the means of private production were eliminated, including the private plots and farm animals that the Chinese peasants had been allowed to keep. It also involved the formation of communes in rural areas in which all farming was done in teams and brigades and wherein the labor and output were shared by all based on their need as well as—or perhaps rather than—their effort. In addition, production of all kinds of locally needed goods, such as farm machines, iron, cement, and fertilizer, was to take place locally regardless of local conditions and realities of comparative advantage. In some cases, actual output of these locally produced goods increased, but the quality of the products was frequently so poor that they could not be used and were therefore discarded. Moreover, when the incentives of private plots and animals had been removed, farm production collapsed owing to the inability of the peasants to grow crops and husband animals for themselves and their families. For example, farm pigs had formerly been an important source of organic manure and mulching materials in the Chinese countryside, and much of this was lost with the disappearance of the peasants' pigs and other farm animals. A severe famine ensued (see chapter 5), and as many as 35 million people starved to death in what can only be described as one of the great catastrophes of the twentieth century. This was in fact a policy-induced famine of enormous proportions and one of the greatest human cataclysms of the socialist period.

Chairman Mao was heavily criticized by other top leaders in China, and the Communist Party Central Committee moved away from the most radical aspects of this leap to socialism. It allowed some restoration of private plots and animals in an attempt to restore farm and food production and to avert a more serious crisis. Yet Mao was not finished, and by 1964 he set about galvanizing the young people of China to counteract what he regarded as the too-conservative members of the CCP and the educated elites of the party, especially those in the party leadership who opposed his idea of continuing revolution. In 1965, in a mass movement involving millions of young people, Mao sought to gain firm control of the CCP by unleashing these young people to renew the revolution through a program known as the Great Proletarian Cultural Revolution, or simply the Cultural Revolution.

Massive social protests and chaos ensued, and the country was plunged into disarray as schools and universities closed, ports were shut down, and factories were boarded up. In this movement, the peasants were left alone, although many urban elites and intellectuals were sent to rural areas for "thought reform" and "rehabilitation." The education of a generation of students was interrupted, and the country's planned economic production was seriously disrupted.

This event also coincided with geopolitical circumstances that pitted China against the United States during the Vietnam War, and China moved many of its new industrial enterprises into remote areas of the interior, such as Sichuan, that were believed to be less vulnerable to attack from the coast. This policy of locating new industries away from the coastal regions and into the deep interior for security reasons was termed the *san xian*, or Third Line, a reference to the movement away from an earlier and more vulnerable coastal First Line (Cannon 1990; Naughton 1988). While such industries may have been more secure against external attack, the irrationality of their location relative to markets and shipping points created highly inflated costs for their products on the way to markets and thus represented an enormous misallocation of scarce capital and resources based on a perceived and perhaps unrealistic appraisal of the security threat.

The more radical phases of the Cultural Revolution began to subside by 1971, but an actual shift to reforms and a restructuring of China's economy did not begin until after Mao's death in 1976, an event that was followed by a leadership succession struggle that culminated in the arrest of Mao's wife, the infamous Jiang Qing, and her radical cronies in 1977. Shortly thereafter, a new group of leaders led by Deng Xiaoping took over and adopted major reform policies for the economy at the Eleventh Communist Party Congress.

In 1978, far-reaching reforms were therefore put in motion that would lead to a remarkable shift in the manner in which China's economy was planned and operated. It is these reforms that have led to a whole new way of doing business in China, for they have sought to take advantage of market forces in allocating resources and determining prices. Yet the Chinese government was careful to implement the reforms gradually, first in rural areas and then in the cities and towns with their extensive network of state-owned enterprises (SOEs). Moreover, it proceeded differently in different regions of the country because some regions responded to the use of market forces much more readily and vigorously than others.

New Economic Reforms

Much has been written about the economic reforms initiated following the death of Mao and the ascendance of the so-called second generation of leaders assembled around paramount leader Deng Xiaoping. They were initially implemented in agriculture through a responsibility system of household production (see chapter 8), and the effect of this system was to return the farmland to the farmers through leasing arrangements and to allow the farmers to take advantage of incentives on raising foodgrain and cash crops as well as on related sideline activities. Output of farm products rose immediately as the farmers responded, and there was a dramatic rise in crop and animal production as well as in farm family income.

However, the new prosperity was not universal; as more enterprising farmers or those farming in more advantageous locations benefited greatly, others languished or lagged behind in the new drive to riches. In conceptualizing how these reforms were proceeding, we may refer to the three alternative systems that Eckstein (1977) identified and postulate that these reforms were attempting to use aspects of a market system to rationalize the allocation of resources and to raise production based on incentives and on a regional specialization of crop production based on comparative advantage. Yet the role of the central state remained an active one, especially in the matter of continuing to levy requirements and quotas on farmers for the production of food grain.

Reforms Extend to the State Sector and Urban Areas

In 1984, a policy decision was made to extend these reforms to the state sector of the economy and to apply them extensively within urban areas. However, this was not intended as a full-blown reform, and it was meant to take effect once the basic production quotas of the SOEs had been met by what were still enterprises of the central state. The SOEs, however, had serious problems. First, such enterprises typically had large numbers of redundant employees, and their efficiency and productivity was low. Many also had enormous debts to state banks or credit institutions, and this debt was increasing, as was long-term financial liability. Moreover, during the first decade of the reforms, these SOEs continued to add substantial numbers of employees, and thus their problems of low productivity and efficiency continued, and the red ink associated with them increased (Lardy 1998).

In the state sector of the economy, SOEs typically operated housing estates for their employees, provided health and hospital services, and operated schools for the children of their employees. They also had pension liabilities in their responsibility to provide housing and health services to all their employees throughout their lives. As was becoming increasingly clear, however, and also more ominous in its long-term outlook, the prospects and fate for most of these SOEs were not sustainable. Yet in a socialist system, it is difficult to face the reality of impending financial failure when so many in the system have become accustomed to the entitlements of their jobs and positions.

Reforms and Spatial Redeployment

In parallel with the growing problems of the SOEs, the CCP had decided to accelerate economic growth in coastal areas and to seek to take advantage of the global market-place. Thus, locations such as the Pearl River Delta, which is proximate to Hong Kong, were allowed to establish SEZs, where the more rigid rules of the central state would be relaxed to encourage foreign investors to commit their resources to production facilities that would take advantage of the very low cost of Chinese labor and related positive factors in construction, land, waste removal, and transportation. At the same time, many local entrepreneurs in China took advantage of the new more relaxed rules and environment for doing business, and a number of township-village enterprises (TVEs) were created. Nominally part of the collective economy and owned by a village or township, many of these TVEs were in fact private in all but name because many were funded by private individuals but were operated under the guise of public ownership by a township or village (this is known as "wearing a red hat"). Such enterprises flourished in the more independent atmosphere of Southeast China, where the entrepreneurial spirit of family capitalism had had a long tradition of success, especially in connection with family relatives and common-surname clansmen in Southeast Asia.

The four initial SEZs (Shenzhen, Zhuhai, Shantou, and Xiamen) were successful, and other regions clamored for equal status; thus, numerous other locations, including Hainan along China's coast, were quickly awarded similar status (see figure 7.2). A kind of freewheeling, capitalist-oriented market socialism spread to other coastal cities. Some developed their own special style and set of products that were distinctive, and the Wenzhou model, so named for a small city on the coast of Zhejiang, became well known for its independent mode of operation as well as its remarkable success. Clearly, a major spatial reordering with a strong coastal orientation of production and distribution was under way, and it was leading China toward much closer links to the global economy.

Growing Regional Disparities and Social Unrest

Yet all was not well. As the success of many of the TVEs grew, the economy heated up to a point of almost unsustainable growth. Moreover, there were now rapidly

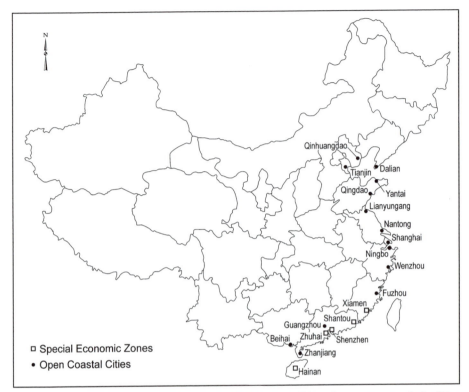

Figure 7.2. Special economic zones (SEZs) and China's open coastal cities (adapted from Phillips and Yeh 1990)

growing disparities between those who were earning lots of money and those who were falling behind. Inevitably, the new wealth quickly attracted the attention of local and higher officials, who insisted on their share in order to approve virtually any transaction. Thus, levels of corruption throughout the bureaucracy grew to egregious heights, and the problem of unequal incomes became more obvious. Economic growth was palpable, and the prosperity it brought was obvious, but so too was the greed of some officials and of those who were paying the bribes. In an emerging new spatiality of production and consumption, regional disparities in income were also becoming increasingly apparent between urban and rural areas as well as between coastal and more remote interior areas.

In the spring of 1989, a broad movement erupted among students that was a mixture of efforts to liberalize the politics of China and to create a society that shared its wealth in a more equitable and fairer way. In fact, a precedent for this movement was the May Fourth movement of 1919, a mass movement led by patriotic students and intellectuals that had the goal and ideal of rejecting foreign imperialism and Japanese intrusion into China while promoting a strong, modern, and democratic China. It is difficult to pinpoint precisely the causes and objectives of the 1989 movement, yet clearly one of the key elements of dissatisfaction was a result of the intolerable levels of corruption and greed that permeated Chinese society and the bureaucracy. The movement resulted in a brief uprising in Tiananmen Square, Beijing, with a subsequent crushing of the uprising by the People's Liberation Army on June 4, 1989. International outrage followed, and there was a reduction in trade and international transactions, which reduced the rate of economic growth in China. Political repression came next, and China entered a period of relative calm (Liang, Nathan, and Link 2001).

Three years later, Deng Xiaoping made his epic trip to South China, during which he unleashed a new wave of relaxed rules and proclaimed that to get rich was good and that the accelerated model of market socialism as practiced in the Pearl River Delta was an admirable thing for China. A modern emperor, in so many words, had put his imprimatur on an open, relatively unrestricted, aggressive, and market-oriented approach to doing business. It was now okay to be a capitalist, although this was described as "socialism with Chinese characteristics."

Deng Xiaoping's trip also validated the value of regional location in relation to the Pearl River Delta and the various scales at which the new production and ties to the global economy were taking place and being formed. China is a vast spatial system, and its different regions were advancing under the new rules at different speeds and by building on different sets of comparative advantages in restructuring their production and distribution systems. Hong Kong and its already established networks for shipping and trading gave the Pearl River Delta an early start and an enormous advantage. Money from overseas Chinese in Southeast Asia as well as Hong Kong and Taiwan was flowing in and being invested in factories producing consumer goods for the domestic and foreign markets. Local and regional Chinese economies were booming as they began to reshape themselves.

Deng Xiaoping's southern trip also resulted in a new wave of reforms, beginning in 1993, and these are the policies that have continued to the present. In conceptualizing this growth since 1993, it is useful to invoke the analysis of Barry Naughton (1995), who has described a kind of "dual-track" system and approach that has sought to maintain some elements of traditional central planning, especially in the energy and key infrastructure sectors, while simultaneously encouraging and allowing the full impetus of market forces to propel rapid economic growth and to assist in making SOEs more efficient. China's recent efforts have placed the most emphasis on the market track, in what Naughton and others have described as "growing out of the Plan."

Significance of Employment Growth and Establishing the Market Economy

One of the significant components of the approach of "growing out of the Plan" is to use the private sector to offset job losses in the state sector of the economy as it is restructured and downsized and to make creation of new jobs one of the highest priorities of the central state. Arguably, then, job creation might be seen as a key government policy goal throughout the reform period. In fact, it could be seen as the key policy goal, given its link to social and political stability and the maintenance of public order as a high priority of the central state. As the reforms proceeded—and it can be argued that they became more flexible in their use of market approaches with the transition to more pragmatic CCP leadership—the role of the private sector grew and became an increasingly more important driving force in labor absorption. In the early reforms of the late 1970s and early 1980s, private enterprise and private entrepreneurs (*geti hu*) were allowed into the marketplace; however, they were subject to many rules. As Han and Pannell (1999) explain, there developed a pattern of geography from the manner in which rules and acceptance of private sector enterprise spread over China, and some areas of the country were much more receptive to the use and growth of the private market economy than others. Especially prominent among the faster and steadier promoters of private enterprise were the coastal provinces from Jiangsu south as well as some border areas, such as Inner Mongolia and Xinjiang.

Yet even while the central state was liberalizing policy on private workers during the first two decades of economic reform, significant growth in employment continued in SOEs, as the central state strove to ensure employment for the large number of new workers entering the marketplace each year. From 1980 to 1994, there was an increase of more than 30 million workers in SOEs even after new policies in the early 1990s had been put in place to reduce such employment. It was only after 1996 that the central state and Premier Zhu Rongji got serious about reforming the SOEs and reducing the redundant employees in these units. Thereafter, there began a severe reduction, and within four years more than 30 million workers had been cut from the state sector in urban areas.

The pattern of employment and the balance between the state and the private sector in China, however, is complex and varied. For example, there is also the collective sector of the economy, and in rural areas the TVEs have been a key component in accounting for rural employment. In urban areas up to the mid-1990s, the collective sector grew substantially but began to shrink markedly thereafter. Meanwhile, the individual private sector grew rapidly during the 1990s as the rules governing it became increasingly flexible and controls on its activities were eased. By 2000, private employment, including various corporate and foreign enterprises in urban areas, had reached more than 50 million, and it was clear that this was a crucial element in the employment equation for China's economic growth and social stability.

Job creation in rural areas also witnessed substantial growth in the private sector; moreover, it was clear that many of the jobs in what were described as TVEs were in fact disguised private jobs. Much private and indeed foreign investment was cloaked under the guise of collective ownership to give it greater political protection in case of a reversal of state policy and a reversion to the more rigorous socialist policies of the past. Investors were ensuring that they would be protected in the event of a shift in political currents that might take China back to a more orthodox period of socialism.

Yet it had become increasingly clear that a return to such policies was highly unlikely given the transition in government leadership to a younger group that was less ideological in outlook and more prone to pursuing policies that would work based on practice and an increased involvement with the global economy. What began in earnest with Deng Xiaoping's southern tour in 1992 and was enshrined at the Fourteenth Communist Party Congress in the same year as a key feature of the socialist market economy came to be regarded as an increasingly liberal and friendly period for the market economy with its emphasis on individual and private commerce. Moreover, as Han and Pannell (1999) have documented, the regions of China where private economic activity was strongest also were associated with the highest rates of economic growth. This established an even stronger imperative for advancing the private economy as an engine of growth and as one that was likely to benefit all of China through its impetus for faster growth.

Premier Zhu Rongji in his final report in 2003 at the conclusion of the ninth Five-Year Plan (in effect from 1998 to 2003) offered the following thoughts on the continuing success of the reforms and opening up of China and the establishment of a socialist market economy:

> The ownership structure was further readjusted and improved. The public sector of the economy grew stronger in the course of readjustment and reform, and efforts to diversify ways of realizing public ownership were successful. The state sector of the economy went through accelerated restructuring, and markedly enhanced its dominance and competitiveness. . . . The collective economy in urban and rural areas made new headway. The joint-stock company sector of the economy expanded continuously. Individually-owned businesses, private enterprises and other non-public sectors of the economy developed fairly fast and played an important role in stimulating economic growth, creating more jobs, invigorating the market and expanding exports. (Premier Zhu 2003, 2)

Premier Zhu went on to applaud the success of the market economy in propelling China's economic growth through its role in allocating resources and in instituting price reforms in such sectors as public services, energy, and transportation. Such a testimonial from a leading technocrat of the retiring administration in 2003 set the stage for continuation of a market approach to economic growth in China and provides a positive outlook for the continuation of economic liberalization and commitment to involvement in the global trading system. China has committed itself to a fast economic growth approach as a means of meeting its labor demands, while it continues to reduce its population growth for future posterity and prosperity.

Regional Development: Spatial Outcomes of Economic Reforms

China is an enormous country and a huge and ever-changing spatial system. This historical reality has presented its people with a continuing challenge, one that its imperial system and bureaucratic structure and apparatus have struggled to counter even as the central state has sought to control the country and keep its peripheral regions under Chinese hegemony. This has been an enduring theme in Chinese history given the rugged surface geography of China and the many formidable physical barriers and impediments it offers to movement. Only in the twentieth century had improvements in transportation and communication advanced to the point where China truly could begin to integrate its many regions and outlying territories in an effective and meaningful manner for governance as well as to make this territory economically contributory to the well-being of the central state. Certainly, one of the key economic tactics of the socialist administration of the People's Republic of China has been to invest heavily in transportation and in this manner to better integrate the country while also seeking to promote a more balanced regional development.

Dr. Sun Yat-sen had a plan for China's development, and he wrote of it in a book that outlined a variety of things needed to advance the country (1953). One of these was promotion of greater spatial integration through a national network of railways that would link all the provinces of China. Now many years after the publication of Dr. Sun Yat-sen's plan, the central state, in its tenth Five-Year Plan, finally completed a major goal of this plan with construction of a long-planned rail link to Lhasa in Tibet, the last remaining province of China with no rail link. It is the recognition of this kind of commitment that helps us understand the enormous challenges facing China in its effort to create a fully integrated modern state and spatial system that can provide for the movement and linkage of goods, people, ideas, and innovations. This is a recent accomplishment and indeed continues apace as China pushes ahead aggressively in building not just railroads but also a new expanded nationwide highway network, more airports and harbors, and more power and telecommunications grids for all its people. Regional development in China, then, is in part a story of building and linking the various regions of China and of seeking to provide a scheme and means of doing this to accelerate economic growth while bringing the advantages and benefits of this growth to all citizens.

Regional Integration in China's History

Three broad periods of spatial development in China's history may be posited (Pannell 1992). The first was a traditional imperial period that lasted for many centuries but that saw China's development focused on the interior of the country and in which typically the impetus for urban and regional growth focused on administrative functions supplemented by economic activities. Over time, these economic activities increased in importance, especially with the Song dynasty (AD 960–1279), and these gradually assumed primacy. Yet China for most of its history has been a nation that focused on its internal development and seemed to stress trade among its regions rather than abroad. These regions, as noted earlier, operated to some extent as discrete entities and were not always well connected or responsive to the center, but the extent to which they were so linked to the center reflected the strength and effectiveness of the central state as a governing and functioning polity. Thus, there was a waxing and waning of the roles of China's regions and its central state over time that paralleled the waxing and waning of dynastic authority and effectiveness (Fairbank 1992; Skinner 1977).

This began to change in the nineteenth century following the arrival of Western colonial powers and especially after the first Opium War (1839–1842), by which the British formally colonized the island of Hong Kong and expanded to a small adjacent territory. This was quickly followed by the establishment of a host of treaty ports

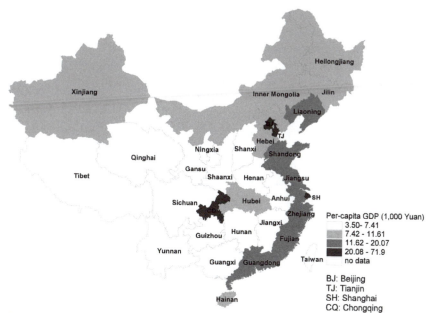

Figure 7.3. Treaty ports and other major cities in China (adapted from Murphey 1974)

that were minicolonies and that served as outposts of an evolving global economy that witnessed the establishment of numerous European and some Japanese colonies (see figure 7.3). The function of these colonial enclaves essentially was to serve the interests of a home country through a mercantilistic system of trade in which natural products from the primary sector of the local economy were sent to the home colonizing country, which in turn sought to return manufactured products to the colony and to maintain terms of trade that were skewed to enrich the colonizing country. Some modest manufacturing was established in many colonies to take advantage of abundant and low-cost labor and to produce products that could be sold in the local economy but would not be a threat to the economy of the colonizing power.

What this meant for China was not only the introduction of new industrial methods and the beginning of modern industrialization but also a remarkable and far-reaching spatial reorientation of its economic and urban structure. In this spatial reorientation and redeployment, the most advanced, dynamic, and rapidly growing urban centers, with clusters of new industries and improved transportation, emerged in coastal or riverine locations where they could connect more easily to an emerging global trading and economic system.

The great port of Shanghai, gateway to the Chang Jiang (Yangtze River) Basin and a quarter of the Chinese population, was the epitome of this new urban center. It quickly became both symbol and archetype for the new, foreign-influenced China—admired for its dynamism and modernity yet despised for its subservience to Western foreign "devils" and seen as a center of wealth and power that was far too "perverted" from Chinese cultural roots and ethnocentrism to be acceptable as truly Chinese. Numerous other treaty ports, such as Canton (Guangzhou), Amoy (Xiamen), Fuzhou,

Ningbo, Nanjing, Hankou (Wuhan), Qingdao, Tianjin, and Dalian, followed suit and attempted to mimic Shanghai. Thus, along China's coast and greatest river were clustered the key economic centers of the late nineteenth and early twentieth centuries, for here were easy links to world shipping and the international marketplace.

Meanwhile, the vast body of China with its huge rural population was not profoundly affected, and rural life continued apace much as it had for many centuries. Unfortunately, this growing population was placing increased stress on the environment to the point that food production was hard pressed to keep up with population growth, and calls for change added to the pressure to modernize. China's old regime ended, and a new era began with the creation of the Chinese republic, soon followed by civil war, World War II, and a communist victory in 1949. The communist victory would lead to a significant spatial reorientation and a new effort to redirect development to the interior.

Regional Development in China

In 1949 when the communists assumed power, Mao Zedong and the new leadership sought to shift the emphasis away from the coast and back to China's interior, the traditional regions of the country's seats of power. This was also an effort to reduce the power and influence of centers of capitalism and global or colonial influence. Thus, cities like Shanghai were seen as sources of funding to be exploited so that capital could be transferred to other cities and regions and in this way fulfill the promise of a nationalist revolution that would restore traditional centers of development and return China to its earlier focus on interior places and development (Wu 1967).

The extensive literature on regional policy and development in China during the first two and half decades of socialist rule (1949–1976) clearly indicates this focus on restoring more regional balance through the transfer of investment to interior locations (for a good review of this literature, see Wei 2000). At the same time, the communist regime built on industrial bases in places like the Northeast, where the earlier investments of Japanese colonialists in a dense transport network linking a cluster of industrial cities had accelerated the rate of industrial production in China. Other centers, such as Beijing, were supposed to become both industrial and administrative in function to justify their proper role in a people's republic as "producer" rather than "consumer" cities. The putative "consumer" role that had previously prevailed in such capitalist outposts as Shanghai and Canton, both former treaty ports and trading centers, was seen as evil.

As noted earlier, this regional policy was accelerated by so-called Third Front industrial development during the Vietnam War era, when key industries and new rail lines were constructed in the deep Chinese interior in such provinces as Sichuan, Guizhou, and Yunnan to protect them from possible attack from the coast or from the Soviet Union. While such investments may appear irrational in pure economic terms owing to their distance from appropriate market centers and low-cost shipping locations, they did have the effect of stimulating development in China's interior and acted as an impetus to provide some corrective to existing regional inequalities.

A number of scholars have written extensively on regional development in China, and there have been a variety of opinions offered as to the trajectory and effectiveness of regional development policy during the past half century in remediating spatial inequalities in income and well-being (Fan 1995; Li 2000; Lin 2000; Marton 2000). Wei (2000) has summarized well the conceptual approaches and contrasted them in the context of mainstream neoclassical theory on regional development and economic growth. As he notes, while there have been varying opinions, the consensus viewpoint suggests that during the Maoist years, there appeared to be a pattern of declining regional inequality as the state directed more of its investment to the interior. After the beginning of the reform period in 1978, however, the pattern was reversed, and there is now more evidence of a growing regional inequality as the coastal regions benefit from state policy to advance their growth. The idea in China, in part derived from Western economic theory, is that the coastal regions should be allowed to grow faster because of their locational advantages and connections to the global economy. Theoretically, as Wei points out, this should follow more or less an inverted-U model of regional development, with the idea of a spatial "trickle down" by which "some regions advance that others may follow"—if put in a more benign context (see also Veeck 1991). How well this is working in China is open to debate, but there is substantial evidence of very rapid growth in some, if not all, of the coastal provinces. Per capita income figures continue to provide a quick regional descriptive picture of this reality (see figure 7.4).

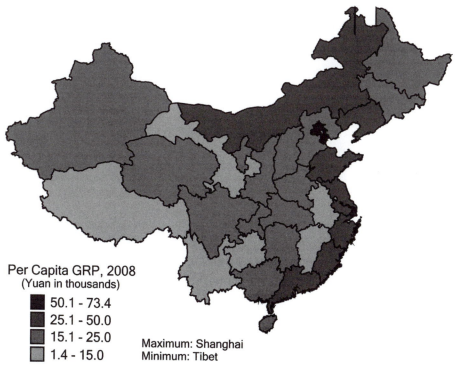

Per Capita GRP, 2008
(Yuan in thousands)

- 50.1 - 73.4
- 25.1 - 50.0
- 15.1 - 25.0
- 1.4 - 15.0

Maximum: Shanghai
Minimum: Tibet

Figure 7.4.   China's per capita gross regional product, 2008 (National Bureau of Statistics 2009)

Promotion of coastal development was first enacted in policies on the establishment of SEZs in the 1970s, and SEZs have since spread from the Southeast to other coastal provinces and then into interior locations. In its seventh Five-Year Plan, China in 1986 laid out a regional scheme that divided the country into three main regions—coastal, central, and western—and these have remained a convenient if sometimes difficult-to-interpret mechanism for describing the manner in which the country has been allowed to develop (see figure 7.5). Wei (2000) in his study tracked and mapped both per capita GDP as well as growth in GDP in the provinces. He concluded that while there were sporadic investments in the interior provinces for various reasons such as security or to promote industrialization during the 1950s, there was a decrease in regional inequalities. In the 1960s and 1970s, however, while there was some industrial investment owing to Third Front development, the impact was spotty, many of the interior provinces did not benefit, and regional inequalities increased.

With the recognition of a new regional scheme in 1986, there was also the recognition that a greater policy commitment to the coastal regions would result in these advancing faster than the interior provinces. Yet this was considered appropriate because China now accepted the use of market mechanisms—such as the principle of comparative advantage and the role of regional specialization of production—that were likely to favor the coastal regions. The regional pattern of growth and well-being

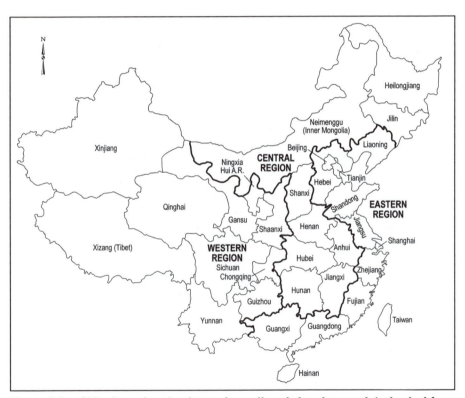

Figure 7.5. China's regional scheme for national development (adapted from Wei 2000 as derived from official Chinese sources)

quickly revealed much more rapid development in the coastal areas, especially of the Southeast but also in Zhejiang, Jiangsu, Shanghai, and Shandong (Cartier 2001). As a result of complaints from other regions and with a concern for satisfying other regional interests, the regulations on foreign investment and other incentives for regional growth were modified. This allowed for faster growth in the interior provinces, with a special focus on remote border regions such as Xinjiang (perhaps for geopolitical as well as economic reasons). These steps appear to have had some success, for per capita income in Xinjiang has grown rapidly (see figure 7.4). Yet this may also reflect rapid growth of the Han Chinese population in that region, especially in the industrial centers with their focus on oil and gas development (Loughlin and Pannell 2001).

As a spatially uneven pattern of development evolved and disparities in regional income and well-being became increasingly visible, the issue of social and political stability in interior regions became more serious. Owing to this concern, by the late 1990s there were increasing calls to develop the western regions of China, and indeed a commitment to increased funding emerged. In March 2003 at the Tenth National People's Congress, Premier Zhu Rongji in his main report on the work of the government gave a review of the program to develop the western region:

> Over the past three years since the introduction of the western development strategy, the government has given a powerful push to the region's development by increasing investment, stepping up transfer payments and introducing preferential fiscal and taxation policies. Work was begun on 36 new key projects, which called for a total investment of over 600 billion yuan. (Premier Zhu 2003, 2)

While it is clear that a much-intensified effort has been made to create a more spatially balanced development policy in recent years, the data continue to demonstrate the enormous advantage of the coastal provinces as seen in virtually all indicators of well-being. Moreover, despite efforts to counter this as described above, the disparity is already great, and the advantages of these coastal locations so linked to world trading and financial systems are only likely to intensify.

Another way to judge the enormous and growing power of the coastal provinces is to note the migration streams in China and to see where these migrants are going. While there is cross migration throughout the country, the main migration streams are from the interior provinces, such as Sichuan, to the east and especially to the dynamic coastal provinces and cities where the jobs in manufacturing, construction, and service trades are located (Fan 2005). These regions are powerful magnets and will no doubt continue to be so for the foreseeable future as they are the leading edge of China's fast economic growth and dynamic structural transformation.

China has had extraordinary economic growth, especially in past two decades, and this has resulted in the lifting from poverty of literally hundreds of millions of people. At the same time, there have been striking spatial consequences and outcomes of this economic growth that have resulted in an uneven pattern of economic growth between different regions of the country as well as between urban and rural areas. Coastal regions and urban areas have prospered, while interior regions and rural areas have lagged somewhat. However, the pattern is not as simple as merely differentiating from the coast and the interior and from city and rural areas, as some deep interior regions, such as Xinjiang, have seen fast economic growth, although there are sharp differences within the region, and some rural areas have done reasonably well, while others have been left behind. Policy makers have grappled with these variations and spatial inequities, and the state continues to pursue policies aimed at reducing such differences, but the outcome of market approaches to economic growth will likely continue to lead to significant differences in economic growth rates among regions and urban/rural locations throughout China.

Questions for Discussion

1. China has experienced very rapid economic growth over the past two decades. What accounts for this growth, and is this high growth rate likely to continue in the short term?
2. How soon do you anticipate that primary-sector employment will drop below the employment rate in the tertiary sector of China's economy, and what will this indicate about the nature of economic shift and change in China?
3. How important is central planning and control in China's economy today? Do you expect this role of central planning to grow or to decline in the future?
4. Is the trajectory of China's dynamic economic growth and change different in character and structure from the manner in which the market economies of Western capitalist democracies have proceeded? If so, what are some of the key differences?
5. China's coastal regions have prospered greatly in recent years, and this has created a spatial inequality in different regions of the country. Is this growing or declining, and how serious is this as a social and political issue?
6. Can the central state in China use policy initiatives to counter the trend of spatial inequality in incomes and economic conditions?

Note

1. In discussing and analyzing the Chinese economy, it is especially wise to be cautious in the use of Chinese economic data. While Lardy (2002) complains about the quality and accuracy of economic data, perhaps reflecting a commonly held view among many economists and other social scientists, Holz (2003) has analyzed in depth some of these data and concluded that this perception is probably not accurate as a general proposition. Holz has concluded that there is probably not a great deal of deliberate falsification of Chinese economic data; at the same time, he agrees that there may well be a substantial margin of error in much of the official Chinese statistical data.

References Cited

Cannon, Terry. 1990. Regions: Spatial inequality and regional policy. In *The Geography of Contemporary China*, ed. Terry Cannon and Alan Jenkins. London: Routledge, 28–60.

Cartier, Carolyn. 2001. *Globalizing South China*. Malden, MA: Blackwell.

Dicken, Peter. 2007. *Global Shift: Mapping the Changing Contours of the World Economy*. 5th ed. New York: Guilford.

Eckstein, Alexander. 1977. *China's Economic Revolution*. Cambridge: Cambridge University Press.

Fairbank, John K. 1992. *China: A New History*. Cambridge, MA: Belknap.

Fan, C. Cindy. 1995. Of belts and ladders: State policy and uneven regional development in post-Mao China. *Annals of the Association of American Geographers* 85 , no. 3: 421–49.

———. 2005. Modeling interprovincial migration in China, 1985–2000. *Eurasian Geography and Economics* 46, no. 3: 165–84.

Han, Sun Sheng, and Clifton W. Pannell. 1999. The geography of privatization in China, 1978–1996. *Economic Geography* 72: 272–96.

Holz, Carsten. 2003. Fast, clear, and accurate: How reliable are Chinese output and economic growth statistics? *China Quarterly* 173: 122–63.

Lardy, Nicholas. 1998. *China's Unfinished Economic Revolution*. Washington, DC: Brookings Institution Press.

———. 2002. *Integrating China into the World Economy*. Washington, DC: Brookings Institution Press.

Li, Siming. 2000. China's changing spatial disparities: A review of empirical evidence. In *China's Regions, Polity, and Economy*, ed. Siming Li and Wing-shing Tang. Hong Kong: Chinese University Press, 155–86.

Liang, Zhang, Andrew J. Nathan, and Perry Link, eds. 2001. *The Tiananmen Papers*. New York: Public Affairs Press.

Lin, George C. S. 2000. State, capital, and space in China in an age of volatile globalization. *Environment and Planning A* 32: 455–71.

———. 2009. *Developing China: Land, Politics, and Social Conditions*. Oxford: Routledge.

Loughlin, Philip H., and Clifton W. Pannell. 2001. Growing economic links and regional development in the Central Asian republics and Xinjiang, China. *Post Soviet Geography and Economics* 42: 469–90.

Marton, Andrew. 2000. *China's Spatial Economic Development: Restless Landscapes in the Lower Yangzi Delta*. London: Routledge.

Monahan, Andrew. 2011. China overtakes Japan as world's no. 2 economy. *Wall Street Journal*, February 14.

Murphey, Rhoads. 1974. The treaty ports and China's modernization. In *The Chinese City between Two Worlds*, ed. Mark Elvin and G. William Skinner. Stanford, CA: Stanford University Press, 17–72.

National Bureau of Statistics. 2009. *Zhongguo tongji nianjian 2009* [China statistical yearbook 2009]. Beijing: China Statistics Press.

Naughton, Barry. 1988. The third front. *China Quarterly* 155: 381–86.

———. 1995. *Growing Out of the Plan*. Cambridge: Cambridge University Press.

———. 2007. *The Chinese Economy: Transitions and Growth*, Cambridge, MA: MIT Press.

Oshima, Harry. 1987. *Economic Growth in Monsoon Asia: A Comparative Survey*. Tokyo: Tokyo University Press.

Pannell, Clifton W. 1992. The role of great cities in China. In *Urbanizing China*, ed. Gregory E. Guldin. Westport, CT: Greenwood, 11–40.

Pannell, Clifton W., and Laurence J. C. Ma. 1983. *China: The Geography of Development and Modernization*. London: Edward Arnold.

Phillips, David R., and Anthony Gar-On Yeh. 1990. Foreign investment and trade: impact on the spatial structure of the economy. In *The Geography of Contemporary China*, ed. Terry Cannon and Alan Jenkins. London: Routledge, 224–44.

Premier Zhu. 2003. Government work report delivered at the first session of the Tenth National People's Congress. March 5. http://www.chinadaily.com.cn/highlights/nbc/news/319zhufull.htm.

Skinner, G. William, ed. 1977. *The City in Late Imperial China*. Stanford, CA: Stanford University Press.

Studwell, Joe. 2002. *The China Dream: The Quest for the Last Great Untapped Market on Earth*. New York: Atlantic Monthly Press.

Sun Yat-sen. 1953. *Fundamentals of National Reconstruction*. Taipei: Chinese Cultural Service.

Veeck, Gregory, ed. 1991. *The Uneven Landscape: Geographical Studies in Post-reform China*. Geoscience and Man Series, vol. 30. Baton Rouge: Geoscience Publications, Louisiana State University.

Wei, Yehua Dennis. 2000. *Regional Development in China*. London: Routledge.

Wu, Yuan-li. 1967. *The Spatial Economy of Communist China*. New York: Praeger.

Yeung, Henry Wai-chung, and George C. S. Lin. 2003. Theorizing economic geographies of Asia. *Economic Geography* 79: 107–28.

CHAPTER 8

Agriculture: From Antiquity to Revolution to Reform

Early Farming Traditions

The origins of agriculture in China are lost in time, but most scholars favor multiple hearths of domestication dispersed throughout what became modern China. One was most certainly within the valleys of the Huang He (Yellow River) and its tributaries. Scholars usually locate another in South or Southwest China, including the areas now incorporated as Guangdong and Guangxi provinces of modern China. There are written records of rice (*Oryza sativa*) cultivation from as early as 5000 BP, but production in southern China most likely predates this. Archaeological remains of seeds thought to be millet and rice found near Hangzhou in Zhejiang province have been radiocarbon dated to 8500 BP. These deposits and other excavations of crop remains provide ample evidence that sedentary agriculture as a way of life in China could be at least eight thousand years old (Sun 1988, 193; Zhao 1994, 49).

Chinese agriculture is popularly associated with rice cultivation, but China's farmers have made many other vital contributions to the global food system. The list of Chinese domesticates or hypothesized domesticates is long and includes tea (*Camellia sinensis*); hemp for seed, fiber, and medicinal use (*Cannabis sativa*); several species of melon (*Cucumis*); foxtail millet (*Setaria italica*); proso millet (*Panicum miliaceum*); soybean (*Glycine max*); adzuki bean (*Vigna angularis*); several types of onions and chives (*Allium fistulosum* and *Allium tuberosum*); buckwheat (*Fagopyrum esculentum*); peaches (*Prunus persica*); apricots (*Prunus armeniaca*); and some types of oranges (*Citrus*) and kiwifruit (*Actinidia*) (Smartt and Simmonds 1995). Early farmers worked not only with grain crops but also with a broad range of dicots, herbaceous shrubs, and trees to produce a considerable array of useful foods and fibers. These crops were cultivated in a growing range of environments as Chinese territory expanded in all directions from its early dynastic origins within the Wei and Huang valleys.

China's farmers also developed or adapted many important agricultural technologies, some in use (virtually unchanged) to the present, such as gravitational and mechanical irrigation, wet-rice production practices, terracing and other land reclamation techniques, anaerobic composting, multiple cropping (the growing of more than one crop per year), and the extensive use of organic materials for improving and maintaining soil fertility. The first recorded use of insects as biological control mechanisms, in approximately AD 340, was the use of a type of ant to control mites and spider infestations on orange trees in Guangdong province (Needham 1981, 13). Irrigation systems evolved in many world regions at the same time as those in China, but few locations matched those of China in terms of the scale of these systems and their extensive distribution throughout the nation. An irrigation system located in Sichuan province, called Dujiangyan, has operated continuously since 400 BC to the present, supplying water to over 500,000 ha of cropland—an area 1.25 times the size of the U.S. state of Rhode Island. Some of these innovations are discussed in greater detail later in the chapter. At the outset, however, it is easy to argue that more than those of any other nation, China's farmers have transformed the landscapes in which they live for the purposes of agricultural intensification by irrigating, manuring, terracing, draining, burning, and deforestation.

Legend has it that the sage-emperor Shen Nong transformed Chinese culture and society through the introduction of agriculture and herbal medicine sometime during the third millennium BC. Another of these sage rulers, Emperor Yu (Yu the Great, 2197 BC), is credited with taming the disastrous floods of the Huang He and inventing irrigation for wet-rice cultivation. As a result, Yu saved the Chinese people from widespread famine. Such apocryphal stories illustrate the salient role of agriculture in the origins and growth of Chinese civilization.

Agriculture, then, has a central place in China's history as well as its present. From ancient times, it has been the nation's very economic foundation. A steady growth in population, combined with natural disasters and refugees from warfare, has constantly pushed China's farmers to expand and farm with greater intensity. As mentioned in chapter 2, much of China's lower-quality arable land, particularly in the north-central and northwest portions of the nation, would not be farmed in a nation with less pressing needs. For China, however, land shortages have been an issue at least since the Tang dynasty (AD 618–907). Even regions with very limited rainfall or extreme slope were brought into cultivation as regional population pressures increased. Over the centuries, the cultivation of marginal land in China, without environmental safeguards, led to countless local ecological collapses that are an important, if overlooked, aspect of the tapestry of China's long history resulting in regional wars, mass migrations, and monumental famines (Smil 1993).

A major effort by the Chinese government beginning in 1998, discussed at length in the next section, is focused on improving conditions of much of this marginal land and for the moment underscores the enduring nature of many of the challenges facing the agricultural sector at the present time. The current effort, known as the Comprehensive Agricultural Development (CAD) Program, is unprecedented in scope of world history, although it has garnered little attention in the Western press. Programs include physically improving land and drainage through grading and subsoiling land and terracing; adding massive amounts of organic material, such as crop straw, and other organic materials, such as manure, to improve fertility and drainage; and promoting the balanced use of inorganic fertilizer after testing on a field-by-field basis. These actions, concurrent with the construction or upgrade of irrigation systems and thousands of small- and medium-scale dams for water storage and conservation, are stabilizing and improving productivity on much of this marginal land that is critical for China's food security in the coming decades.

The Agricultural Sector in Historical Context

THE COMMUNE ERA

Any discussion of contemporary agriculture in China must begin with the remarkable commune system initiated by Mao Zedong in the early 1950s under which all farmers and their families were assigned to live and work on one of these large, collectively managed farms. At their peak, commune farms incorporated 90 percent of China's arable land. In 1973, there were approximately fifty thousand communes in China. Communes ranged in size from 25 to 130 km² and averaged fifteen thousand members. The communes were envisioned by Mao to be self-reliant in all ways—not only for food production but also in terms of the provision of housing, education and health care, and even retirement benefits. The commune system was not merely a theoretical experiment in social engineering. It was a fundamental aspect of Mao Zedong's vision of a strong and self-reliant rural China that would in turn serve as the foundation for the modernization of all of China.

Although Mao Zedong's egalitarian vision of a cohesive, classless rural China organized under the commune system initially appealed to the peasants supporting the revolution in the late 1940s and early 1950s, their enthusiasm faded over time, and food production collapsed. Decades of hardship and deprivation, including the tragic Great Leap Forward (1958–1960), eroded most farmers' collective dreams of prosperity couched in equity. In the mid-1980s, one farmer from Haimen county in Jiangsu province summed up the commune era to me in this way: "Under Mao, we were equal—equally poor."

In general, the commune era was one characterized by a limited diet, ration coupons, and regular food shortages in both rural and urban China. Cooking oil, sugar, salt, cloth, grain, pork, soy sauce, eggs, liquor, and beer were but some of the staples included in a massive rationing system in which individual communes, cities, counties, and provinces all issued distinct coupons that could be used only within these jurisdictions. Fruits and off-season vegetables were seldom included in these systems because before 1980 such products were so scarce that there was no reason to pretend that there was enough for everyone to have even a taste. Up until 1984, urban workers would have fruit only a few times a year, most often when their work units (*danwei*) would buy a truckload, and then distribute bags of apples or pears at New Year's as a bonus. A middle-aged farmer we interviewed in 2008 in western rural Jilin province recalled her first orange eaten in the early 1980s to be very bitter until her daughter told her she was not to eat the rind. The people in her village had never seen one until the reform era began in 1978. Under the commune system, central planning, poor organization and transport, and the lack of a profit motive severely restricted the productive potential of the land and kept China's farmers terribly poor.

Still, it must be recognized that there were important successes during the commune era, notably in improvements to farm mechanization, irrigation, organic agriculture practices, multiple cropping systems, and plant breeding programs. The famines of the war years (World War II and the civil war) were truly horrible, and older people can still vividly remember the servitude and hardship before 1949—particularly during the Japanese occupation and the civil war. The commune era was a lot better than life under Japanese occupation. More to the point, however, the successes of the commune system were too few and too far apart and touched the lives of too few people too lightly. In short, nothing associated with the communes seemed to work very well despite so many good intentions and an undeniable nobility of purpose.

THE RESPONSIBILITY SYSTEM (*BAOGAN DAOHU*) AND MARKET REFORM

The Responsibility System was introduced in December 1978 and broke up the communes by returning land to the "responsibility" (control) of individual farmers through land-use contracts based on family size and local land availability. This daring change introduced by Deng Xiaoping has altered rural China in countless ways that are still playing out. Concurrent with land reforms were a series of changes that once again allowed farmers to determine what crops could be grown and how crops and farm inputs were purchased and priced. While some price controls remain for grains, the essence of the current system relies on prices for all products determined by market demand and farmers' own decisions—as in capitalist nations.

Under the Responsibility System, each farmer contracts a specific amount of land from the village or township authorities, initially for five to seven years but typically now for thirty years. Allocations were based largely on family size but also by the amount of available land and the voluntary use of other types of land contracts negotiated with local officials or other farmers. China's land is owned collectively by all Chinese people, present and future, so, since 1949, farmers cannot own land. Now, however, farmers can subcontract land from other farmers or from public land controlled by the village and negotiate the terms of these exchanges on a case-by-case basis. Farmers who commit to only growing grain can get larger land allotments and many types of technical support for free or subsidized process as well.

The reforms started with legislation in December 1978 are ongoing and constantly adjusted to meet new problems—and their results are ever changing as well—representing another challenge to China's rural planners. Change under the responsibility system came slowly at first. Even in the early 1990s, most farmers in any given village or county grew the same crops and had similar lifestyles and similar incomes. Farmers in 2010 are far more diversified, producing different cash crops on different amounts of land, depending on the extent of their contracts. Their activities are also more diverse, including crop farming, husbandry, aquaculture, and forestry as well as the off-farm manufacturing work discussed extensively in chapter 9. Incomes and educational levels among the farm families even in a single village now vary greatly, presenting very different socioeconomic conditions from those of the more homogeneous commune era or even the early reform era to 1995.

The benefits of the responsibility system are unquestioned, but by the late 1990s, it was clear that more was needed and that new problems associated with the small scale of China's farms, a lack of infrastructure oversight (lost with the repeal of the commune system), and implementation of the new commercial market system have developed. China's farmers are now helped by many new government programs that are discussed in the next section when the CAD program is introduced, but a single example might help set the stage and show how far China's farmers have come. In 2009, during a survey of farmers in western Jilin, one farmer reported taking a great but ultimately profitable chance by growing a special variety of red pepper used to dye silk that he shipped to Jiangsu province on the coast after seeing a request for the crop posted by a company in Shanghai on the Internet. Getting prices from special government call sites on cell phones or looking up prices for inputs or commodities on the Internet from home is now a common practice for many of China's farmers—even in poor areas of West and Central China.

AGRICULTURAL MODERNIZATION EFFORTS IN THE NEW MILLENNIUM

Currently, China can boast of one of the world's largest and most ambitious bioengineering research efforts, while the nation's agricultural, aquacultural, and husbandry sectors are now among the world leaders in research and development. Still, the sector suffers greatly from uneven development (Dong, Song, and Zhang 2006; Huang et al. 2002; Jin et al. 2010) in terms of both products and regions. On the one hand, China's farmers compete well in global markets and are price setters for an impressive list of internationally traded labor-intensive products. On the other hand, many other millions of China's farmers in interior regions remain trapped in endemic poverty because of degraded environments, limited land, undeveloped markets, undeveloped transportation systems, and poor technical training. This is the challenge currently facing China's agricultural planners: How can the entire farm sector advance to ensure sufficient supplies of strategic crops such as food grains while raising incomes and quality of life for *all* the nation's farmers?

China's CAD program was developed and introduced in 1988 in response to these concerns in light of the specific conditions facing the nation's agricultural sector in the early to mid-1980s (Kueh 1995). At the time, both population and arable land loss were increasing, threatening the nation's food security and forcing the nation to spend increasing amounts in foreign exchange for food grain and feedstuffs. Lester Brown (1995), a well-known U.S. expert, posited at the time that China's grain imports could potentially collapse the global grain market. In addition, agricultural production capacity remained weak in many parts of the country, in part because of an excessive amount of low-yielding farmland relative to total arable land. Local governments lacked the funds and tools to reverse the trend. In 1998, low-yielding cropland accounted for more than 1 billion mu of a total land area of less than 2

**Photo 8.1. CAD-funded program in north-
ern Ningxia Hui Autonomous Region in
Northwest China that provides funds to line
irrigation ditches with concrete to conserve
irrigation water (2010) (Gregory Veeck)**

billion mu. At the same time, disaster-affected cropland (flooding, drought, and insect
plagues) accounted annually for about 20 percent of total sown area. Further, a high
proportion of arable land was degraded, and soil erosion was unchecked in many
areas. In addition, agricultural technology and farm extension services and outreach
lagged behind similar factors of production in other sectors, such as industry. Finally,
agricultural infrastructure, often ignored since the commune era, was aging and often
of low quality, and farmers had little incentive and less capacity for effecting changes
to these conditions (Prandl-Zika 2008).

From 1998 to 2007, the total funding of the CAD program was 320.3 billion
yuan. Of this, 99.2 billion came from the central government, 76.8 billion from
participating local governments, 34.3 billion from bank loans, and 110 billion from
farmers and other minor sources. Since inception, more than 520 million mu (34.66
million ha) of low-yielding land has been improved through inclusion in the CAD
program. As a consequence, average annual grain yield increased by 1.5 to 2.25 tons/
ha through the transformation of low-yielding farmland. New and upgraded irrigated
farmland rose by 480 million mu (32 million ha).

Improving the quality of farmland was just the beginning. Historically, the most distinctive characteristic of Chinese agriculture in relation to Western agriculture is not in the particular crops that are grown but in how these crops are produced. Until recently, crops were supplied with a remarkable diversity of organic nutrients that permitted the production of two or even three crops on the same land, year in and year out, without impacting soil fertility and drainage characteristics. The production of more than one crop per year remains an important characteristic of China's modern farm sector under the CAD program. Throughout much of China, roughly south of Beijing, two or more crops are planted sequentially whereby winter crops of winter wheat, rapeseed, or winter barley planted in the later summer or fall are followed by summer crops such as rice, corn, or soybeans. The southern provinces, such as Guangdong, grow three crops, often a winter crop and two rice crops. With improved drainage, irrigation, and soil fertility, multiple cropping systems are more reliable, with fewer losses to drought and flooding and higher yields due to improvements to cropland and the use of better varieties designed to "fit" local conditions.

For centuries, China's farmers were also renowned for their careful collection and processing of organic materials for use on their fields. To some extent, these traditions continue, but in more developed regions, the time and labor costs associated with these sound ecological practices have reduced their popularity. Throughout China, chemical fertilizers are now much more important than organic fertilizers, and evidence of overuse is easily observable in the eutrophication of lakes, canals, and rivers (see chapter 2). Massive increases in fertilizer use is the primary reason both for China's higher yields and great increases in absolute production (figure 8.1).

On average, China applies about 260 kg/ha of chemical fertilizer as compared to about 103 kg/ha in the United States. Reliance on chemical fertilizers also impacts soil quality and structure because the benefits of the use of organic fertilizers, such as improved drainage, better soil structure, better water retention, and less soil compaction, are lost when farmers use only inorganic fertilizers. Under the CAD program, efforts are under way to increase the use of organic fertilizers once again while improving chemical fertilizer and farm chemical application methods so that "less is more."

An additional benefit of these CAD-directed efforts is that production costs are reduced and net returns increase with fewer cash inputs. Also as a consequence,

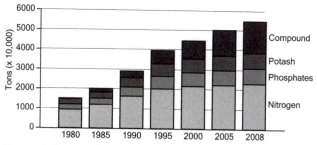

Figure 8.1. Fertilizer use in China by type, 1980–2008. Compound fertilizer is a premix of nitrogen, phosphorous, and potassium fertilizers blended to meet the needs of local soils (National Bureau of Statistics 2009, 453)

chemical residues on grain, fruits, and vegetables, a major source of concern and anger among China's urban consumers, are reduced. Pollution of grain, vegetable, and
horticultural products, however, still remains an important and challenging problem.
In part because of these real concerns, the market for organic and "green food" in
urban China has grown rapidly in recent years, and major CAD-related programs are
currently being developed to help farmers develop the higher-value, higher-quality
products for this market.

Modern Realities and Challenges

Understanding the relationship between a healthy agricultural sector and the history
and progress of the Chinese people is vital for a clear picture of contemporary China.
The ancient saying *Zhu yu feibao wu gu weibao* (The most precious things are neither
pearls nor jade but the five grains) underscores the importance of good harvests and
domestic food security. Implied in this ancient saying is the recognition that if China's
agricultural sector cannot effectively meet the growing demands of its people, severe
social, political, and economic problems far beyond the farm sector will invariably develop (Huang 1998, 166–70). The most recent official census began on November 1,
2010, but the Chinese Academy of Social Sciences estimated the nation's population
on July 1, 2009, to be 1.34 billion. China's population is expected to peak at around
1.46 billion around 2030 and slowly decline thereafter. Feeding this many people is a
challenge, and while imports can be purchased to meet domestic shortages, they must
be purchased with foreign capital reserves that could be put to better use. Meeting this
strategic goal of near self-sufficiency in grain (more than 95 percent of year-on-year
demand) is particularly challenging given that available arable land is estimated at 1.22
million ha, or only 0.085 ha per capita (1.28 mu), approximately 40 percent of the
world average (Yan et al. 2006).

Currently, China's agricultural planners are directing efforts in several directions.
Two are central to this summary: 1) the implementation of a wide range of policies
and actions that promote food security, particularly with respect to grains in light of
the nation's growing population and changing diet (CAD), and 2) the introduction
of policies that promote greater rural–urban equity and that enhance farmers' income
potential. Simply put, both of these are very expensive and challenging targets and
sometimes come into conflict at the policy level. Grain production policies focus on
improvements to farmland quality, rural infrastructure, farm inputs, and biotechnology (Xu and Bai 2002). Income generation policies center on creating accessible
market conditions, developing new higher-value products, providing better market
information and marketing systems, upgrading farmers' education and farm extension
training, and implementing subsidy payments and tax relief programs. Commitments
to both underscore an assumption that significant government investments, including
the burden of a complete elimination of taxes for farm families in 2004 and massive
subsidies for crop inputs such as quality seed, fertilizer, and farm equipment, will
continue for many years to come. This reliance on farm subsidies to ensure acceptable
incomes and maintain targeted production levels is a well-trod road—traveled to the

present by all nations of the European Union, the North American nations, Japan, and South Korea, among others—and has always proven expensive.

As is shown in the next section, China has essentially met its demand for grain to the present, but the price supports and subsidies used to meet these targets will grow increasingly expensive and may prove unpopular with urban consumers, as is the case in other nations.

It should also be recognized that despite ongoing structural shifts in the economy and a gradual decline in the share of the labor force in the sector, more than 306.5 million people remain directly employed in the Chinese agricultural sector (National Bureau of Statistics 2009, 111). In addition, the agricultural products processed in China's cities and exported to the rest of the world (see chapter 9) increased dramatically in the past decade and are vital for those employed in the massive agroprocessing industries that have developed (National Bureau of Statistics 2002). Increasingly in a post–World Trade Organization (WTO) era, China's farmers are linked economically to the domestic and global food system. In 2009, China exported US$34.86 billion in agricultural products with a growing portion coming from processed fruits, vegetables, essential oils (mint and sesame), honey, seafood, and medicinal plants (National Bureau of Statistics 2009, 726). The farm sector, then, is of strategic importance not only because of national food security and exports but also because it must continue to provide acceptable employment opportunities for both the massive rural workforce and the food industry manufacturing workforce—still considerably larger than the total population of the United States. If both food and employment are not provided, even greater uncontrolled migration to the cities and social instability will be the inevitable result. The success, then, of the agricultural sector in continuing to meet these goals for China's people is a critical requirement for continued economic, political, and social stability in the new century.

Food Production in Modern China

Setting cost aside, the considerable efforts devoted to ensuring grain security, including the 1978 Responsibility System and market reforms and the 1998 CAD program, have proven successful in terms of both increased gross production and very significant increases in unit area yields for all major crops (World Bank 1997; Zhou and Tian 2005). One indicator of the efficacy of these changes in policy is the fact that total grain production has increased for six straight years beginning in 2004. From 2005 to 2008, total grain production increased from 484.02 million tons to 528.5 million tons. The mean annual average rate of increase for this four-year period was 3.14 percent. It is expected that total grain output for 2010 will be a new record, suggesting the "over 500 million metric tons" trend will continue. Table 8.1 summarizes trends in China's total grain production over time as well as indicating corresponding proportional changes to the major grain crops: corn, wheat, and rice.

Perhaps as important as the absolute increases in all three major grains as well as for aggregate beans and tubers is the interesting shift in shares represented by rice and corn (maize). While wheat output has remained largely unchanged at about 25 percent

Table 8.1. Grain Production and Proportional Shares by Crop, 1991-2008

Year	Total "Grain" (1,000 tons)	Cereal (1,000 tons)	Cereal as % of All "Grains"	Rice (1,000 tons)	Wheat (1,000 tons)	Corn (1,000 tons)	Beans (1,000 tons)	Tubers (1,000 tons)	Rice (%) of All Cereals	Wheat (%) of All Cereals	Corn (%) of All Cereals
1991	435,293	395,663	90.90	183,813	95,953	98,773	959,530	987,730	46.5	24.3	25.0
1992	442,658	401,696	90.75	186,222	101,587	95,383	1,015,870	953,830	46.4	25.3	23.7
1993	456,488	405,174	88.76	177,514	106,390	102,704	1,063,900	1,027,040	43.8	26.3	25.3
1994	445,101	393,891	88.49	175,933	99,297	99,275	992,970	992,750	44.7	25.2	25.2
1995	466,618	416,116	89.18	185,226	102,207	111,986	1,022,070	1,119,860	44.5	24.6	26.9
1996	504,535	451,271	89.44	195,103	110,569	127,471	1,105,690	1,274,710	43.2	24.5	28.2
1997	494,171	443,493	89.74	200,735	123,289	104,309	1,232,890	1,043,087	45.3	27.8	23.5
1998	512,295	456,247	89.06	198,713	109,726	132,954	1,097,260	1,329,540	43.6	24.0	29.1
1999	508,386	453,041	89.11	198,487	113,880	128,086	1,138,800	1,280,863	43.8	25.1	28.3
2000	462,175	405,224	87.68	187,908	99,636	106,000	996,360	1,059,998	46.4	24.6	26.2
2001	452,637	396,482	87.59	177,580	93,873	114,088	938,730	1,140,877	44.8	23.7	28.8
2002	457,058	397,987	87.08	174,539	90,290	121,308	902,900	1,213,076	43.9	22.7	30.5
2003	430,695	374,287	86.90	160,656	86,488	115,830	864,880	1,158,302	42.9	23.1	30.9
2004	469,469	411,572	87.67	179,088	91,952	130,287	919,518	1,302,871	43.5	22.3	31.7
2005	484,022	427,760	88.38	180,588	97,445	139,365	974,451	1,393,654	42.2	22.8	32.6
2006	498,042	450,992	90.55	181,718	108,466	151,603	1,084,659	1,516,030	40.3	24.1	33.6
2007	501,603	456,324	90.97	186,034	109,298	152,300	1,092,980	1,523,005	40.8	24.0	33.4
2008	528,709	478,474	90.50	191,896	112,464	165,914	1,124,641	1,659,140	40.1	23.5	34.7

Note: "Grains" in China include estimates for shelled soybeans at a one-to-one ratio and whole tubers (sweet potatoes and potatoes but not cassava or taro) at a five-to-one ratio.
Source: National Bureau of Statistics (2009, table 12-15) and calculations by the authors.

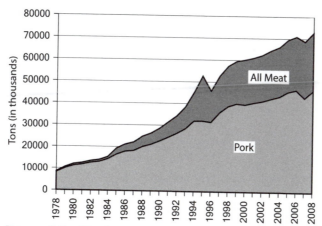

Figure 8.2. Meat production in China, 1978–2008 (National Bureau of Statistics 2002, 406; *Xin Zhongguo Wushi Nian Tongji Ziliao Huibian* 2000, 34; National Bureau of Statistics 2009, 471)

of total cereal production over the past eighteen years, rice production has slowly lost just over 6 percent of total share to corn, which now accounts for 34.7 percent of total cereal production at 165.9 million tons—this reflects changes in the Chinese diet, as much of this additional corn (and soybeans) is used to produce meat, most of which is sold to increasingly prosperous residents in China's cities, although some is actually exported. As a consequence of cheaper (and more available) feedstuffs, meat production (and consumption) has skyrocketed in the reform era with meat output rising from only 19.26 tons in 1985 to 72.8 million tons in 2008 (figure 8.2) (National Bureau of Statistics 1994, 352; 2009, 471). Of course, as meat production has increased and the scale of the operations has expanded to achieve economies of scale—particularly in the eastern and northeastern provinces—groundwater and surface-water pollution from manure-tainted runoff has increased as well.

The current target of the Chinese government is to increase grain production to 540 million tons by 2020 while doubling per capita income over the twelve years from 2008 to 2020. To reach this goal, the Chinese government raised investments to agricultural development by 30.3 percent in 2007, 37.9 percent in 2008, and another 20.2 percent in 2009. In these three years alone, total government investment in agriculture increased by 88.4 percent over 2006—much of this money dedicated to improving marginal land, improving production technologies, and subsidizing farmers.

China's CAD programs initiated in 1998 have paid off. Aggregate grain yields (output per unit of land) increased 24.1 percent from 1991 to 2008. Other important economic crops had even greater increases, with cotton yields increasing by 33.0 percent, peanuts by 34.9 percent, and rape by 33.9 percent (table 8.2). These figures are simply remarkable given the modest increases in yields for most other nations of the world for the same period.

As noted earlier, the emphasis by national planners on grain security conflicts to some extent with farmers' profit-maximization strategies that could be achieved via

Table 8.2. Changes in Unit Area Yields (kg/ha) for Major Crops, 1991–2008

Year	Cereals	Cotton	Peanuts	Rapeseeds	Flue-Cured Tobacco
1991	4,206	868	2,189	1,212	1,709
1992	4,342	660	2,000	1,281	1,687
1993	4,557	750	2,492	1,309	1,654
1994	4,500	785	2,564	1,295	1,491
1995	4,659	879	2,687	1,415	1,584
1996	4,894	890	2,804	1,366	1,750
1997	4,822	1,025	2,592	1,479	1,809
1998	4,953	1,009	2,943	1,272	1,740
1999	4,945	1,028	2,961	1,469	1,797
2000	4,753	1,093	2,973	1,519	1,763
2001	4,800	1,107	2,888	1,597	1,732
2002	4,885	1,175	3,011	1,477	1,792
2003	4,873	951	2,654	1,582	1,768
2004	5,187	1,111	3,022	1,813	1,889
2005	5,225	1,129	3,076	1,793	1,956
2006	5,310	1,295	3,254	1,833	2,072
2007	5,320	1,286	3,302	1,874	2,044
2008	5,548	1,302	3,365	1,835	2,133

Source: National Bureau of Statistics (2009, table 12-16, 467).

crops where China has comparative advantage, such as the cultivation of labor-intensive fruit, vegetable, fungi, and medicinal plants for growing domestic consumption and export. This is the fundamental dilemma challenging China's farm sector. Again, the nation wishes to be self-sufficient in grain while also raising farm incomes and improving living conditions in rural areas. Government policy and intervention are significantly less influential with respect to both gross production and yields for the latter group of products, but the higher prices for these products ensure that production of these products have increased significantly and that farmers enjoy higher incomes. Of course, exports of these products help expand and stabilize these markets (figures 8.3 and 8.4). All this reflects a maturing agricultural sector, producing more products for consumers in more places than ever before. Consumers in the nations of the European Union and in North America routinely purchase fruit juices from China as well as honey, mint products, grass seed, mushrooms, pepper- and garlic-based products, potted plants such as orchids, cut flowers, canned meats, and frozen and canned fish, shrimp, and crayfish.

Common to all agricultural production systems, however, are the benefits of vastly improved extension and farming technologies. Because of these improvements, there has been a significant reduction in the proportion of China's arable land base used to meet food security goals. In 2008, 54 percent of China's sown (not arable) land was devoted to producing cereals. When soybeans and tubers are included in "total grain area," as is the convention in China, the share increases to 67 percent. In contrast, in 1991, over 72 percent of sown area was devoted to cereal production, and when tubers

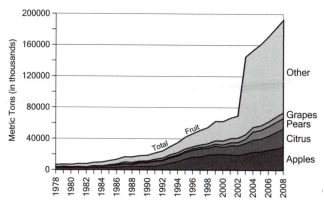

Figure 8.3. Fruit production in China, 1978–2008 (National Bureau of Statistics 2009, 466)

and soybeans are included, the figure rose to 82 percent of China's total sown area. As a result of the CAD program and related efforts, over the past twenty years, more than 20 percent of the nation's arable land has been freed up for use, growing more lucrative economic crops or fruits and vegetables. That this has occurred is evidenced by the remarkable increases in the production of these products, shown by the data depicted in figures 8.3 and 8.4. Meeting growing demand while reducing the share of land resources used for grain production reflects China's successful "agricultural revolution" and provides some idea that the commercialization of the farm sector will continue well into the future.

As befits a large country, China's environmental diversity is paralleled by a remarkable range of agricultural systems. On the largest scale, China can historically be divided into two main regions: a water-rich South relying heavily on rice production and a water-poor North that produces the lion's share of China's wheat, barley, corn, and other coarse grains (figure 8.5). In recent years, however, these crop regions have

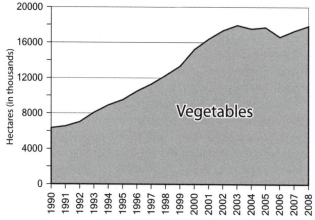

Figure 8.4. Changes in farmland devoted to vegetable production in China, 1978–2008 (National Bureau of Statistics 2009, 462)

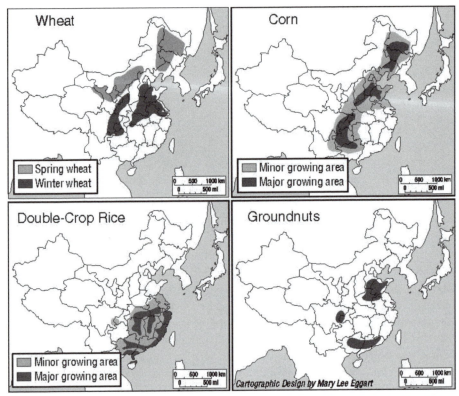

Figure 8.5.   Crop distributions of selected crops (derived from data in the *China Agricultural Yearbook* 2002)

blurred in important ways as changes in technology and consumer demand have resulted in changes in crop production. Most important, the amount of area sown to rice in North China has increased rapidly because of the development of new short-season varieties of rice that can be grown as far north as the Russian border. This would be equivalent to growing rice in northern Canada, which is not where most people expect rice to be grown.

The regional pattern of rice cultivation in China today also illustrates the effect of market reform, the CAD program, and different income levels derived from different levels of commercialization on China's farm systems (figure 8.6). The amount of land sown to rice within traditional rice-growing provinces in South and Southeast China has declined as farmers have switched to more lucrative aquacultural products, fruit, vegetables, and specialty crops (bamboo, mint, and grass seed). In Northeast China, however, irrigated rice offers much greater returns than the traditional crops of corn, soybeans, and spring wheat. So, despite potential water shortages and ecological damage in the future, rice production in the Northeast has exploded as new short-season varieties are introduced. Farmers here, even paying more for water than in the past, can generate more net income from rice (almost double that of corn or soybeans), especially because imports of corn and soybeans from the United States and other nations (a result of post-WTO economics) hold domestic prices for these crops down.

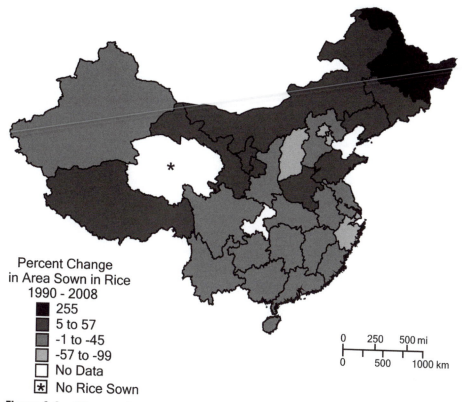

Figure 8.6.   Changes in area sown to rice in China, 1990–2008 (National Bureau of Statistics 2002; National Bureau of Statistics 2009)

China's Agricultural Regions

China's nine major terrestrial agricultural regions are depicted in figure 8.7. It is interesting to note that the distinction made earlier in chapter 2 between North and South China, along the axis of the Qinling Mountains, remains as a dividing line across several of the regions. The following sections introduce these regions in greater detail.

THE NORTHEAST

Long called Manchuria in the West, the Northeast region (Dongbei) is composed of three provinces: Heilongjiang, Jilin, and Liaoning. The region is centered on the great Songliao, or the Northeastern Plain, which is rimmed by mountains on the west (Greater Hinggan), north (Lesser Hinggan), and east (Changbai). The plain is generally low and rolling, and the northern reaches are swampy. In the 1950s, the cultivation of this vast expanse of relatively unpopulated land held great promise as a new base for agricultural production. Except for its southern flank, the Songliao Plain

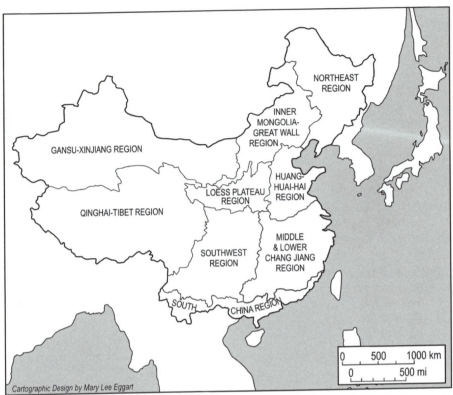

Figure 8.7. Agricultural regions of China (adapted from the *National Economic Atlas of China* (Hong Kong: Oxford University Press, 1994))

has been farmed intensely for only the past sixty-five years. Cold weather and periodic droughts limit the crops that can be produced here, but irrigation and special short-season varieties ameliorate these problems. The most important crops include spring wheat, corn, soybeans, potatoes, and sugar beets. As noted earlier, when water is available, specially bred short-season rice is an increasingly important crop.

The massive alluvial plains of the Northeast were not farmed intensively until after 1949 when the government relocated largely ethnic Han farmers from poor areas in Shandong and other coastal provinces. Consequently, population density is lower, per capita arable land area is greater, fields are larger, and mechanization has proceeded more smoothly. In the three provinces of the Northeast (Liaoning, Jilin, and Heilongjiang), large-scale farm equipment like that used in the United States is common. In fact, U.S. firms such as John Deere have enjoyed strong sales here over the past quarter century. Many of the largest farms in the Northeast are the state farms that are still controlled and managed collectively by the central government, the provincial government, or the military. As discussed in chapter 2, the conversion of northeastern wetlands to farmland, especially on the Sanjiang Plain, is controversial and threatens a very important and diverse ecological region. In the long run, many ecologists believe that this land conversion is not ecologically sound.

The lower portions of the Northeast Plain in Jilin and Liaoning provinces produce corn, soybeans, rice, and potato. Government programs, backed up by hundreds of millions in government and international loans (Asian Development Bank and World Bank) have recently been directed at improving the efficiency of larger-scale grain production and beef and dairy operations in the region. CAD efforts include soil testing to help farmers use fertilizer more efficiently, improved drainage and irrigation systems, the use of hybrid seed, and improved access to farm equipment. In short, the use of modern farming methods will improve farming incomes by raising yields, conserving water resources, and lowering production costs, and national planners anticipate ever greater output from the region. Long-term environmental impacts of this "agricultural intensification" (especially rice production that lowers water tables faster than replacement) remain to be seen.

NEI MONGOL AND THE AREA NORTHWEST OF THE GREAT WALL

Composed primarily of grasslands devoted to herding, the dry and windswept Nei Mongol (Inner Mongolia) Grassland region in recent years has seen a significant increase in field crop cultivation, especially of corn, sunflower, and potato. In the eastern portion of the region, commercial forestry is now discouraged with extensive logging bans and other restrictions, while the arid western portion, excluding the deserts, is the site of the most ambitious reforestation effort in China's history—but for ecological purposes, not commercial forestry. Reforestation is intended to create extensive windbreaks, which will help stabilize the grasslands and reduce the massive volume of windblown soil that rains down on Beijing, Tianjin, and the other large cities of the North China Plain throughout the spring months. It will also protect the last small areas of old-growth forest in North China.

In the recent past, particularly during the Great Leap Forward (1958–1960) and the Cultural Revolution, efforts were made to convert Inner Mongolian grasslands to field-crop production. The plowing up of grasslands in Inner Mongolia during the Cultural Revolution, especially during the Take Grain as the Key Link campaign, turned 391,000 ha of good pasture into desert and resulted in more than 748,000 ha being invaded by sand. Severe environmental damage resulted as the area affected by desertification increased dramatically. Only recently have more reasoned strategies been applied, with much of the land being returned to pasture or managed with more sustainable practices. In the next century, husbandry in the region will probably increase to meet the growing demand for meat, including beef, lamb, and mutton. Dairy operations are fastest growing—and in 2009, Inner Mongolia accounted for the largest share of the nation's milk production. Of course, the wool and cashmere from the huge sheep and goat herds from this area as well as from the Far Northwest supply a growing high-end textile industry increasingly dependent on exports to North America, Japan, and Europe. China is the largest producer of cashmere in the world, with much coming from this remote region.

THE HUANG-HUAI-HAI REGION (EAST CHINA PLAIN)

This region is defined by the drainage basins of the Huang He (Yellow River), the Huai, the Hai, and their great deltas. Long one of China's most densely populated and productive agricultural regions, it is currently troubled by severe water conflicts and declining grain production. Traditionally, most land here was double cropped with winter crops of wheat, barley, and rape and summer crops of corn, peanuts, soybeans, and cotton. Again as in the Northeast, post-WTO imports of corn, soybeans, and wheat have forced production shifts to less traditional products whenever possible. Since the early 1990s, farmers here have converted as much of their land as possible to fruit, vegetables, and other specialty crops that offer higher returns. The subsidies used to promote grain production are vital in the region. Much of China's investment in developing capital-intensive export agriculture is made in this region and other agricultural regions along the coasts. The region has also long been one of the most important for swine production, and recently its western edge has also become important for commercial sheep production for wool but also to meet the changing tastes of urban Chinese consumers.

THE LOESS PLATEAU

The Loess Plateau takes its name from *loess*, the German term for thick deposits of windblown soil originating thousands of miles away on the steppes of Central Asia. Loess is distributed throughout many areas of northern China, but the greatest concentrations are found in this portion of North-Central China along the middle reaches of the Huang He. The Loess Plateau covers an area of about 300,000 km² and accounts for about 65 percent of all the deposits of loess in China. In fact, the Loess Plateau is the largest continuous deposit of loess in the world. It is the yellow color of the loess, eroding into the rivers and streams of the region for thousands of years, that gives the Huang He its characteristic yellow color and name (see photo 8.2).

There is a poignant lesson to be learned from the badlands of the Loess Plateau. In early historical times (Han dynasty, 206 BC–AD 220), the Loess Plateau was thickly forested. Then, much of the plateau was a very important agricultural region that provided grain, lumber, and livestock to an expanding Chinese Empire. Times have changed. Today, travelers visiting the region can ride in a jeep or truck for miles without seeing more than a few scattered trees or some small, widely dispersed shrubs, and even these are recent, stemming from various reforestation campaigns over the past fifty years. Visitors will mostly see a seemingly endless panorama of barren yellow hills, gullies, and ravines.

At the present time, erosion on the Loess Plateau is viewed as one of China's most significant environmental problems. It is estimated that an average of 1 cm of loess is lost from the surface of the entire plateau each year. Ninety percent of the sediment load (1.6 billion metric tons/year) of the Huang He is derived from the Loess Plateau despite its accounting for only 40 percent of the total area of the drainage basin. Not only does this massive silt load dramatically affect river navigation, seasonal flood

Photo 8.2. Loessal terrain in Shanxi province in North-Central China. With rainfall limited to 500 to 700 mm. per year, erosion can occur with only slight rainfall. The loess lacks a conventional soil horizon and is very prone to erosion once the surface cover has been removed (1991) (Gregory Veeck)

events, and urban and industrial water quality for the middle and lower portions of the Huang He, it also represents the loss of a precious soil resource that this impoverished portion of China can ill afford to lose. In part because of its severely degraded environment, the Loess Plateau has been one of China's poorest regions for centuries and remains so at the present time.

In part because of the difficult terrain, there are few good roads, meaning that some villages are isolated for much of the winter. This limits commercial links to the outside world. In turn, tax revenues are low, limiting health care and educational opportunities. The people of the isolated rural villages of the Loess Plateau face a harsh existence. There is an expression, *huichiku*, which means to be able to bear hardship (literally, to eat bitterness). It is indeed a harsh and bitter life the farmers of the Loess Plateau must endure. Almost one-third of China's poorest counties, based on per capita income and gross national product, are found in this region.

Since 1949, considerable efforts have been directed at stabilizing the cropland of the Loess Plateau, but far greater efforts are needed. The Develop the West program initiated in 1999 has targeted this region for central government investments in agricultural infrastructure and terrace construction to improve conditions and reduce poverty. As a result, the government has promoted fruit production, especially of apples, and the Loess Plateau has become China's major apple-producing region with a significant amount of product entering the international market as juice concentrate. In 2009, China exported about 43 percent of the world's total volume of apple juice concentrate to the international market, largely because of increased production in the region.

THE MIDDLE AND LOWER REACHES OF THE CHANG JIANG

This region is truly the historical agricultural core of China. Centered on the Chang Jiang (Yangtze River) and its many tributaries and large lakes, most of this region is a low-lying, well-watered alluvial plain. Double cropping of winter wheat (or winter barley), rape, and rice is most common, although nonirrigated areas are given over to cotton, corn, peanuts, and other dryland (nonirrigated) summer crops. Traditionally, yields for these crops are always among the highest in China because of the rich alluvial soils.

The coastal areas of Jiangsu and northern Zhejiang provinces are at the center of a commercial agricultural revolution that has developed in the past quarter century. As in the Huang-Huai-Hai region, vegetable, fruit, flower, and plant production in the lower Chang Jiang valley has skyrocketed. Often supported by township-village enterprise processing plants for freezing, salting, preserving, and drying farm produce, the demand for these products by urban consumers and international markets has dramatically altered the farm economy in these areas. Vegetables, flowers, mulberry (for silkworm food), tea, and fruits—often tied to export opportunities—are the crops of choice, except on larger-scale specialized grain farms where farmers are allowed to subcontract larger areas if they agree to produce only grain.

As industrial expansion occurs along the Chang Jiang from Shanghai to Wuhan, prime agricultural land is being lost. Further, water for irrigation is often polluted because of point-source emissions from factories. Zoning regulations are constantly being rewritten and upgraded in an effort to mitigate these problems. However, environmental problems continue to the present, in part because local agencies will not (or cannot) enforce the regulations. The harsh fact is that industry is more profitable than farming (even vegetables), and local officials are reluctant to slow rural industrial growth because this would threaten their status, lower tax revenues, or weaken local support. Loss of high-quality cropland is greatest in this region because of urban/

Photo 8.3. A picture of the Grand Canal taken from a hill in Hangzhou City, Zhejiang province (1987) (Gregory Veeck)

industrial expansions. Farmers in the past decade routinely protest land confiscations, and new laws were enacted in 2008 to help stop these "land grabs" to preserve the nations arable land base and address the concerns of farmers. Some of these protests have turned violent in recent years and will probably continue until land is protected more effectively and industrial pollution is mitigated. Lake Tai, discussed in chapter 2, has become a "battleground" for environmental activists demanding better enforcement of environmental regulations and an end to industrial water releases to the lake.

THE SOUTHWEST REGION

The Southwest includes two main subregions: Sichuan province and the significantly higher Yunnan-Guizhou Plateau. The former includes some of the best cropland in China; the latter possesses some of the worst. The Sichuan Basin has long been a vitally important granary for China, and its agricultural productivity in part accounts for the locations of the great cities of Chengdu and Chongqing. With the area's abundant water resources, irrigated rice is produced whenever possible, with high-yield corn and soybeans on the higher slopes. Too far inland to be influenced by imported grain, the region promises to continue to be China's most agriculturally diverse region. The Sichuan Basin also has one of the greatest concentrations of pigs in China because corn and swine production are complementary production strategies.

Despite ample water resources along the rivers, however, the remaining portions of the region remain mountainous and poor, with agriculture less developed than in all other parts of southern China. This is a result of the extremely rugged terrain, a very poor road system, and seasonal (January through March) water shortages in the northern portion of the region. The Three Gorges Dam was built, in part, to store more water for irrigation as well as to generate power and control flooding on the Chang Jiang. Associated with the Develop the West program initiated in 2000 but extending for another ten years (2020), massive investments in thousands of small and medium dams and associated water management systems have also improved crop production systems and mitigated some of the terrible devastation of droughts throughout the region. Except for the Sichuan Basin, good-quality soils are often only found in the narrow alluvial valleys that cut through the sharp mountains and limestone landforms. The terrace systems that are employed in the steep terrain give the region a distinct beauty. In fact, many visitors to China argue that the famous karst landscapes of the Yunnan-Guizhou Plateau constitute the most beautiful natural scenery in China (see chapter 2).

Recent efforts to stabilize slopes and reforest the region to establish a third forestry base have been complicated by this rugged terrain. There are few good roads, little rail transport, and few navigable rivers. Rail and road building throughout the region has been stepped up since 2004 (see chapter 9). This will improve incomes but may also result in further environmental degradation of these fragile ecosystems. Tea and bamboo plantations have expanded over the past twenty years and offer higher returns than subsistence grain production. The Southwest is also a major citrus-producing region but faces competition from other areas of South China and, ironically, from lower-cost imports from Southeast Asia. Again, with investments to improve transportation and expansions of stable terraced areas, this region could make much greater contributions to China's national food supplies. These developmental monies must come from the outside—from international loans and central government grants, however, as there is little that can be expected from local sources.

THE SOUTH CHINA REGION

Although it comes as a surprise to many people, China's southern region is composed mostly of low mountains and hills (90 percent of the land area), with alluvial plains found along the coasts. The major basin is formed by the Xi Jiang (West River) and its tributaries, and the only large alluvial plain found in the entire region is found where the Xi Jiang and its distributary, the Zhu Jiang (Pearl River), empty into the sea. The Pearl River Delta has long been a famous agricultural region. Historically, most cropland here was double or triple cropped. Mulberry (for silkworms) and tea remain important nongrain crops, and bamboo production for the flooring and furniture industries has also expanded rapidly. Very intensive land use is made possible not only by abundant water resources and a long growing season but also by the fertile paddies created by organic fertilizer applications over many centuries. This intensive manuring resulted in the creation of rich anthropogenic soils where poor lateritic soils once dominated the natural landscape. The region is China's major source of cane sugar, citrus, aquacultural products, tropical fruits, and spices, such as white pepper. In a sense, postreform changes in the Pearl River Delta agricultural system exemplify the contemporary problems associated with land conflicts in all of China's important agricultural areas (Wong and Guan 1998). Because it is adjacent to Hong Kong, public and private investors from China and abroad have flocked to the cities and towns of the delta to set up manufacturing plants with links to Hong Kong, Taiwan, and other nations. As these cities and their factories expand dramatically, an ever-increasing amount of arable land is being lost despite the stringent land-use laws on the books intended to protect cropland. As in the Chang Jiang valley, conflicts over land use pit farmers and agricultural planners against more powerful local and regional supporters of urban and industrial growth. The productive land in the delta that led historically to its high concentration of commercial centers is now held hostage by the continued expansion of these centers in the modern era (Lo 1989). The Pearl River Delta, with some of China's most productive triple-cropped farmland—long famous for the production of rice, mulberry, tea, fruit, fish, crabs, and shrimp—is now the center of the most important new industrial region in Asia (Cartier 2001; Nickum 1995, 73–77). The fate of this famous agricultural region, in light of these pressures, remains to be seen.

THE GANSU-XINJIANG REGION

In absolute size, the massive region incorporating China's Far Northwest is second only to the great Qinghai-Tibetan Plateau. It is the driest region of China, with most of the region receiving less than 200 mm of rainfall per year. Only the Gansu Corridor on the southeast edge of the region can support rain-fed agriculture. Irrigated oasis agriculture has been important along the margins of this arid region for several thousand years, and the region has played a central role in China's history despite being so far from China's core. Its irrigation systems are fed by seasonal snowmelt from the east–west-trending mountain chains that surround the basins. These are the oases (and qanats) that served as way stations for several of the ancient Silk Road routes (see chapter 3). The breakup of the Soviet Union in 1990 provided new opportunities for agricultural trade with the new Central Asian republics; these linkages are new, and trade remains limited today. In recent years, the oases have expanded rapidly, with a commensurate increase in grapes, melons, and other specialty crops. Some environmentalists argue that expansion has been too extensive and express concern that the groundwater is being depleted quickly. Since 1990, deep-well irrigation systems have been continuously expanded, tapping the rich aquifers below the basins. When irrigation water is available, the cloud-free sky affords excellent radiation, and the land, through human effort, is quite productive. Recently, irrigated cotton production and sugar beets have vied with grapes, Hami melons, dates, and other traditional oasis crops. This is China's newest cotton base, capitalizing on the long, hot days that are ideal for irrigated cotton. There is little doubt, however, that this expansion—without proper irrigation technology—threatens long-term groundwater resources. The potential here for future water shortages threatens a massive ecological collapse such as the one that affected the Aral Sea region in Kazakstan. Another growing concern is salinization of cropland as irrigated areas are expanded as evaporation from irrigated cropland concentrates salts to levels that eventually render the cropland useless. Along

Photo 8.4.   Kazak herder in the Tian Mountains in Xinjiang Autonomous Region (2003) (Gregory Veeck)

the lines of cotton, rice cultivation is also increasing along the river valleys of the eastern edge of the region, resulting in growing problems with salinization and fierce water conflicts with downstream water users.

Traditional herding activities continue in more remote or nonirrigated areas of the region, but there are growing herds of cattle competing with the more traditional herds of sheep, goats, horses, and camels. Recent efforts to modernize herding practices have been somewhat successful, and new drought-resistant perennial forage crops have altered the traditional migration patterns of the herders. Of course, population throughout the region is very sparse with the exception of the old oasis cities of Lanzhou, Ürümqi, Kashgar, Turpan, and Hami.

THE QINGHAI-XIZANG REGION

Much of this great plateau is over 4,500 m high, and the entire region is cold and dry most of the year. Nomadic herding of sheep, yaks, and horses remains important. In the valleys, cut by mostly seasonal rivers, spring wheat and barley is grown. Other important crops for subsistence farming include numerous varieties of potatoes and vegetables. In terms of farm extension work, efforts remain focused on improving local food supplies. It is reasonable to assume that the region will never fully meet local food needs.

Forest Resources and Production

Forests obviously serve several important roles in any nation, and China is no different. Oak, Masson pine, Chinese fir, birch, and larch are the dominant species in Chinese forests and account for 71.31 million ha, or about 50 percent of total forested area (Butterworth and Lei 2005). Commercial lumbering operations and tree crops are vital to the Chinese national economy, but so are forests with ecological or environmental significance. These include shelter-belt forests to slow wind erosion, upper watershed forests that limit fluvial erosion rates and retain moisture in agricultural areas, and forests in ecologically diverse regions such as Hainan Island and the Southwest, where many species of flora and fauna will be lost if greater investments are not made. Recent estimates of China's forested land as a proportion of total land area indicate that massive spending on reforestation is paying off. China's State Forestry Administration estimates that 18.21 percent of national land area is now covered by forests, up significantly from the 1990s when cover hovered around 12 to 13 percent (Butterworth and Lei 2005).

Presenting a clear, concise summary of existing forest resources is more difficult than it would first appear. In part, this is true because so many forested areas in China are small and fragmented and often represent land with multiple uses (Forestry Department 1995).

Efforts are under way to reverse deforestation and to promote the development of commercial forest industries in new areas that are distinct from remaining old-growth forests (Huang W. 1998).

As some indicator of the great impact that the centuries of human occupation have visited on China, Zhao (1994, 41) estimates that in preagricultural times, as much as 40 percent of China's territory was forested. And even though efforts are under way to reverse the massive deforestation that has taken place, the centuries of abuse have exacted a great toll in many areas. Denuded landscapes in many regions, including the Loess Plateau, the Northeast, the Southwest, the South, and the middle-lower Chang Jiang, are common. Virtually all the agricultural regions of China need reforestation, and efforts to address this problem have received considerable investment in the past decade (Edmonds 1994; Smil 1984).

Historically, Northeast China has had the country's most productive and extensive timber holdings. This area is now less important for the forest industry, as tighter national control and stronger environmental protection laws have slowed old-growth logging in the region and stimulated the planting of commercial forests and shelter belts, particularly in eastern Inner Mongolia. Timber production from managed forests in the southern forest region (northern Guangdong, Guangxi, Fujian, and Jiangxi provinces) exceeded production in the Northeast only after 1993—in part because of these regulations. Although wood quality in the Northeast is higher, southern China now accounts for 43.9 percent of forest area, while Northeast forests and forests in the Southwest accounted for 26 percent each in 2005 at the time of the last national forest census (Butterworth and Lei 2005). China imports massive amounts of milled wood and round wood (trees) from Russian Siberia. These imports have allowed China to place more of its own forested area under environmental protection as state forests and biopreserves.

As China's planners work to shift the forest-products industry southward, competition from other land uses has become an obstacle in achieving this goal. Further, much of the forested land in the southern region, particularly in Jiangxi and Fujian, is covered by fast-growth durable species, such as Masson's pine and cunninghamia, but market demand and returns for these woods are quite low. Commercial operations wishing to contract this land are deterred because the low market value of its current (secondary) forest cover will not finance reforestation. Still, with abundant rainfall and the virtual absence of old growth (except in some isolated mountain areas of Zhejiang and Fujian), physical and legal conditions for the establishment of new or replanted forests are better in southern China than elsewhere. The region also has the added advantage of proximity to markets and vastly improved transport systems that will keep shipping costs low.

Efforts have been under way for the past twenty years to establish a third commercial forestry region in the Southwest, primarily in Yunnan and Sichuan provinces. This is a cause for both celebration and concern. Coverage accounts for 26 percent of national forested area, and much of this increase is due to an expansion of commercial forests. Many of China's most ecologically diverse and important forests are in this last region. High-quality hardwoods, bamboo, and other exotic species have been logged in these areas for centuries, often for the construction of temples and elite residences in the East. Until recently, little was done to reforest and to introduce appropriate commercial species into this remote region. With greater investment in transportation planned for the next decade, the area is becoming increasingly important for high-quality tropical wood. Already, areas planted in tropical tree crops, such as rubber, cocoa, coffee, and citrus, have expanded dramatically in the CAD era. It is hoped that a balance will be struck between the commercial uses of forestland and reforestation efforts for environmental and ecological purposes.

There are considerable opportunities for expansion of the Chinese forest-products industry, although new sources of investment capital and new production bases must be developed. There is virtually unlimited demand for paper and wood products in China's rapidly growing economy. Imports of wood and paper products of all types in 2007 totaled over US$22.5 billion (National Bureau of Statistics 2008, 715). In part, these massive imports—matched with similar domestic production—can be explained by the fact that China has the largest furniture industry in the world, and much of this wood must still be imported because of wood quality issues. At present, China's forest resources remain inadequate, and modern management techniques are being introduced as quickly as financially possible. In the meantime, China is a major global importer, especially with neighboring Russia, Vietnam, Burma, and Laos.

Trends in Aquaculture

One of the most distinctive aspects of China's agricultural sector has been its long reliance on ponds, canals, and rice fields for the production of fish, turtles, freshwater crabs, eels, shellfish, shrimp, and crayfish. China maintains one of the largest fishing fleets in the world, but its freshwater aquaculture has increased steadily in the reform era. The share of freshwater aquaculture to China's total fish and seafood production increased from 16.4 percent of 4.6 Mt of product when reforms began in 1978 to 53 percent of a staggering 49 Mt in 2008 (see figure 8.8) (National Bureau of Statistics 2009, 473). Still, this means that 47 percent of "catch" comes from the world's oceans and China's coastal seas. China's neighbors as well as other members of the international community have repeatedly called for more careful stewardship of the

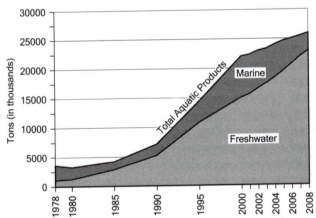

Figure 8.8. Fisheries and aquacultural production in China (National Bureau of Statistics 2009, 473)

oceans, and there is hope that much of the current ocean catch will be replaced with coastal and lake aquaculture. This incredible more-than-tenfold increase in total catch is dwarfed, though, by the proportional increase in aquaculture as a share of total production.

This rapid increase in production can be credited to the increased incomes of many urban Chinese during the reform era in conjunction with a rapidly expanding export market. Even as production has skyrocketed, demand and prices have remained strong because urban residents prefer fish and shellfish to most other sources of protein, especially after years of scarcity. Further, the labor-intensive nature of the production of pond-raised fish, shrimp, crayfish, and eels gives China's farmers a comparative advantage over producers in the West. This in turn has justified the use of better, more sophisticated—and more expensive—production technologies that result in greater efficiency and fewer losses to disease. Aquacultural experts from China's agricultural science institutes and universities have played a critical role in these improvements, and China is now one of the world leaders in aquacultural breeding and production systems. There are, however, serious and warranted concerns regarding the environmental and health implications associated with China's aquaculture industries because of the improper use of medicines, including antibiotics. Greater regulation is needed, but the small scale of producers makes enforcement of regulations and testing of products difficult. Groundwater pollution complaints associated with areas of high production are common. Frozen fish and seafood exports are the number one Chinese food import for both the United States and the European Union.

The Balance Sheet: Environmental and Economic Challenges in the Farm Sector

Based on many criteria, the success of recent efforts to "modernize" China's farm sector cannot be disputed. While it is important to realize that there remain pockets of poverty in many rural areas, particularly in North-Central China, Southwest China, and Tibet—places poorly integrated with the national economy or with limited opportunities because of harsh environmental conditions—the past two decades have generally been prosperous ones for many of China's farmers, especially those near cities or in the East. Montalvo and Ravallion (2010) estimate that the share of China's population that was officially classified as below the "poverty line" (<$1.00/day) dropped from 53 percent in 1981 to 8 percent in 2001–2002.

Location always matters. Farmers living near the east coast have access to more options than those in the rest of the country because of the coast's more diverse markets, including those for export crops and products. As a consequence, farmers living in proximity to large cities or within the coastal provinces have the highest rural incomes, while the more remote, less commercialized, interior provinces report the lowest incomes (see figure 8.9). In this sense, the inequity of space plays out in yet another way in the contemporary Chinese landscape with higher rural incomes in coastal areas.

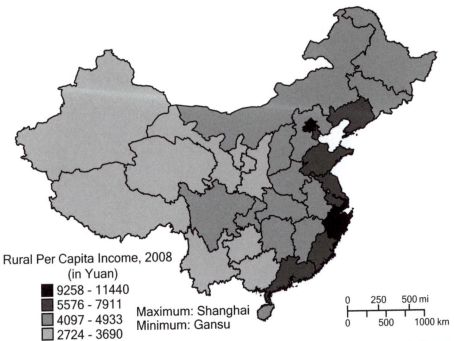

Rural Per Capita Income, 2008 (in Yuan)
- 9258 - 11440
- 5576 - 7911
- 4097 - 4933
- 2724 - 3690

Maximum: Shanghai
Minimum: Gansu

0   250   500 mi

0   500   1000 km

Figure 8.9. Rural per capita incomes by province for China, 2008 (National Bureau of Statistics 2009, 339)

Of course, there are many problems with the current agricultural sector that China's rural planners still must correct or work to mitigate. As in any nation, new problems arise all the time, more so as China's farmers are now integrated into the global food system as never before—impacted by global economic conditions as never before. Still, there are also problems that are unique to China. Landownership remains a complicated and controversial matter. Even with thirty-year leases, farmers are often reluctant to make improvements on land they do not own—especially when they worry that it will be taken from them by local governments looking to expand urban residential areas or establish a new factory. Farm-scale issues, common to many developing nations but also important in developed nations such as the European Union, Japan, and the Republic of Korea, are difficult to address given constraints of population and a limited arable land base. The constant redistribution of small plots under contract by farmers limits effective mechanization and raises costs for grains above those of land-extensive nations such as the United States, Canada, Argentina, and Brazil. Most farms in China are too small to permit either economies of scale or the pecuniary advantages associated with bulk purchases of farm inputs or the sale of farm products. Although the actual amount of land available to each household under the *Baogan Daohu* system varies by location, a rule of thumb is that each household will have only around 5 to 10 *mu* (.34 to .68 ha). Recent efforts to promote farmer's cooperatives are one response to the problems associated with small-scale producers, but the effective development of these institutions is still in its infancy.

Despite massive efforts to improve conditions, most believe that pollution problems related to agriculture and aquaculture (see also chapter 2) continue to increase,
in part because of the more intensive use of more types of farm chemicals, veterinary
and aquaculture medicines, and inorganic fertilizers but also because of the limited
expertise and understanding of many of China's farmers regarding the proper handling
of these products. As farm-related policies have grown decentralized, despite extensive
legislation regarding farm pollution, the central government has limited power to
address these issues just when there is a greater need for oversight. Local government
agencies must be provided the expertise and means to better regulate and train farmers.

In most cases, the extent and impacts of environmental problems is still not clear.
In contrast to readily available data related to crop production, studies suggesting an
increase of environmental problems, particularly in poorer regions, have been available only in the last ten years, and testing is often incomplete and unfunded in many
areas. As in all nations, including the United States, detailed local data on erosion; soil
salinization; lowered water-table levels; surface-water pollution by nitrates, phosphates,
and other farm chemicals; and soil exhaustion in China are incomplete and often
unsystematically collected. More money must be spent on environmental assessment.
Still, Chinese researchers are among the first to recognize that many of the country's
agriculture-related environmental problems have increased in severity, particularly
since the reform era began (Lin, Cai, and Li 1996). If more sustainable approaches
to crop production are not developed, the ensuing loss or pollution of cropland will
exacerbate food shortages and require that greater and greater amounts of capital be
used for food imports.

Further, reliance on the market cuts two ways. For the individual farmer, there
are clear dangers in a market-based farm sector. While grain production remains
protected by minimum prices guaranteed by the government, prices for nongrain
crops fluctuate daily, and markets are more distant than those for grain. Not only
is transporting perishable vegetables, flowers, or fruit to market more expensive, but
choosing the wrong market location, one that has a temporary surplus of the crop
a farmer is selling, for example, drops the price dramatically. Given a transport system burdened by bottlenecks and cronyism, long-distance sales and transportation
of fruits, vegetables, or flowers also has a potential for greater spoilage if a product
ends up at the wrong market on the wrong day. Still, for many of China's farmers,
the bottom line is that potentially higher profits justify the risks. Older farmers and
farms operated by women because of the seasonal out-migration of young males are
particularly challenged. It is difficult for farmers in their fifties or sixties to "catch up"
to the new realities of the farm sector.

For China's farmers, the past thirty-three years has been a remarkable journey,
transforming many of them from commune workers to entrepreneurs to price setters
in the global marketplace. China's farmers will continue to seek out production opportunities that incorporate exports to other countries. Exports stabilize domestic prices
and open up new niche opportunities for domestic producers and consumers. Trade
disputes related to agricultural exports will increase as other nations come under pressure to protect their own domestic fruit, flower, and vegetable producers. Perishables
are more profitable, more risky, and so more contentious in every nation. Although

the international grain trade is often fractious, the fights that are developing related to the growing global trade in other farm products that impact a greater number of small producers in more nations will become more pronounced in the coming decades.

Questions for Discussion

1. How has the Chinese agricultural sector changed in the past twenty-five years?
2. Why is China's agricultural sector important for the nation's future economic progress?
3. What are some important economic and environmental problems facing China's agricultural sector?
4. Looking at figure 8.9, how might you explain the differences in mean rural per capita incomes as they vary across the provinces?
5. What are some important implications related to the differences in urban and rural incomes that are present in China today?
6. What do you think are some potential implications of China exporting labor-intensive agricultural crops to other nations (both developing and developed)?

References Cited

Brown, L. 1995. *Who Will Feed China? Wake-Up Call for a Small Planet.* Worldwatch Environmental Alert Series. London: Norton.

Butterworth, J., and Z. Lei. 2005. *People's Republic of China Solid Wood Products China's Sixth Forest Inventory Report.* USDA Foreign Agricultural Service Gain Report CH5027. Washington, DC: Global Agricultural Information Network.

Cartier, C. 2001. *Globalizing South China.* New York: Blackwell.

China Agricultural Yearbook. 2002. [Zhongguo Nongye Nianjian]. Beijing: China Agriculture Publishing House.

Dong, X. Y., S. F. Song, and X. B. Zhang, eds. 2006. *China's Agricultural Development: Challenges and Prospects.* Burlington, VT: Ashgate.

Edmonds, Richard Louis. 1994. *Patterns of China's Lost Harmony: A Survey of the Country's Environmental Degradation and Protection.* London: Routledge.

Forestry Department. 1995. *General: Principles, Difficulties, and Achievements with China's Forestry Development.* Country Report. http://www.fao.org/docrep/w7707e/w7707e02.htm (accessed August 19, 2010).

Huang, J. K., S. Rozelle, C. Pray, and Q. Wang. 2002. Plant biotechnology in China. *Science* 295, no. 55 (January 25): 774–777.

Huang, W. 1998. Lumbering halted to save endangered natural forests. *Beijing Review* 41, no. 51: 8–11.

Huang, Y. P. 1998. *Agricultural Reform in China: Getting Institutions Right.* Cambridge: Cambridge University Press.

Jin, S. Q., H. Y. Ma, J. K. Huang, R. F. Hu, and S. Rozelle. 2010. Productivity, efficiency and technical change: Measuring the performance of China's transforming agriculture. *Journal of Production Analysis* 295, no. 55 (January 25): 774–77.

Kueh, Y. Y. 1995. *Agricultural Instability in China, 1931–1991.* Oxford: Clarendon.

Lin, J. Y. J., F. Cai, and Z. Li. 1996. *The Chinese Miracle: Development Strategy and Economic Reform*. Hong Kong: Chinese University Press.

———. 2003. *The China Miracle: Development Strategy and Economic Reform*. Hong Kong: Chinese University Press.

Lo, C. P. 1989. Population change and urban development in the Pearl River Delta: Spatial policy implications. In *The Environment and Spatial Development in the Pearl River Delta*. Beijing: Academic Books and Periodicals Publishing House.

Montalvo, J. G., and Martin Ravallion. 2010. The pattern of growth and poverty reduction in China. *Journal of Comparative Economics* 38: 2–16.

National Bureau of Statistics. 1994. *Zhongguo Tongji Nianjian, 1994* [China statistical yearbook, 1994]. Beijing: China Statistics Press.

———. 2002. *Zhongguo Tongji Nianjian, 2002* [China statistical yearbook, 2002]. Beijing: China Statistics Press.

———. 2008. *Zhongguo Tongji Nianjian, 2008* [China statistical yearbook, 2002]. Beijing: China Statistics Press.

———. 2009. *Zhongguo Tongji Nianjian, 2009* [China statistical yearbook, 2009]. Beijing: China Statistics Press.

Needham, J. 1981. *Science in Traditional China: A Comparative Perspective*. Hong Kong: Chinese University Press.

Nickum, James E. 1995. *Dam Lies and Other Statistics: Taking the Measure of Irrigation in China, 1931–91*. Honolulu: East-West Center.

Prandl-Zika, V. 2008. From sustainable farming towards a multifunctional agriculture: Sustainability in the Chinese rural reality. *Journal of Environmental Management* 87, no. 2: 236–48.

Smartt, J., and N. W. Simmonds. 1995. *Evolution of Crop Plants*. 2nd ed. Essex: Longman Scientific and Technical.

Smil, Vaclav. 1984. *The Bad Earth: Environmental Degradation in China*. Armonk, NY: Sharpe.

———. 1993. *China's Environmental Crisis: An Inquiry into the Limits of National Development*. Armonk, NY: Sharpe.

Sun, J. Z. 1988. *The Economic Geography of China*. Hong Kong: Oxford University Press.

U.S. Department of Agriculture. 1995. *China's Forest Products Market*. Washington, DC: Agricultural Trade Office of the Foreign Agricultural Service.

Wong, C. Y. A., and L. J. Guan. 1998. The application of remote sensing techniques to analyze the dike-pond resource sustainability in the Zhujiang Delta. *Zhongshan Daxue Xuebao*, October, 64–69.

World Bank. 1997. *At China's Table: Food Security Options*. Washington, DC: World Bank.

Xin Zhongguo Wushi Nian Tongji Ziliao Huibian [Comprehensive statistical data and materials for fifty years of new China]. 2000. Beijing: China Statistics Press.

Xu, Z. H., and S. N. Bai. 2002. Impact of biotechnology on agriculture in China. *Trends in Plant Science* 7, no. 8 (August): 374–75.

Yan Y., J. Z. Zhao, H. B. Deng, and Q. S. Luo. 2006. Predicting China's cultivated land resources and supporting capacity in the twenty-first century. *International Journal of Sustainable Development and World Ecology* 13, no. 3: 1–16.

Zhao, S. Q. 1994. *Geography of China: Environment Resources Population and Development*. New York: Wiley.

Zhou, Z. Y., and W. M. Tian, eds. 2005. *Grains in China: Foodgrain, Feedgrain, and World Trade*. Burlington, VT: Ashgate.

CHAPTER 9

China's Industry, Energy, Trade, and Transportation in a Global Context

The landmark economic reforms of December 1978 began in agriculture, but the industrial sector has been the great engine of China's economic growth for the past three decades. China's major challenges—ensuring adequate energy supplies and sustainable development, building essential infrastructure, improving environmental protection, countering hyperurbanization, controlling population growth, and maintaining stable foreign relations—are all influenced by the long-term health of the industrial sector. Reforms in the Chinese industrial sector include many challenging issues: closing unprofitable plants, improving product quality, ensuring environmental protection, cutting massive state subsidies to unsuccessful firms and factories, streamlining banking and export regulations, promoting exports, and, finally, complying with international trade agreements that play an increasingly important part in China's role as the "factory to the world." These are tall orders, particularly as the nation rapidly transitions from an industrial sector controlled largely by government bureaucracy to one now mostly dependent on market forces.

This chapter documents the most important challenges facing China's industrial, energy, trade, and transportation sectors and also includes related material on strategic energy and industrial resources. A brief history of China's industrial history is given with the hope that it fills out the discussion begun in chapter 7 of the nation's changing spatial economy over the past century, including new patterns of trade and transportation.

Domestic and Global Implications of Chinese Industrialization

China's expanding role in international politics is closely related to the complex industrial issues that Chinese planners must address in the coming decades. Global trade in light manufactures originating from China and the complex questions related to this trade will constitute seminal economic and political debates for both workers and consumers throughout the world for at least the next fifty years (see chapter 7). China's entry into the World Trade Organization on December 11, 2001, along with a thirty-year-long surge in manufactured goods, at once provide considerable challenge and opportunity for other nations. The value of China's exports increased a staggering 147 times over the years from 1978 to 2008 (US$9.75 billion to US$1.43 trillion). Over the same period, imports—especially of energy, ores, and industrial technology—also increased dramatically, from US$10.9 billion to US$1.13 trillion. The majority of China's exports, 95 percent, are industrial manufactures. As fundamentals of comparative advantage and supply and demand have taken root in China's reformed, market-oriented economy, the export of labor-intensive Chinese manufactured goods has moved to the center of trade debates in many nations, rich and poor. Labor groups and governments of many nations have reacted quickly to try to stem the flow of imports from China that undercut production and sales of domestic equivalents, but with little effect.

China's industrial products constitute more than household goods, shoes, clothing, toys, and furniture. The country's entry into high-end manufacturing has been surprisingly swift. The production of more sophisticated products, particularly electronics and computer components, has brought Chinese firms into fierce competition with those of more industrialized nations—and much earlier than expected. China's diverse and balanced export strategy has firms of many different "ownership" types selling virtually any product you might imagine (see http://www.Made-in-China.com for some idea of this great range of goods). A decade ago, Chinese industrial planners pinpointed "four main high-technology targets [for manufacture]: computers, information technology, pharmaceuticals, and new materials," and this focus is paying off (Walcott 2002, 352).

If agriculture was the key link to China's well-being in the past, there is no question that the promotion of sustainable growth in the industrial sector is now far more important (figure 9.1). Despite agriculture's successes (see chapter 8), the farm sector—even with highly processed exports—is simply not profitable enough or large

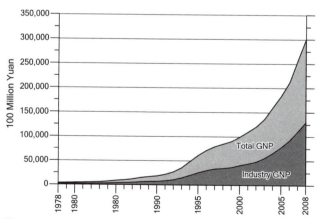

Figure 9.1. China's gross national product and share from industry, 1978–2008 (National Bureau of Statistics 2009, table 2-1, 37)

enough to drive China's overall national development. Already, growth in the industrial sector, including the expansion of light industry into rural areas, has increased wages and living standards far above those possible even in the 1990s. Throughout the nation, then, industrial growth has generated significant increases in tax revenues for reinvestment in public welfare while siphoning off surplus labor from the farms. These increased returns to both labor and capital have profoundly influenced Chinese society. Most Chinese working steadily in industry now live better, earn more, and dream about better lives for themselves or their children than even twenty years ago. Higher wages and greater productivity allow for the expansion of the service sector and the growth of the tax base, which spreads the benefits of industrial growth far beyond the factory gate.

But critics of China's many industrial policies are also correct; there are still many daunting problems facing China's industrial sector that must be addressed—and quickly. Much of China's industrial output once came from state-owned enterprises, but many of these factories were inefficient and could survive only with massive subsidies from the government. Beginning in the mid-1980s, much of this support has been withdrawn from all but a few strategic industries, such as ore mining and processing and energy production; many firms have gone bankrupt and their workers left largely to fend for themselves. In traditional cities with a significant share of state-owned enterprises, unemployment may be as high as 20 to 25 percent. These unemployed or underemployed workers face constant uncertainty, never knowing in advance if they will be able to find enough part-time work to feed their families. As in all nations, such problems have resulted in significant social unrest and grave concerns about the future for older members of China's working class. Further, energy and labor costs are rising—especially along the east and southeast coast—often forcing firms with slim profit margins to close. Some foreign firms seeking lower labor and production costs have even gone so far as to relocate from China to Vietnam.

In addition to pressing economic concerns, the many environmental problems associated with the very rapid growth of industry and manufacturing remain. The diversified industrial sector that has emerged in many rural areas throughout China has led to a proliferation of pollution problems, but water pollution issues are particularly tragic given the large number of rural residents dependent on surface water for cooking, drinking, and hygiene and the types of chronic illnesses caused by many waterborne industrial wastes (Economy 2004; Smil 2000). Laws and regulations are in place; the major issues revolve around more effective monitoring and enforcement.

While water quality in urban areas is improving in many eastern cities as factories move from densely populated areas, air pollution in China's major cities has never been as bad as it is now—and it is getting worse as more and more people buy automobiles—an issue discussed later in the chapter. Reconciling industrial growth and environmental protection is now widely recognized at the highest levels of Chinese government as a prerequisite for sustainable economic expansion. More money is being spent on mitigation of pollution than ever before, so there is reason for hope, but the extent and severity of the many problems are daunting.

In the late 1990s, citizens' groups throughout the nation began demanding a more rigorous and successful government response to industrial pollution. There are more than 2,200 nongovernmental organizations focused primarily on environmental protection at the present time. It certainly seems the government is increasingly willing to listen to complaints from these increasingly visible groups. The terrible pollution that occurred over the past thirty years in Lake Tai, located along the border of Jiangsu and Zhejiang and home to the major industrial cities of Suzhou, Wuxi, and Yixing, offers an excellent example. Runoff from agriculture and point-source releases from factories over a half century made shallow Lake Tai, long famous for its history and beauty and a major tourist destination as well, one of the most noxious water bodies in the nation. Local activists and citizens groups in the cities surrounding Lake Tai began demanding improvements in the late 1990s. To draw attention to the problems, there were many local demonstrations; videos were made and distributed; government agencies, including the Chinese Communist Party (CCP), were petitioned for help; and letters and invitations for tours were sent to all major media outlets to make more influential people aware of the problems. While all this took years, there are clear signs that these efforts have paid off—at least for Lake Tai. Some factories were closed, local officials were punished, and more rigorous bans with larger fines have been instituted. Without such concerted promotion, policies related to pollution control are often difficult to enforce, especially at the local level, for a variety of reasons. Leaders in rural and suburban areas, traditionally places with weak environmental monitoring systems and limited environmental protection staff, must address these problems with very limited resources and mixed local support. Shutting down polluting factories increases unemployment and is usually unpopular in the short term despite the fact that the polluting factories may cause long-term health problems, first and foremost, for local residents.

Background: Industry in Traditional China

China's industrial history is, in relative terms, rather short. Manufacturing in traditional China concentrated largely on the same products that had been produced for millennia. The technical transformations that revolutionized Europe's social, economic, and political relations in the 1700s did not occur in China until much later. The reasons for this are complex. There was no shortage of technical/"scientific" understanding and competence, but inventions with potential industrial applications were not applied in China as they were in Europe and later in North America and Japan. Fundamental understanding of chemistry, physics, and metallurgy in classical China rivaled or exceeded that of Europe in the same period. This knowledge led to many commercial applications in Europe (invention to innovation), but a domestic "industrial revolution" where inventions became innovations with commercial applications did not occur in China.

Shortages of capital and strong central government control over manufactured goods such as salt, cloth, ceramics, and metal goods may partly explain these different trajectories. During the late dynastic era, most existing industries in China, such as silk and porcelain manufacture, salt production, tea processing, and food processing (rice mills or oil-pressing plants), were owned by local elites but closely monitored and taxed by departments within the imperial court bureaucracy. Exports of the most lucrative products, such as silk, tea, and porcelain, were controlled by the court through licensed firms, often maintained by lucrative relationships for government officials. This system of strict if sporadic control may have limited local initiative (Braudel 1986, 586). An early summary by Lucian Pye (1972, 82) argues for the limiting effect of arbitrary taxation and internal taxes and tariffs (*liken*) on capitalist-style industrial expansions. Wallerstein (1976) adds additional factors, including strict court regulation, a largely agrarian economy, and a highly stratified social system that limited the rise of a merchant class (in contrast to European guildhalls) to the mix.

Before 1949, rural manufacturing efforts in most areas of China often concentrated on small-scale local agroprocessing of consumer products such as rice- and sorghum-based alcohol, soy, rough cloth, farm implements, fermented foodstuffs, and noodles. A major turning point came after the first Opium War (1839–1842) with the advent of a subsequent era known as the "unequal treaties" period, which is discussed in chapters 5 and 10. As foreign governments and businesses occupied coastal areas after the 1840s, output of manufactures such as textiles, tea, glass, and porcelain grew considerably, but profits were collected by foreign investors rather than reinvested to improve factories or production efficiency. As a consequence, the quality of manufactures and the way they were produced in China remained unchanged for much of the period. The dominant role of foreign powers in China's early industrial history was remarkable. After the first decade of the twentieth century, more than 80 percent of Chinese shipping, 30 percent of cotton-yarn spinning, 90 percent of the rail network, and 100 percent of iron production was under foreign control.

The greatest concentration of Chinese industries, including textiles, emerged within the Chang Jiang valley and "Jiangnan" (southern Jiangsu and northern Zhejiang provinces). These regions, together with Shanghai, accounted for more than half of China's industrial output even as late as the 1930s. Notably, the renowned prosperity of Jiangnan's agricultural economy provided capital that nurtured investments in the mechanization of traditional textile industries. This region remains among the most prosperous in all of China (Wei and Fan 2000).

This eastern concentration of capital, technology, and low-cost transport represents an important legacy to the present from the unequal treaties period in that the industrial base by 1949 was spatially concentrated in a few coastal cities, in Jiangnan, and in ports on the Chang Jiang. In practical terms, all these regions were first controlled by foreign capitalists and then nominally by the Japanese until the end of World War II. In the 1930s, the cities of Shanghai, Tianjin, Qingdao, Hangzhou, Suzhou, Beijing, Nanjing, and Wuxi accounted for around 94 percent of the total industrial output of China exclusive of the areas the Japanese controlled in the Northeast (Manchuria or Manchukuo).

This second industrial base, located in what is now called the Northeast, is a legacy of Japanese occupation from 1905 to the end of World War II. Resource-poor Japan invested heavily here through the puppet state of Manchukuo, concentrating on ore and fuel extraction and ferrous metal production to support the rapid expansion of the Japanese Empire.

In summary, China's prerevolutionary industrial base was extremely small, technologically backward even for the era, and spatially concentrated in the coastal regions. This spatial concentration of industrial production and associated expertise, credited to foreign interests and advantageous export and import locations on the southern and eastern coasts, remains a critical component of contemporary discussions of China's industrial geography and resource use at the present time. Coastal concentrations of skilled labor, capital, managerial expertise, research universities, and foreign connections all helped ensure that many of China's factories would never stray too far from the coasts even in the Maoist era. Even under the market-oriented policies of the present, overcoming the "inequality of space" remains a major challenge and justifies the massive investments in transportation systems described later in the chapter.

Stages of Industrial Development after 1949

During the past sixty years, China's leaders have followed several different approaches to industrial development. These approaches reflect changes in China's domestic political patterns and changing relations with foreign countries. Eight fairly distinct periods of industrial development between 1949 and the present have been identified. The mercurial changes in policy responsible for these periods have had many noticeable effects on the industrial landscape. Again, location plays an important role in the current situation as well as in the past. Location matters. The fundamental imbalance between the location of most of China's strategic resources in the western and central portions of the nation and the location of China's historic manufacturing bases in the Northeast and on the eastern seaboard should remain foremost in the minds of readers.

REHABILITATION AND RECOVERY (1949–1952)

Immediately after the Chinese communist revolution (1949), virtually all factories still in operation after twenty years of Japanese occupation and civil war were nationalized, in keeping with the communist vision of the ideal state as manifested in the Soviet Union. While the efficacy of this policy is now commonly criticized, the then Soviet Union, in retrospect, provided the only viable model in accordance with communist ideology. China was a largely agricultural and commercial society, torn asunder by decades of war and unrest, and its industrial base was small, concentrated, and poorly integrated with the national economy. China was faced with a possible U.S.-supported invasion by the nationalists who were based on Taiwan, and it had only a tenuous grasp on its outer regions, where local warlords continued to have considerable influence. The state had to rapidly increase or initiate the production of all strategic materials needed for the defense and the consolidation of the nation. Given these political realities, China's planners acted logically in trying to rapidly expand domestic output by substituting labor for scarce capital and technology to meet immediate needs. Growth rather than efficiency was the critical impetus for many of the state's policies.

Further, as Western diplomacy effectively sealed China off from the rest of the world, Mao Zedong quickly realized that China must be self-sufficient. By 1952, the capacity of China's industrial output had returned to prerevolutionary levels, and the nation was ready to undertake long-term programs to develop the economy.

THE FIRST FIVE-YEAR PLAN (1953–1957)

Following the Soviet model and encouraged by Soviet advisers before its relations with the Soviet Union soured in the mid-1950s, the Chinese central government took the dominant role in determining how China's industrial expansion was to proceed. Massive state investments in key industries such as metallurgy, oil and gas, mining, and transport quickly followed. The collective and commune system used for agricultural production (see chapter 8) was quickly introduced in urban areas as well—for all types of factories. Large state-owned factories were designed to be as self-reliant as possible. Workers lived on the factory grounds or in housing blocks close by. Elementary schools and kindergartens were also located on the grounds of each factory so that young children could walk to school and visit with parents at lunch. All major factories issued scrip (rather than cash) that could be exchanged at factory stores for necessities such as soap, toilet paper, and seasonings (salt, vinegar, soy, and sugar). The performance of the industrial sector improved during the first Five-Year Plan, partly as a dividend from peace. Existing plants were used more intensively, and new industrial projects were constructed with priority going to the construction of heavy-industry infrastructure, including iron and steel mills, machine tool and die plants, construction material plants, and mining- or metallurgy-related processing plants. Over 80 percent of the nation's total industrial investment during this period went to heavy industries. In an effort to counter the effects of pre-1949 investment patterns, approximately 55 percent of total investment and approximately 75 percent of all monies invested in plant construction were allocated to inland areas—in an effort to improve general economic conditions in these areas. It is ironic that in 2011 a similar effort, the Develop the West program, has been under way for a decade with very similar goals. At any rate, with greater investment, many industrial centers in the interior grew rapidly, including Baotou, Wuhan, Lanzhou, Taiyuan, Xi'an, and Liuzhou.

THE GREAT LEAP FORWARD (1958–1960)

Few goals of the Eighth Congress of the CCP, convened in September 1956, were as touted or as radical as the introduction of a national plan for superaccelerated industrialization. China's leadership estimated that three five-year plans would be needed to industrialize the nation. As a consequence of these plans, the disastrous Great Leap Forward was introduced in 1958 with the slogan "Twenty years in a day." In addition to the introduction of the commune system borrowed from agriculture, industrial growth was theoretically to be ensured through the use of "backyard" furnaces for the local production of pig iron, which then could be further processed in centralized mills or cold worked into farm implements and construction steel. The output from most of these operations turned out to be of very poor quality—often useless. The Great Leap Forward placed a special emphasis on the development of local industries as this was in line with Mao's desire to upgrade the level of economic growth in rural areas and especially in the western provinces.

The Great Leap Forward was an unmitigated disaster, and all Chinese, urban and rural, grew bitter and frustrated in the face of extreme hardship and mass starvation in many areas. Estimates range up to 30 million deaths as a result of the program and the resulting famines, a number that speaks louder than any to follow about the dangers of poorly devised government policies.

RECOVERY AND READJUSTMENT (1961–1965)

In response to the tragedy—perhaps the worst in China's long history–fundamental policy changes were required and implemented. Agriculture, once again described as "the foundation of the national economy," was assigned top priority for development, as the need for reliable food supplies for all, urban and rural, was now clearly understood. Industrial growth was based on much more pragmatic and cautious planning, with a heavy emphasis given to those sectors that could support agricultural development. The entire economy recovered slowly but steadily, and consumer goods, especially those for peasant consumption, were produced in greater quantity to mollify frustration in the rural areas. This recovery was based mainly on the more effective use of existing facilities rather than on the construction of new projects.

By 1964, the economy had made sufficient progress to prompt Premier Zhou Enlai to announce at the Third National People's Congress that the economy had recovered and that it was time to modernize the nation based on the "Four Modernizations" of industry, agriculture, national defense, and science and technology. The expression "Four Modernizations," however, was seldom mentioned in the following decade because of various upheavals caused by the Cultural Revolution. Still, it is important to realize that the basic blueprint put forth in early Four Modernizations documents reappeared in 1975 and became national policy shortly after the death of Mao Zedong in 1976.

THE CULTURAL REVOLUTION (1966–1969)

China's economic development has always been closely linked to domestic political events, an association that, in hindsight, has caused clear problems. Economic decisions always have political contexts, but the rapid-fire shifts in government policies, foci, and investment certainly inhibited the nation's efficient industrialization. The Cultural Revolution brought profound and often violent changes to Chinese society, almost freezing it in time and certainly limiting industrial growth and technical advance.

Fearful of the political consequences of any action, officials were often scared into inactivity, and efforts to improve industrial output and efficiency were delayed. Further, the outbreaks of xenophobia that occurred during the Cultural Revolution gave rise to an almost complete self-imposed national isolation. Foreign trade stagnated, and the importing of foreign equipment was strongly discouraged, increasing the technology gap between China and the industrial nations of the world. Most Chinese colleges, research institutes, and universities were closed for at least some portion of the Cultural Revolution. This resulted in later shortages of industrial scientists and technicians and a lack of managerial expertise, leading to further inefficiencies and widespread losses at most manufacturing concerns.

REORIENTATION (1970–1976)

During this complex and poorly understood period, a precarious balance of power existed between pragmatists led by Zhou Enlai and radicals led by the now infamous (and later purged) Gang of Four, which included Mao's last wife. The former group ultimately gained the upper hand, but again valuable time was lost. Ultimately, with the consent of an ailing Mao, Zhou Enlai engineered a more Western-friendly foreign policy. One important reason for this change was China's desire to obtain Western technology to accelerate economic development—especially industrial production. On the whole, national economic growth registered impressive gains during this period, and by 1974 the nation's industrial output had more than doubled from that of 1967. Growth in output for heavy industry fueled most of this domestic recovery. For the first time since 1949, credit was used in foreign trade to purchase advanced, if expensive, Western technology with the goal of catching up to the developed countries. The living standard of the average Chinese person began to improve during this period but not at the pace of the post-1978 period.

THE FOUR MODERNIZATIONS (1976–2000)

The goal of the Four Modernizations program was clear. By the end of the century, China was to be at the forefront of the world's industrial nations in a number of areas. Per capita gross national product was to reach US$1,000, based largely on industrial expansion. The return to Four Modernizations policy is clearly an important benchmark in China's economic history and heralded the adoption of many strategies that were later championed by the resilient Deng Xiaoping. Undoubtedly, this period laid the foundation for all that has followed. Once the reforms were started in rural areas and agriculture, they quickly spread to the industrial sector. While not all goals were met, the economy was energized, and the open-door policy once again brought foreign capital, technology, and expertise to the nation. Schools were dramatically improved, and postsecondary opportunities expanded. This brief but important era was the "launchpad" of China's "economic miracle."

ONGOING REFORM IN THE INDUSTRIAL SECTOR
(2001 TO THE PRESENT)

The manufacture of products that people actually want to buy keeps the workers working and spending their wages in a growing economy. A mere thirty years ago, China had legendary inefficiencies in industrial planning, production, and distribution and hence produced a very limited mix of products that few with any other options wanted to buy. Gross output had gone up most years since 1949, but the quality of most products was inferior to those traded on international markets, and prior to the December 1978 reforms, no efforts to improve them seemed to work. In retrospect, scholars generally agree that reliance on state subsidies and centralized planning isolated factories and their management from the frustration of consumers and other downstream users of their products. The poor quality of manufactured goods in the early years of "new" China was initially accepted as yet another cost of true political and economic autonomy.

Times certainly have changed. Currently, consumers are protected by government-sponsored complaint hotlines and websites, product warranties, and even "lemon laws" that prosecute firms that produce shoddy goods. Some Chinese products compare well to the best in the world. In 2011, dozens of high-quality, low-cost Chinese products are found in every American home. And the United States is hardly unique in its consumption of Chinese goods, as thousands of Chinese products are now industry leaders throughout the world. International trade has not only provided China with profits and higher wages but has also required competitiveness, modern manufacturing technologies, sophisticated product marketing and advertising, up-to-date distribution systems, and modern managerial expertise. Most important, as China's consumers, particularly those in the cities, have grown more prosperous, domestic demand for processed food and manufactured goods of all types has also exploded, clearing the way for greater and greater levels of production (Veeck and Burns 2005).

Energy Production and China's Consumption of Strategic Resources and Implications

Simply put, China cannot be the "factory to the world" without using massive amounts of raw materials—derived either from domestic or international sources. It is ironic that news reports about China's environmental problems or massive purchases from global commodity markets pay scant attention to the fact that if foreign consumers were not buying, the Chinese would not be manufacturing. Like many land-extensive nations, the country is well endowed with the mineral resources vital for economic development. Reserves of 148 kinds of minerals have been surveyed and assessed, and except for fossil fuels, China has sufficient domestic supplies of most strategic resources needed for continued industrial growth. Unfortunately, these resources are often in the wrong places in terms of processing and manufacturing plants and consumers, and the cost of mining, processing, and moving domestic supplies is often more expensive than purchasing imports (Shen 1999). As a consequence, a few "key" strategic metals and ores are imported in large volumes. China has become a price setter in markets for copper, aluminum, iron ore, steel, and many other metals simply because the nation is consuming so much so quickly.

Steel production and consumption present a useful example. The majority of the largest steel mills are located in the East: in Liaoning (Northeast China), in the cities of the North China Plain provinces, and in the cities of Wuhan (Hubei) and Shanghai. In addition, however, there are still too many small, inefficient steel mills that must be consolidated (figure 9.2). Further, China's housing and transport booms are greatest in the eastern provinces, and, as a consequence, there is a massive demand here for construction-grade steel. About 95 percent of China's iron ore deposits are classified as low-grade ores, with an average iron content of 32 percent, and only 5 percent of reserves are classified as high-grade ores, with iron content over 60 percent. The major ore-producing areas are found in Yunnan (Panxi area), Hubei (centered on Wufeng), Jiangsu-Anhui (Ningwu area), Hebei-Shanxi (Hanxing and Wutai areas), Hebei (Jidong areas), and Liaoning (Anben area) (*National Economic Atlas of China* 1994, 15).

Applying the most advanced technology, much of this ore could be processed profitably, but at present, a significant share of China's building boom relies on grow-

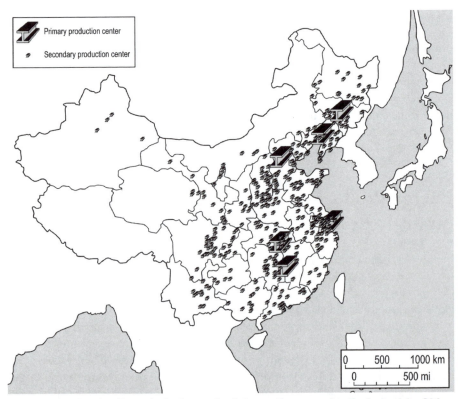

Figure 9.2. Production of pig iron, steel, ferroalloys, and rolled steel in China (after *National Economic Atlas of China* (Hong Kong: Oxford University Press), 1994)

ing steel imports, particularly of "scrap steel" or prepared steel products for construction (I-beams, "reinforcing bar," and sheet steel). Still, China's steel production ranked first in the world in 2010, the fifteenth year at this rank, but its production costs in many cases remain higher than those of most other major international producers. Steel plate shipped from South Korea or Japan to Shanghai, Tianjin, or Dalian can be significantly cheaper than the same product produced domestically and shipped downriver from Wuhan. Demand for scrap steel, largely from China, has increased the world market price 4.6-fold since 1991, illustrating not only China's insatiable demand but also the effect of its consumption on world prices. On the bright side, when prices are this high, the old industrial regions—the "rust belts"—of Europe and North America are being slowly beautified, stripped of scrap from ancient factories, junkyards, and derelict rail cars and freighters.

To continue the story, coal—vital for processing the iron and steel as well as for power generation and many other uses—is distributed extensively throughout many regions of China, with the most important concentrations found in the northern and northwestern provinces or autonomous regions of Shanxi, Shaanxi, and Inner Mongolia as well as the southwestern province of Guizhou. Again there is a "mismatch" between the location of much of that coal and where it is needed. Traditionally, Shanxi province, with the greatest deposits of coking and noncoking bituminous coal, has been the center of China's coal industry—so much so that it is known as the "coal province," but again, the power plants for electric generation and manufacturing remain concentrated in the East (figure 9.3). In 1990, the date of the last comprehensive survey of coal resources, China's reserves were estimated to be 954.4 billion metric tons (Mt), with an annual output of 1.08 billion Mt, ranking it first in the world (*National Economic Atlas of China* 1994, 74). Consumption for 2009 was estimated at 1.419 billion tons, or 43 percent of total world consumption. This is a great deal of coal. Coal accounted for 70.3 percent of total energy production in 1978, and this figure increased to 76.7 percent in 2008. Logically, coal production accounts for most of China's electric power generation, with hydropower making up a large portion of the remainder. Reflecting growing imports of other fossil fuels, coal accounted for 68.7 percent of China's total energy consumption in 2008.

The energy sector is further complicated by the fact that many, not all, firms remain state owned. This means many decisions are made for reasons beyond the "balance sheet." As pressure to reform state-owned mining operations mounted in the early 1990s, planners had to decide which locations would qualify for loans to upgrade equipment to become more efficient and which mines would be closed for either inefficiency or the safety of miners. The latter became a "hot-button" topic in 2010 after a series of tragic mine accidents showed that many stringent regulations were being systematically ignored by individual nongovernment mine operators. Decisions about shutting down mining or smelting operations combine central planning and market economics and as a consequence are both economically and politically difficult. Many coal-mining areas are among China's poorest, and mining jobs—while dangerous—pay better than staying on the farm (figure 9.3). Illegal mining operations spring up because of unemployment, leading to more mine accidents and more unrest.

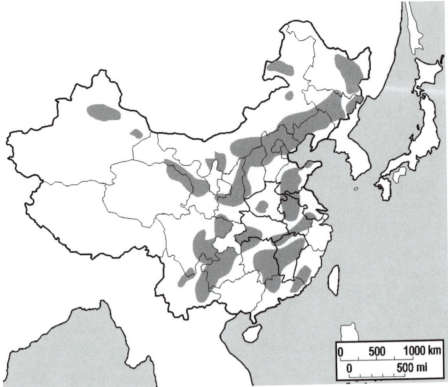

Figure 9.3.   Major areas of coal production in China (after *National Economic Atlas of China* 1994)

While coal is widely recognized to cause major pollution problems and excessive greenhouse gas emissions, the nation has massive reserves of coal in contrast to much more limited oil and natural gas deposits and is expected to depend primarily on coal well into the future. This should sound familiar to many readers. China is now the world's largest emitter of greenhouse gases, according to recent reports (Wilbanks 2008). With greater industrialization, these problems will only get worse. Of course, as the nation industrializes, demand for crude oil also grows rapidly with the expansion of the transportation network and a dramatic increase in automobile ownership. There are also health dividends to be considered in shifting from cheaper coal to other, more expensive, energy alternatives (which at least in the short term will mean more imports). For rural Chinese, respiratory diseases are the third leading cause of death (16.9 percent). In cities, respiratory disease is ranked fourth and accounts for 11.86 percent of all deaths (National Bureau of Statistics 2009, 907). Many of these deaths can be credited to the extremely polluted air in China's cities and to the use of high-sulfur soft coal in rural homes, which are tightly sealed during the winter to conserve heat.

While China's demand for energy grows rapidly in step with its economic growth and, as noted, it is now the world's largest emitter of greenhouse gases, there is positive news. The energy intensity with which China is consuming energy has improved greatly in recent years. Energy intensity is measured as the amount of primary energy consumed per unit of gross domestic product (GDP) and can be measured annually. This is, in fact, a measure of the efficiency with which energy is being used in an economic system. In China's case, the energy intensity has declined dramatically in recent years. For example, in 1980, China's energy intensity was measured at almost 1,700 tons of oil equivalent (toe) per million U.S. dollars of GDP. By 2007, this had been reduced to 274 toe per million U.S. dollars of GDP, a level not greatly above the energy intensity level of the United States and a truly remarkable achievement for a country that is still developing its industrial and energy infrastructure (Sagers and Pannell 2008).

Since 1993, China has been a net importer of crude oil, importing 19.9 Mt that year, much from the Middle East. This source remains important, but China is diversifying. It is now the world's third largest importer of crude oil after the United States and Japan and imports more than half of what it annually consumes. Recently, China has been focusing more attention on African sources of oil, especially Angola and Sudan. China is seeking to expand its sources of supply and to pursue these new sources, which have not been previously locked up by the Western-owned international major oil companies, such as Exxon and British Petroleum. In 2008, Angola was the second major supplier of crude oil to China after Saudi Arabia and accounted for a larger share than Iran. Iraq and Russia are other major suppliers (U.S. Energy Information Administration 2010). In 2009, China's largest state-owned oil company, CNPC, made a successful joint bid with British Petroleum to develop a portion of the Rumaila oil field, Iraq's largest.

There are significant foreign policy issues that are linked to the search for oil and other strategic minerals, and China has been especially active in Africa in providing aid as well as low-cost infrastructure development projects ranging from all modes of transportation development to energy projects (hydropower dams, electric grid systems, power plants, and oil refineries and pipelines), urban housing and administrative buildings, hospitals, schools, and agricultural development. China has also concluded deals with Venezuela, Libya, and a number of other suppliers in Africa and the Middle East as it ensures reliable sources for its rapidly growing consumption of oil. The search for strategic resources goes hand in hand with Chinese diplomatic and aid efforts while China also extends its marketing for low-cost Chinese consumer and other products throughout the world.

There are still many important unknowns regarding China's domestic oil reserves, and exploration is ongoing, with assistance from many multinational firms, including Royal Dutch Shell, Exxon, Marathon, ConocoPhillips, and British Petroleum. Joint ventures have played a dominant role in this exploration, especially in the Far Northwest and in the South China Sea fields. The presence of these firms reflects their belief in the growth potential of the China market and the government's desire to expand production as quickly as possible. Exploration within the South and East China seas has both economic and geopolitical overtones, however. Many other nations contest all or part of the Chinese claims in the South China Sea, particularly in the areas around the Spratley Islands (see figure 4.3 in chapter 4). While all recognize the need for a peaceful resolution of these claims, as China's economic and naval power grow, it will be a formidable force in projecting its claims for offshore petroleum and other resources in the neighboring seas. Where and how China acquires its oil, whether domestically from the Far Northwest or contested offshore locations or from imports, will be a major foreign policy issue.

In 1975, about half the nation's crude oil came from the Daqing field in Heilongjiang province (figure 9.4). The Shengli fields in Shandong province and the Dagang field in Hebei province combined to contribute another quarter of all production in 1975. The great Daqing field in the Northeast remains one of China's largest oil fields in 2010, and while its share of total production has declined considerably as new fields are developed, 40 million tons of oil were produced in 2009, and the manager of the field, PetroChina, estimates that with the use of new technologies, this level of production will continue until 2020. Shengli continues to produce, and a new field nearby (Qiaodong) was announced in 2009.

New fields are needed. Some of the most promising recent expansions include several major fields in Xinjiang Autonomous Region and Inner Mongolia (Chaogewunduer field). In 2008, crude oil output from Xinjiang reached approximately 40 Mt, equivalent to that currently produced by the Daqing field but less than anticipated. This may explain China's recently signed long-term agreement with Kazakhstan to jointly develop a massive pipeline to transmit oil from the Caspian Sea region to refineries in China—a sort of geopolitical energy insurance policy. Another pipeline

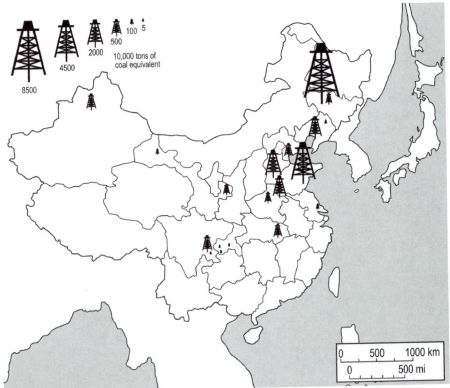

Figure 9.4. Production of crude oil and natural gas in China (after *National Economic Atlas of China* 1994)

is planned to connect Russian fields to the Xinjiang refineries at a cost of more than US$2 billion. These projects will take many years to complete, but such international joint projects reflect China's new position in terms of international relations.

Natural gas in 2008 accounted for only 3.8 percent of China's energy consumption, but efforts are under way to increase this, largely through growing domestic output. Natural gas production is largely located in the West with major fields in the Tarim and Junggar basins of Xinjiang, in Qinghai province, and in several major fields in Sichuan province. As with coal, major consumers are in the East, and again massive investments in pipeline development are required to transport the gas from these distant sources to the East. There is a new pipeline (the East–West Gas Transmission Project) from Sichuan to Hubei province that should increase both supply and demand. Eventually, this will link to another new pipeline from Xinjiang (Tarim Basin). Offshore locations that are thought to have considerable gas reserves include the Donghai and Chunxiao gas fields in the East China Sea. Currently, the largest producing field is the Yacheng 13-1, with proven reserves of 3 trillion cubic feet, with the gas shipped largely to the cities of Guangdong province (Cordesman 1998, 52). The target market for these fields will be the cities of East China, including Beijing, Tianjin, Shanghai, and Hangzhou.

Hydroelectric power is an area that has both great potential for China and strong support in the central government. Hydropower, along with nuclear and wind power in 2008, accounted for 9 percent of China's energy production and 8.9 percent of its consumption (National Bureau of Statistics 2009, 243). The Three Gorges project, discussed in chapter 2, is just one of dozens of major hydroelectric projects that have been built, and four of the largest twenty-five hydropower dams in the world are in China. Out of 680 gigawatts of explored resources, the Ministry of Electric Power estimates that 380 gigawatts can be harnessed for the generation of electricity. One estimate by Biello (2010) is that China has 26,000 dams of all types. There are genuine concerns, however, regarding the environmental impact of hydroelectric dams—in China as everywhere else in the world. Nonetheless, China has pursued a very ambitious program of dam construction for power generation, water management, and irrigation. In the past twenty years, China has built several hundred major hydropower plants, with several dozen more currently under construction. Twelve large multipurpose dams are planned for the Jinsha River (source of the Chang Jiang River) in West China alone, illustrating the importance of hydropower in the coming years. Lower power costs will also prove a powerful incentive for industrial relocation or expansion, bringing industries to the southwestern provinces of Yunnan, Guangxi Autonomous Region, and Sichuan to create a new "manufacturing base" specifically because of low-cost power. In 2010, Ford Motor Company announced the construction of a new plant in Sichuan. Many other firms, international and domestic, are relocating for the same reason.

The construction of multipurpose dams (water management and power) in Southwest China is particularly complicated, as many of these rivers flow into Southeast Asia, raising geopolitical issues in addition to technical and environmental ones. Few issues related to foreign diplomacy are as complicated as those associated with shared natural resources. Simply put, water from these rivers is vital for the economies of these nations, and China must establish that the large-scale projects currently under construction and planned for the future will ensure continued access as the nation works to develop the region.

Wind power has received considerable attention in China, and there are currently at least several hundred thousand wind turbines operating in China—mostly in remote areas. The first large wind farm in China was established in Shandong in 1986, and it has been upgraded to a capacity of 14 megawatts with the help of Danish development funds. More recently, more extensive wind-turbine fields have been established in several locations in Xinjiang, Gansu, and Inner Mongolia. In addition, there are many local programs that provide subsidized wind-turbine technology to farm households that are "off the grid." While there are no concrete numbers available with respect to coverage, there is no doubt that hundreds of thousands of households in West China derive their electricity from the wind. A typical setup, including a turbine and storage batteries, costs about $350 after subsidies. The overall contribution of wind power in China, as in most nations, is minor at the present time but holds promise for the future, and the local use of small turbines has had a dramatic impact on the quality of life for many households in remote areas where grid-sourced power is unavailable.

Completing this census, we note that China has significant uranium reserves and has embarked on a very ambitious plan to expand its nuclear power production. As of September 2010, China has twelve nuclear reactors online with another twenty-four under construction and at least four or five more waiting for permits (World Nuclear Association 2010). Initially, critical components for all these reactors were imported from France, Canada, Russia, and the United States. Chinese scientists are working to develop their own domestic expertise, however, and soon hope to produce major reactor components domestically. China is currently negotiating to build two nuclear reactors in Pakistan.

Industry consumed 71.1 percent of China's total energy in 2008, as production of all types of manufactures increased at an unprecedented pace for sales to both domestic and international markets (figure 9.5). Of course, this great explosion of industrial production did not occur evenly throughout the nation, with coastal provinces having the greatest concentration of manufacturing and thus manufacturing wages (Wei 1996) (figure 9.6). Manufacturing will continue to require the "lion's share" of energy, but this may begin to change in the coming years as more and more increasingly affluent Chinese purchase cars and build larger living spaces—all requiring greater energy resources (Dargay, Gately, and Sommer 2007).

Increased prosperity will not only increase overall energy consumption but also shift China's demand toward petroleum products, especially gasoline. Improvements in energy efficiency have been made and will continue, but conservation will provide only a small fraction of future needs unless there are radical changes in basic industry and power-generating technologies. A study by the Rand Corporation estimates that China will be importing approximately 60 percent of its oil and 30 percent of its natural gas by 2020 (Downs 2000). Domestic production of oil and natural gas has risen, but domestic production of crude oil has increased slowly in recent years while demand has grown rapidly. This has resulted in rapid increases in the import of crude oil, and China today imports slightly more crude oil (more than 4 million barrels per

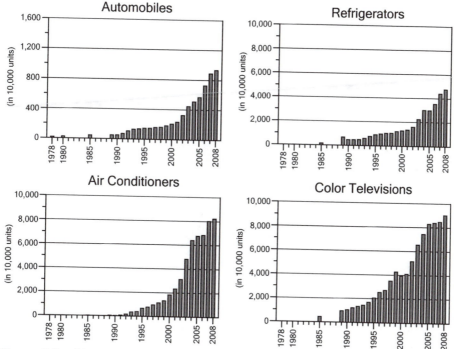

Figure 9.5.   Changes in output for select industrial products, selected years. Note that all these products were produced in small volumes since 1978 but below the 10,000-unit thresholds used for these graphs. The key point is the growth in these durable goods after 1990. (National Bureau of Statistics 2009, 473–75)

day) than it produces, and these imports will only increase in the future (U.S. Energy Information Administration 2010). While total domestic energy production rose from 627 Mt of standard coal equivalent in 1978 to 2,600 Mt in 2008, increasing more than four times in thirty years, much of this was, as noted earlier, based on increases in coal production.

Trade and China's Future Economic Growth

China's trade involves both domestic and international trade, and for much of China's history, domestic trade among the various but sometimes spatially discrete and functionally separate regions of the country made up the great bulk of trade and the exchange of goods and money. Domestic trade was traditionally composed of both local and regional exchange, and the main exchange of goods was at the local level, where farmers and petty merchants traded commodities for tools, clothing, and other needed items, virtually all of which were derived from or related to the rural economy (Brandt 1989). Regional exchange involved the movement of grain from the grain-surplus regions of Central China to political control centers of the north, as one key example, but there were many patterns of movement of bulk commodities drawn from

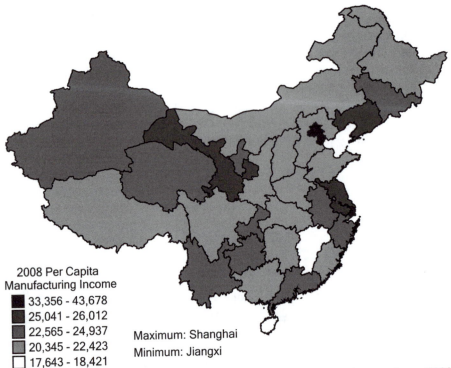

2008 Per Capita
Manufacturing Income
- 33,356 - 43,678
- 25,041 - 26,012
- 22,565 - 24,937
- 20,345 - 22,423
- 17,643 - 18,421

Maximum: Shanghai

Minimum: Jiangxi

Figure 9.6.   Average annual wage for manufacturing workers by province, 2008 (National Bureau of Statistics 2009, table 4-25, 155)

the primary sector along with selected flows of precious goods, salt, and even money. Such flows, however, had little effect on the daily lives of the peasants who made up most of China's population.

Foreign trade for most of China's history was very limited and involved the sale and export of costly goods such as silk, porcelains, fine handicrafts, and teas that were much in demand outside China. China's traditional posture, as noted above and in chapters 3 and 4, was one focused on internal development, and for much of its history it was largely closed to foreigners and to trade. China took the position for much of its history that it did not need foreign products. Western interest in China, however, began early and in earnest as explorers and traders from Europe penetrated beyond South and Southeast Asia, and the Portuguese set up a small enclave on the southeast coast of the Pearl River estuary at Macao as early as 1557 (chapter 12). Trading with czarist Russia began in the seventeenth century, but all this trade was regulated, and the Chinese continued to seek to control its trade and contact with the West and to keep foreigners out. In South China, a guild system, the *cohong,* was established at Canton to oversee and control this trade, and it worked reasonably well as a regulating and control mechanism that tightly restricted the access of Western traders.

The British, who were very interested in the China trade and who were looking for something to sell China to help pay for the Chinese goods—especially tea—that were increasingly in demand in Britain and Europe, seized on opium, grown in India, that could be purchased cheaply and sent to China, where it would fetch a good price to balance the terms of trade. The Chinese resisted, and in 1839 the British initiated hostilities in the First Opium War, which led to China's defeat. This military conflict was followed by the Treaty of Nanking in 1842, a humiliation for China that ceded the port of Hong Kong to the British and opened certain Chinese cities, designated as treaty ports, to foreign commercial and diplomatic activities and in which foreigners were given immunity from Chinese laws. As foreign powers rushed to China to share in the new access, the number of these treaty ports expanded rapidly, and by the turn of the century they numbered almost one hundred. As a result, foreign trade grew rapidly, and China began to import a variety of foreign manufactured goods as well as foodstuffs while exporting valuable commodities such as silk, tea, and other foodstuffs and goods from the rural economy. By the end of the nineteenth century, China's key trading partners were Hong Kong, Japan, the United States, and Great Britain, a pattern that to some degree remains in the early twenty-first century.

As China entered the twentieth century, its foreign trade was strongly associated with the coastal regions of the country and the Chang Jiang (Yangzi River), and the role of foreigners with their special privileges was prominent. In part it can be argued that this was based on a mercantile pattern of trade and exchange that was established to benefit the various imperial powers that controlled the treaty ports. It was thus seen as both exploitative and humiliating for the Chinese. At the same time, this trade brought with it new ideas and innovations for doing business and also the beginning of new technical processes related to modern industry and production. In this sense, it assisted in China's initial steps toward the modernization and industrialization that began in the late nineteenth century and accelerated through the twentieth century.

Trade in China and Its Role, 1949 to the Present

Trade and especially foreign trade during the Maoist period was modest and restricted, based on the idea that a powerful socialist state should be self-reliant and not dependent on others. The principle was termed *autarky*, and it implied a strong sense of going it alone and not allowing or encouraging reliance on any other countries. Thus, trade was used mainly to meet real scarcity in items that could not be produced at home. Moreover, in China, local areas were supposed to be self-sufficient and were discouraged from looking outside for goods or services. Thus, the principle of autarky was both a domestic and an international policy. The result of this was to keep trade to a minimum, and China traded mainly with other Soviet-bloc states during the first two and a half decades of CCP control, although it maintained a considerable exchange with its colonial neighbor, the territory of Hong Kong.

In doing this, China rejected the notion of comparative advantage, with its associated outcome of regional specialization of production, among factors of production such as land, labor, capital, and technology. According to this idea, such production would allow different regions to produce the goods they could most efficiently produce and then to exchange such goods in trade; theoretically, this would benefit all producers. But after the economic reforms of 1978, China largely shifted its conceptual approach to accepting the idea that there are differences in comparative advantage that can be used to allow regional specialization of production, although a policy remains in effect that requires all regions to produce food grain.

As early as 1972, China began to expand its trade with Western economies as its political relations improved with the United States and other countries such as Japan. Its trade began to grow from almost nothing in the mid-1970s, and China's whole concept of trade and its role in economic growth and development has shifted substantially since then. This is best seen in the rapid and almost astonishing growth of China's value of trade as a share of GDP. In the early years, it was as small as 4 percent of GDP, whereas by the early twenty-first century, it had risen in value to more than 30 percent of GDP even as China's economy had grown rapidly. What is now clear is that China has studied the success of other East Asian nations that have used trade and especially exports to power their economic growth engines (table 9.1). The key model is Japan, but the examples of Taiwan and South Korea are also very compelling in demonstrating how export-driven strategies of trade, buttressed by supportive industrial and economic policies, have driven rapid economic growth and development in the last three decades of the twentieth century.

Volume and Trade Partners

China's foreign trade has grown extraordinarily rapidly in the past three decades. This is illustrated dramatically in table 9.1, which provides data on the growth in China's total trade from 1978 (US$20.6 billion) to 2008, when the total reached US$2.563 trillion, a more than one-hundred-fold increase. China is today one of the top three trading nations in the world. This is a remarkable achievement that affirms the claim made above that China has indeed elected to use an export-driven model of economic growth to propel its development as it enters the twenty-first century. Only once in the past twenty years has there been a deficit in China's terms of trade, and in 2008 China

Table 9.1. China's Total Value of Imports and Exports, 1987–2008 (US$Billion)

Year	Total Trade	Total Exports (FOB)	Total Imports (CIF)	Balance
1978	20.64	9.75	10.89	−1.14
1980	38.14	18.12	20.02	−1.90
1985	69.60	27.35	42.25	−14.90
1990	115.44	62.09	53.35	8.74
1993	195.70	91.74	103.96	−12.22
1995	280.86	148.78	132.08	16.70
2000	474.29	249.20	25.09	24.11
2003	850.99	438.23	412.76	25.47
2005	1,421.91	761.95	659.95	102.00
2008	2,563.26	1,430.69	1,132.56	298.13

Source: National Bureau of Statistics (2009).

had a surplus of US$298 billion. It is clear that trade offers enormous advantages for China. This trade engine in turn is based on China's recent success as a manufacturing center for many products, a matter to discuss when we examine the commodity composition of its trade.

This rapid growth in trade has led China to focus most of its activity and exchange on its immediate neighbors in East Asia as well as on other leading market economies, such as those of the United States, Germany, and Britain. In 2008, China's leading trade partner in Asia was its regional neighbor Japan; the value of this trade was US$267 billion (see table 9.2), and China had a US$33 billion deficit. The United States was China's leading trading partner in 2008 with a total of US$334 billion in trade. China, according to its trade figures, had a substantial surplus with the United States amounting to roughly US$170 billion, although the United States claimed the number was in fact a larger amount.[1]

Another key trading partner is Hong Kong, with whom China has an enormously favorable balance or surplus amounting to more than US$175 billion, and much of this is reexports, that is, goods manufactured in China and sent out through Hong Kong for shipment to other final destinations, including the United States. Two other key regional trade partners are South Korea and Taiwan. In both cases, as seen in table 9.2, China has a substantial trade deficit, because China is buying the products that it needs at competitive prices from regional producers with low transport costs. China's trade deficit with Taiwan in 2008 amounted to more than US$75 billion, and it may be that China is also pursuing political goals by making itself a critical buyer of goods from the Taiwan economy. The evolving trajectory and terms of the China–Taiwan trade is a topic for continuing study because it can provide clues into the emerging political as well as economic relationship between these two parts of China (see chapter 13).

The Commodity Composition of China's Foreign Trade

When we think of the composition of China's trade, it is useful to think of the *factor content* of the products China imports and exports, that is, the share of the value of the

Table 9.2.   China's Foreign Trade with Key Trading Partners, 2008 (US$Billion)

Country	Exports	Imports	Total Trade Value
United States	252.38	81.36	333.74
Japan	116.13	150.60	266.73
Hong Kong	190.73	12.91	203.64
South Korea	73.93	112.14	186.07
Taiwan	25.88	103.33	129.21
Germany	59.21	55.78	114.99
Russia	33.08	23.83	56.91
Singapore	32.31	20.17	52.48

Source: National Bureau of Statistics (2009).

product that is related to the various factors of its production, such as land, labor, capital, and technology. This is a conventional means of analyzing trade structure and can be done quantitatively in detailed trade studies. Although such an analysis exceeds the scope of our discussion, it is a useful way to think about China's trade and its overall economic structure as we go about assessing the role of trade in China's development.

In 1980, half of China's exports derived from the primary sector (raw materials), and only about 12 percent were manufactured goods. This would suggest that a strong share of what was then a relatively modest amount of exports was derived from land as a factor of production.

In 2008, only about 5 percent of China's exports derived from the primary sector, while 95 percent were produced in the secondary sector and were described as manufactured goods. What we can infer from this is that China has now shifted to a much greater use of its labor and capital as factors of production in the content of its exported goods. If we were to examine more specifically the nature of the various products among these manufactured and exported goods, we would discover that many of them have substantial labor content as a share of their total value, and thus we could conclude that China is making good use of its greatest comparative advantages: low-cost labor. It is this approach both to manufacturing and to its related trade in a global trading system that has enhanced China's remarkable recent rise as a powerful competitive force among the world's industrial powers and that is causing considerable concern among many nations, such as Mexico and India, and among China's Southeast Asian neighbors, which also seek to share in the global market for low-cost textile goods, garments, footwear, sporting equipment, machinery, and other products with a significant labor content in their total value.

On the import side, the pattern is similar, although the changes are not as striking. For example, in 1980, about 35 percent of China's imports were primary-sector goods, including petroleum, and 65 percent were manufactured goods. In 2008, 68 percent of China's imports were manufactured goods, including a significant share of transportation equipment, and only about 32 percent came from the primary sector, again including petroleum. Thus, China continues to need manufactured goods, but today these are more sophisticated and higher-end manufactures, such as chemicals, plastics, vehicles, aircraft, and high-tech machinery. At the same time, the country also continues to import some foodstuffs and related primary-sector goods as well as petroleum, an import commodity that is increasing as China continues to struggle in its search for petroleum self-sufficiency while its consumption grows rapidly in step with its growing automobile usage.

The "China market" continues to be both a dream and a frustration for businesspeople and manufacturers. The idea of "oil for the lamps of China" remains a goal for commercial operators and entrepreneurs everywhere, but the reality is that the Chinese market has proven difficult to break into. Moreover, China's recent rise as a major manufacturing power has made this dream more unrealistic, for it is increasingly clear that few external producers of manufactured goods will be able to undersell China in its ability to produce a vast array of goods, including increasingly more sophisticated industrial and electronic products, at low cost. The next two or three decades are likely to see China enhance its position and power as a great manufacturing center and formidable trade competitor. China's recent entry into the World Trade Organization should help other countries gain official entry to its markets, but China's increasing competitiveness in the many products it manufactures will allow it to maintain a strong position as it meets the challenge of a more open domestic market for its vigorous foreign competitors, such as Japan, the United States, Germany, and Britain. It now has the capacity to satisfy much of its domestic demand for goods and services, and it continues to develop its internal technical capacity and the quality of its workforce to meet the challenges of globalization.

Spatial outcomes related to China's rapidly growing economy will continue to yield new "spatialities" of production, consumption, and distribution. Continuing investment in transportation as well as in commercial and industrial infrastructure is proceeding at a frenetic pace in the coastal as well as the interior regions of the country. Hong Kong and Shanghai are two keys to the international commercial, manufacturing, and shipping marketplace, while Beijing, as China's capital, has a very strong international presence in banking, finance, and research and development. Meanwhile, there are literally scores of other cities, both coastal and interior, that are also advancing their goals and roles in the marketplace and assessing their positions and their chance to be key future players as China extends its reach into the global economic system.

Transportation: A Key to China's Continuing Economic Growth

The creation and existence of an effective and efficient modern transportation system is indeed a necessary, if not sufficient, condition for the economic development and modernization of a nation in the twenty-first century. This is perhaps nowhere more obvious than contemporary China, which, because of the continental size of its vast territory, must be brought together to function effectively as a national polity and space economy. Since the reforms of 1978 and especially since the early 2000s and acceleration of the market forces in the economy, enormous investments have been made in all modes of transportation and communication in China (Lin 2002). This has been in parallel with the very rapid and continuing growth of the economy, which has experienced an explosion in the movement of goods, people, and information.

In August 2010, news reports of a sixty-mile-long "traffic jam" on China's National Expressway #110, running from Beijing through Inner Mongolia and ultimately southwest to Tibet, had people around the world shaking their heads in disbelief. It took eleven days to clear this single traffic jam and get conditions back to "normal"—normal being when traffic moves along at thirty to forty miles per hour (Coonan 2010). The massive snarl resulted in hundreds of millions of dollars in lost productivity and spoilage and generated a huge public outcry (Oster 2010). The August traffic jam was newsworthy solely because of its extreme size and duration, but similar traffic jams are the norm at congestion points throughout China. The rail system, second now only to that of the United States in total length, is similarly congested with trains often leaving within three to ten minutes of each other on the same tracks. The World Bank estimates that losses due to poor infrastructure and transport systems cost China a full 1 percent of annual GDP. Shipping costs from western China to the east coast can be 50 percent higher compared to equivalent distances in other land-extensive nations, such as Australia, Canada, or the United States. Transportation planners in China—under fire by government officials, business and industry leaders, and angry citizens' groups to "do something"—are working to improve the system as quickly as possible, and expansion of the gigantic system is now one of the government's highest investment priorities.

Ironically, most agree that the current transport crisis has its origins in the very successful economic reforms adopted in December 1978 that dramatically increased manufacturing, commerce, and international trade while also raising incomes, resulting in more passengers able to afford to travel more often. Even as late as 1978, the best routes of the interprovincial rail and expressway system were located largely within the eastern and southeastern provinces. However, as greater volumes of goods, raw materials, and people moved between East China and the rest of the country, it was clear that the system needed expansion as quickly as possible. Even with massive investment since the late 1980s, the national freight transport system remains insufficient to meet the demands of the nation's expanding economy. Similarly, the passenger rail system, while making strides unprecedented since the expansion of the U.S. rail system in the early twentieth century, also remains inadequate in terms of the tremendous growth in demand. Rail passenger-kilometers (a value of 1 for every kilometer traveled by a single passenger) increased more than sevenfold from 109,320,000,000 passenger-kilometers in 1978 to 777,860,000,000 passenger-kilometers in 2008. The increases for freight are even greater (figure 9.7).

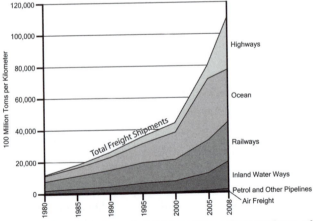

Figure 9.7. Growth in freight ton kilometers by type of conveyance, 1980–2008 (National Bureau of Statistics 2009, table 15-9, 614)

Urban and regional suburban transportation systems are also overwhelmed. Most regional networks serving megacities such as Beijing-Tianjin, Shanghai, Guangzhou, or Chongqing commonly experience "traffic paralysis" for a minimum of three or four hours each morning and at night during "rush hour." In an effort to resolve or at least mitigate these problems, investments in rail, road, and subways and light rail in the past decade have been the greatest in the nation's long history. The most recent program, made in conjunction with China's 2008 National Economic Stimulus Package, invested $586 billion for new infrastructure and earthquake recovery, including $189 billion for railway upgrades and expansions and more than $100 billion for road construction. This does not include local investment—the source of funds for most local and regional roads and highways. Similar investment of this magnitude over the next decade will improve infrastructure and lower the shipping costs from central and western regions for goods while putting hundreds of thousands to work who may have lost employment during the 2008–2010 global economic downturn.

Despite this great investment, there are many types of transportation-related infrastructure projects that require funding, and, as in every nation, the distribution of infrastructure projects is as much political and economic. Typically, every Five-Year Plan (chapter 7) includes a range of investments distributed throughout the nation to minimize charges of favoritism. For just the period from 2006 to 2010 (the last Five-Year Plan), for example, the following national projects were funded: six new passenger-only long-distance rail lines, five electric intercity lines, five upgraded rail lines, fourteen new expressways, new container facilities for twelve existing ports, deep-water dredging for the Shanghai Container Port, dredging for the entire "Grand Canal" from Beijing to Hangzhou, and the expansion of ten airports.

Seldom is the relationship between economic development and transportation infrastructure as clear as in China. The vastness of China's territory and the ruggedness of the terrain in the central and western regions make the construction of a national transportation network a great challenge, as evidenced by the significantly lower densities of rail in these regions (figure 9.8). The nation must find ways to build infrastructure fast enough to meet the growing demand brought about by the expanding manufacturing sector, which is increasingly dispersed throughout the nation rather than concentrated along the coasts as in the previous century (for a good summary of emerging spatial patterns, see Fan 1995). Of course, the nation has lower labor costs than developed nations, but as the price of labor increases in the eastern provinces, firms increasingly look to the interior for bargain-priced labor. This means shifting factories to interior provinces where transportation remains poor but where operating costs, labor, and power are significantly cheaper. Continued growth of the economy, especially over the great spaces of western China, depends then on the development of superior transport systems linking interior regions to the coasts, allowing the lowest possible costs for moving both goods and passengers.

Of course, the expansion of transport routes into distant places such as Xinjiang and Tibet (Xizang) has significant implications for those already living in these regions. Commerce and connectivity will be improved, resulting in more jobs and higher incomes. But these new transportation systems allow eastern capital and entrepreneurs' easy access to less developed interior regions as well. Social relations and local

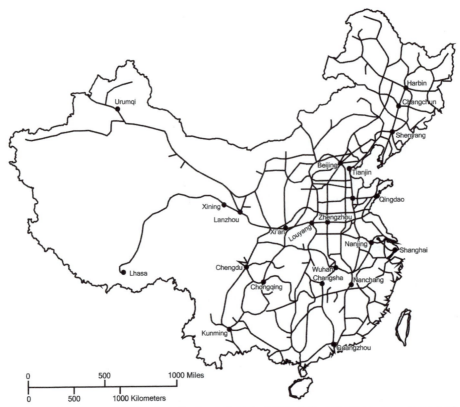

Figure 9.8.   Map of China's railway network, 2009 (after similar maps found at http://www.travelchinaguide.com/china-trains/railway-map.htm and http://www.chinatraintickets.net/china-trains)

autonomy will be altered in the face of this greater integration into the national system. Greater investment and incursion by eastern China concern many Tibetans and other minority groups living in the west, and eastern workers are not always welcome.

As some measure of how fast these systems are expanding, from 1980 to 2008, railway lines increased by 50 percent, highways increased by 319 percent, aviation routes by 1,279 percent, and pipelines by 570 percent (table 9.3). The frequent "bottlenecks" that develop at the present time—despite these rates of expansion—highlight an important consideration: Will any transport system ever be adequate for China, or will demand constantly overwhelm supply no matter how quickly or how often the system is expanded? In the United States, there are just over 765 registered vehicles per 1,000 citizens—this has remained fairly steady for some years—suggesting that some equilibrium has been reached. In China, at present, this figure is only 128 vehicles per 1,000 citizens, and already the road and highway systems of most major cities are overwhelmed. How many more consumers, once possessed with the means as China's economy continues to grow and the middle class expands, will choose to buy personal vehicles regardless of occasional inconvenience? Sitting in your car for an hour or two in traffic still beats standing packed on the bus or waiting in line to board for the same time or longer.

Table 9.3. Major Indicators for China's Transportation System: Selected Years (10,000s of Kilometers)

Year	Length of Railways (including electrified rail)	Length of Highways (including multilane expressways)	Navigable Inland Waterways	Regular Air Routes	Petroleum and Gas Pipelines
1980	5.3 (0.2)	88.8 (0.0)	10.9	19.5	0.9
1985	5.5 (0.4)	94.2 (0.0)	10.9	27.7	1.2
1990	5.8 (0.7)	102.8 (0.1)	10.9	50.7	1.6
1995	6.2 (1.0)	115.7 (0.2)	11.1	112.9	1.7
2000	6.9 (1.5)	140.3 (1.6)	11.9	150.3	2.5
2005	7.5 (1.9)	334.5 (4.1)	12.3	199.9	4.4
2008	8.0 (2.5)	373.0 (6.0)	12.3	246.2	5.8

Source: National Bureau of Statistics (2009, table 15-4, 612).

China's Road Network

China's road network, especially the national expressway system, is growing faster than rail or waterways. The national expressways are the crown jewels in a system that is still largely composed of provincial-, county-, and township-funded and -maintained roads. In 1992, there were only 652 km of expressways. This increased to 65,000 to 70,000 km at the end of 2010 and should increase by another 20,000 km to 80,000 km by 2025 (Li et al. 2010). The expressways are well maintained and well regulated and have significantly reduced travel times from city to city, especially in the East. Unfortunately, most of the road system is composed of lesser-quality roads. In 2008, there were 3,669,900 km of lower-quality national and provincial "highways"—many of which are actually only two lanes (one lane each way).

The benefit of roadways (over rail) includes their low construction costs, convenience, and ease of access. They can be built nearly anywhere—so as local firms and economies have expanded in many areas, more locally financed highways follow. Some of these are well constructed; others are poorly designed, use shoddy materials, and wear out in a year or two. An interesting recent initiative involves loans provided by Beijing and provincial governments to local township (subcounty) governments to improve "farm roads," the lowest level of road in the network. Recent research has shown that improvements to these local roads, which are now usually packed earth or gravel, will allow greater mechanization in agriculture, and farmers will have improved access to more distant commodity markets, improving competition and allowing the farmers fair prices from wholesalers (Li et al. 2010).

In 1980, slightly over 6 percent of total ton/kilometers of freight were moved on highways. This jumped to over 29% of total ton/kilometers of freight in 2008—a sharp rise even over 2007. As new rail lines are brought online—especially those connecting the West to the East—this figure should once again decline, but road networks will continue to expand as quickly as possible at least for the next several decades because of their low cost and the ability of concerned local governments to absorb the cost. However, there is little doubt that China is looking ahead and seeking lower-cost future options to move both people and freight as quickly as possible. While recognizing that the national roadway system must be upgraded, widened, and expanded, it is clear that the long-term preference is for proportionally greater investments in light rail, subways, rail, high-speed rail, and pipeline systems rather than investing in road systems.

Experts argue that there are already too many cars, with estimates ranging from 45 million to 50 million in 2009. Thirty-five million of these are privately owned with the remainder owned by government agencies or government or private companies (Gordon and Sperling 2009). Construction of new highways is occurring with remarkable speed, but in almost any large city in China, one gets the feeling that no matter how many new roads and expressways are built, they will always be overwhelmed by the ever-growing number of cars that use them. The city of Beijing, for example, adds 1,500 cars to its roads every day.

The paradox, of course, is that, while congestion is very bad for business and represents a direct cost to the economic system by way of lost time and labor and health impacts, building transportation infrastructure, such as rail lines, roadways, vehicles, pipelines, electricity pylons, and transmission towers, is very good for business, both domestic and international. In December 2008, China finally surpassed the United States as the largest manufacturer of motor vehicles. In 2008, the United States built 8.7 million cars, trucks, and other four-wheeled vehicles, while China produced 9.5 million (U.S. Department of Energy 2009). The automobile industry in China is similar to that of the United States in that the wages are relatively high, the work is (usually) steady, and the profits are much greater than many other manufactures. Chinese enthusiasm for cars has many international implications as well. It is possible that by 2015, General Motors will build (and sell) more cars in China than in the United States.

When we first visited China in the early and mid-1980s, it was a country of surprisingly few private cars and millions of bicycles. Increasingly, this is not the case, as cities have grown too large for bicycle commuters to make the trip from home to work, not to mention that pollution on the streets is too noxious and the streets are too dangerous. In a sense, then, the bicycle is being driven from the field by prosperity. Further, while public transportation systems are far more comprehensive and convenient than those in the United States or many other nations, they are also very crowded, and commuting times can be some of the most stressful of a workday. Chinese consumers will continue to purchase cars for the same reasons as everyone else. The personal automobile is convenient, quiet, and private and sometime serves as an important status symbol as well.

Railroads Revisited

As in many other nations, planners increasingly argue that despite the current trend of increasing auto ownership, China cannot "afford" a long-term love affair with the car and the "open road." As a consequence, the nation has made massive investments in alternative transport systems, particularly rail and light rail but also high-tech pipeline systems that can move liquids and even solids such as coal in slurry form so as to limit the growing freight burden on the nation's motorways. Rail networks in West China are improving dramatically, and most lines are now joined with the eastern grid and end at state-of-the-art marine container shipping terminals. Rail expansions also include interurban systems, such as regional commuter rail, subways in high-density areas, and urban light rail, as well as standard long-haul freight trains, express rail (bullet trains), and even several magnetic levitation train systems that are in operation or currently under construction. The Chinese government has promised to spend more than $1 trillion expanding and improving the rail system from its current 78,000 km to over 110,000 km by 2012 (figure 9.8).

The construction of rail lines, subways, and urban light rail systems will continue for decades, but the expansion of these systems is well under way. Whether any expansion, no matter how great, can meet total demand will remain an intriguing question, but there is little question that China's transportation system will continue to improve and expand. All of China's cities with populations over 5 million now have multiple rail stations increasingly linked to commuter rail systems into a system that is already unprecedented in scope with respect to both extent and capacity. Almost all major cities have subways or light rail networks that move millions of people about these cities every day. Many of these new urban light rail systems employ cutting-edge, low-energy technologies, and, as a consequence, China has emerged as a leader in high-speed,

Photo 9.1. The express train from Beijing to Tianjin, which went into operation in 2008, travels the 117 km between the two cities in thirty minutes at speeds up to 350 km (217 miles) per hour. Several other high-speed dedicated routes are in operation, including one from Wuhan to Guangzhou. In 2011, the Beijing-to-Shanghai high-speed rail line will be completed, significantly reducing air and road congestion (Clifton Pannell)

high-technology rail systems. The world's fastest long-distance passenger train (the Harmony) averages 354 kilometers per hour (220 miles per hour) on the 1,062-km trip (660 miles) from Wuhan to Guangzhou. At the present time, there are another 8,000 miles of high-speed tracks and trains planned for construction, but already as costs escalate, as might be expected, there are disagreements regarding the shares that the provinces must pay in terms of the central government's contribution. The planned route from Shanghai, down the mountainous and crowded coast to Fuzhou in Fujian province, for example, has been temporarily put on hold as the financial agreements are ironed out more completely. Given the massive sums already invested, it seems certain that construction will be completed on all routes already started. An interesting and somewhat ironic aside, with implications for nations such as the United States and Canada looking to upgrade passenger rail, is the fact that China's air carriers have already expressed concern about the unfair subsidies going to the rail sector and the possible effects of the these modern efficient rail routes on the revenues for air carriers on these routes.

The rail link to the capital city of Tibet, Lhasa, from Xining in Qinghai—the Qingzang Railway—is, without exaggeration, one of the engineering marvels of the world. It extends 1,956 km (1,215 miles), of which 530 km are constructed over permafrost and at such great elevations (maximum 5,072 meters above sea level) that the cars must be pressurized for the comfort of passengers. Opened to all in July 2007 after a year of testing and trials, the train has come to signify the great potential and aspirations of the Chinese people in a way that the Three Gorges Dam did for the previous generation. It is interesting that the seventy-eight locomotives for the line were supplied by the U.S.-based multinational General Electric and that the 361 pressurized passenger cars were manufactured by Bombardier Sifang Power (Qingdao) Transportation Ltd, a joint-venture firm formed for this purpose by Bombardier of Canada and the South China Rolling Stock Corporation.

In 2010, China proposed a high-speed, high-tech railway from North China to Europe. Such a line, romantically called "The new Silk Road," would radically alter the spatial economy of Inner Asia. Although such a system is probably decades away, the possibilities are remarkable to consider and shed some light on the plans and ambitions of the nation at the present time. Given the pressing need represented by China's still-growing population and the nation's expanding role in global manufacturing and trade, it is sure that the nation's transportation system will soon be the largest and most integrated in the history of the world and will get only better.

Final Thoughts

As might be expected, winners and losers have emerged from the ongoing and dramatic shakeup of China's industrial sector, resulting, in part, from the rapid shift to a market economy. It seems a time of feast or famine. Stories of newly made millionaires vie for space in Chinese newspapers with articles on the sad plight of hundreds of thousands of laid-off factory workers. For young, college-educated, urban Chinese, this could be the best of times. Careers in real estate, industry, manufacturing, sales, and marketing potentially offer career satisfaction, independence, and the possibility at least of great riches. On the other hand, for many of the folks pushed off the factory floor because of the closing of inefficient factories and firms or global downturns, the future is far less rosy and increasingly uncertain.

China's industrial expansion will continue into the foreseeable future with both positive and negative implications. On the plus side, for many of China's people, the transformation of China's industrial sector has systematically resulted in a significantly higher standard of living. This is true not only in China's cities but in many places in rural China as well. The "new China" anticipated with fierce dedication by Mao Zedong in 1949 now is a reality—and in large part this is due to new policies associated with the industrial sector.

Still, the market can be a harsh taskmaster, and many firms throughout China have gone bankrupt because of dated technology, poor management, high transportation costs, or the production of obsolete products. Problems related to unemployment and underemployment must be addressed, and social services for the unemployed and underemployed must be both extended to rural areas and strengthened in the cities. As in every nation, firms with very low profit margins may overlook environmental controls and exacerbate local pollution problems. China's leaders clearly realize that "development at all costs" is not sustainable, and environmental monitoring is being "stepped up."

Dependence on global markets for mass production of consumer and industrial goods for export has been an important part of China's strategy for economic growth both for the labor absorption capacity of export-oriented industries and for the valuable foreign exchange these exports earn. On the one hand, this focus on exports generates a substantial trade surplus on which to create wealth for China while at the same time allowing China to purchase extensive energy and raw materials as well as advanced industrial and transportation products from abroad. The terms of trade have largely favored China, and this trend appears likely to continue for the foreseeable future as China seeks to employ as many workers as possible in its factories while also creating more wealth on which to build its future. On the other hand, as many other nations have learned over the past several hundred years, greater integration with the global economy exposes domestic economies to economic shifts that are beyond the control of national planners.

Questions for Discussion

1. How has China's early industrial history influenced present conditions and the spatial patterns of the nation's industrial base?
2. Describe the relationship between the requirements of China's industrial sector and manufacturing exports and the nation's developing foreign policy goals.
3. The relationship between China's transportation system to industry and trade is complex. What are some of the important challenges related to transportation, and how do these relate to the overall economy?
4. Why are local environmental problems in China difficult to address, and what do you think should be done to improve conditions in local areas?

5. China has massive coal reserves. Why are these insufficient for further economic growth?
6. How might you comprehensively explain the patterns depicted in figure 9.6 (map of manufacturing wages)? What past and present conditions are responsible?

Note

1. Trade figures vary according to the source of the information. Each country typically seeks to report a maximum import quantity based on cost, insurance, and freight while minimizing the value of its exports to gain the maximum advantage in bilateral or multilateral trade discussions. Consequently, trade data may not be consistent among various sources.

References Cited

Biello, David. 2010. The dam building boom: Right path to clean energy. *Yale Environment 360.* http://e360.yale.edu/content/print.msp?id=2119 (accessed September 28, 2010).

Brandt, Loren. 1989. *Commercialization and Agricultural Development: Central and Eastern China, 1870–1937.* Cambridge: Cambridge University Press.

Braudel, Fernand. 1986. *The Wheels of Commerce.* Vol. 2 of *Civilization and Capitalism, 15th–18th Century.* New York: Harper and Row.

Coonan, Clifford. 2010. The ten-day traffic jam driving China mad. *The Independent.* http://www.independent.co.uk/news/world/asia/the-tenday-traffic-jam-driving-china-mad-2061184.html (accessed August 25, 2010).

Cordesman, Anthony H. 1998. *The Changing Geopolitics of Energy—Part VI: Regional Developments in East Asia, China, and India.* Washington, DC: Center for Strategic and International Studies.

Dargay, Joyce, Dermot Gately, and Martin Sommer. 2007. Vehicle ownership and income growth, worldwide: 1960–2030. January. http://www.econ.nyu.edu/dept/courses/gately/DGS_Vehicle%20Ownership_2007.pdf (accessed August 25, 2010).

Downs, Erica Strecker. 2000. *China's Quest for Energy Security.* Santa Monica, CA: Rand Corporation.

Economy, Elizabeth. 2004. *The River Runs Black.* Ithaca, NY: Cornell University Press.

Fan, C. Cindy. 1995. Of belts and ladders: State policy and uneven regional development in post-Mao China. *Annals of the Association of American Geographers* 85, no 3: 421–49.

Gordon, Deborah, and Daniel Sperling, 2009. Surviving two billion cars: China must lead the way. *Environment 360.* Editorial. http://e360.yale.edu/content/feature.msp?id=2128 (accessed August 25, 2010).

Li, T. A., F. Shilling, J. Thorne, F. M. Li, H. Schott, R. Boyton, and A. M. Berry. 2010. Fragmentation of China's landscape by roads and urban areas. *Landscape Ecology* 25: 839–53.

Lin, Shuanglin. 2002. China's infrastructure development. In *China's Economy into the New Century: Structural Issues and Problems,* ed. John Wong and Ding Lu. Singapore: Singapore University Press, 291–324.

National Bureau of Statistics. 2009. *Zhongguo tongji nianjian 2009* [China statistical yearbook 2009]. Beijing: China Statistics Press.

National Economic Atlas of China. 1994. Hong Kong: Oxford University Press.

National Economic Atlas of China. 1994. Bk. 2. Hong Kong: Oxford University Press.

Oster, Shai. 2010. China traffic jam could last weeks. *Wall Street Journal.* August 24. http://
online.wsj.com/article/SB10001424052748704125604575449173989748704.html
?mod=WSJ_hpp_MIDDLETopStories (accessed August 25, 2010).

Pye, Lucian W. 1972. *China: An Introduction.* Boston: Little, Brown.

Sagers, Matthew J., and Clifton W. Pannell. 2008. The clean energy dilemma in Asia: Observations on Russia and China. *Eurasian Geography and Economics* 59: 391–409.

Shen, Xiaoping. 1999. Spatial inequality of rural industrial development in China, 1989–1994. *Journal of Rural Studies* 15, no. 2: 179–99.

Smil, Vaclav. 2000. China's environment and natural resources. Chap. 14 in *The China Handbook: Prospects onto the 21st Century*, ed. Christopher Hudson. Chicago: Glenlake, 188–97.

U.S. Department of Energy. 2009. Fact of the week: Fact # 600 China produced more vehicles than the US in 2008. December 7. http//www1.eere.energy.gov/vehiclesandfuels/facts/2009_ fotw600.html (accessed February 18, 2011).

U.S. Energy Information Administration. 2010. Country analysis brief, China, 2009. http://www.eia.doe.gov/cabs/China/oil/html.

Veeck, Ann, and Alvin C. Burns. 2005. Changing tastes: The adoption of new food choices in post-reform China. *Journal of Business Research* 58: 644–52.

Walcott, Susan M. 2002. Chinese industrial and science parks: Bridging the gap. *Professional Geographer* 54, no. 3: 349–64.

Wallerstein, Emmanuel. 1976. *The Modern World Capitalist System: Capitalist Agriculture and the Origins of the European World-Economy in the Sixteenth Century.* New York: Academic Press.

Wei, Yehua. 1996. Fiscal systems and uneven regional development in China, 1978–1991. *Geoforum* 27, no. 3: 329–44.

Wei, Y. H., and C. C. Fan. 2000. Regional inequality in China: A case study of Jiangsu province. *Professional Geographer* 52, no. 3: 455–69.

Wilbanks, Thomas J. 2008. The clean energy dilemma in Asia: Is there a way out? *Eurasian Geography and Economics* 59: 379–90.

World Nuclear Association. 2010. Nuclear power in China. September. http://www.world
-nuclear.org/info/inf63.html (accessed September 28, 2010).

CHAPTER 10

Urban Development in Contemporary China

Youqin Huang

Having been an agrarian society for centuries, China is in the midst of an urban revolution that has had and will continue to have profound impact on the world. Although it is not easy to measure the momentum precisely (owing to the complex and changing system of urban definitions; see, e.g., Chan 2007; Chan and Hu 2003), the latest official statistics show that by 2008, urban population in China had reached 607 million, or 46 percent of the national total (National Bureau of Statistics 2009b). In absolute numbers, China has the largest urban population in the world, and its urban population is larger than the total population of any other nation in the world except India. There are 655 cities in China, and 122 of them have more than 1 million residents and another 118 have more than half a million residents in 2008 (National Bureau of Statistics 2009b). In addition, there are 19,234 designated towns (*zhen*), the other type of urban places in China.[1] With about 225 million to 250 million rural-to-urban migrants every year, China is also one of the most rapidly urbanizing regions in the world.[2] By 2030, there will be more than 900 million people living in cities, transforming China into a mainly urban society. With its unprecedented speed and scale, the urban transition in China is a world-historical event that is transforming not only China's economy, urban landscape, and culture in profound ways but also the global economy with massive exports of products "Made in China."

Urban development in China has long been considered unique in balancing social equality and economic efficiency. Although socialist Chinese cities were formerly crowded and poorly serviced compared to cities in developed countries, they were virtually free of many of the urban problems that were widespread and seemed unavoidable in other developing nations, such as high crime and unemployment rates and acute inequality (Whyte and Parish 1984). At the same time, China has achieved rapid industrialization and economic growth. Since 1978, economic reforms have injected new energy into Chinese cities, especially those in the coastal regions, which have contributed to the country's spectacular economic growth (see chapter 7). Many scholars have asked whether Chinese cities offer an alternative model for urban development in developing countries. In this chapter, we examine the dynamics of urban development in China, focusing on the post-1949 era. We specifically consider the role of government policies and ideology and how they affect the social and spatial structure of Chinese cities.

A Brief History of Chinese Urban Development before 1949

China has a long and elaborate history of urban development. Urban centers in China probably first appeared on the North China Plain during the Shang dynasty (1766–1122 BC), and a well-structured urban system—with a national capital, a number of regional centers and provincial capitals, and a network of county seats and commercial towns—was put in place that lasted for more than two thousand years (Ma 1971; Skinner 1977). There were about 3,220 cities and towns in China in the late eleventh century, and several cities had populations of more than 1 million (Ma 1971; Steinhardt 1999). About 6 to 10 percent of the Chinese population at the time lived in cities, but in absolute numbers, there were more city dwellers in China than anywhere else in the world before the mid-nineteenth century (Whyte and Parish 1984).

Traditional Chinese cities were mainly political and administrative centers with limited commercial functions (Ma 1971; Skinner 1977). The size of a city was often determined by its status in the administrative hierarchy, and this often corresponded to the rank of officials living in that city (Chang 1977; Sit 1995). However, during the Song dynasty (AD 960–1279), commerce grew significantly, especially in the port cities, because of the expansion of maritime trade (Ma 1971). Later, a Chinese diaspora further contributed to the rapid growth of port cities because of their strong economic connections with Southeast Asian countries (Cartier 2001a). In AD 1077, there were 170 large commercial centers in China, but as Ma (1971) has argued, in spite of the increasing importance of commerce in China, the rise and fall of its cities was still closely tied to political factors. For example, the site change of a capital usually led to a significant decline of the old capital city and a corresponding rise of the new (Steinhardt 1999). In addition, traditional Chinese cities served multiple functions associated with military, transportation, communication, religious, cultural, and intellectual activities (Mote 1977). Many Chinese cities prospered during the period of the tenth to the nineteenth century, and one foreign visitor reported that Nanjing in the late sixteenth century surpassed "all other cities in the world in beauty and grandeur" (Ricci 1953).

Despite China's long history of urban development and the existence of its many great metropolises, no single city dominated Chinese civilization as Rome dominated Roman history (Mote 1977). Chinese civilization did not see the city as superior to the countryside; thus, there was no need to build one great city to express and embody an urban ideal. Despite their complex bureaucracy and urban sophistication, traditional Chinese cities had a rural component in both their physical and their social organization, demonstrated by an urban architecture made up mainly of one- or two-story buildings in courtyard style that were rarely distinguishable from those in the countryside. An organic urban–rural continuum was achieved in China because of the freedom of social and geographic mobility between the two (Mote 1977).

Arguing from China's vast territory and diverse social and physical environment, Skinner (1977) suggests that there was not a single integrated national urban system in China but rather several self-contained regional systems defined by physiographic features such as river valleys and mountain ranges. Other scholars divided Chinese cities into coastal port cities and inland commercial and administrative centers. This dichotomous urban system was further strengthened in the middle of the nineteenth century when China was defeated by a newly industrialized Britain in the Opium Wars (Murphy 1970). As a result of this defeat, many of China's coastal and river cities (known as treaty ports) were forced to open up to foreign trade (see figure 7.3 in chapter 7). Hong Kong and Macau were the first foreign enclaves to be occupied by Britain and Portugal, respectively, and were only recently returned to China's sovereignty (in 1997 and 1999). Treaty ports such as Shanghai, Nanjing, and Qingdao were also forced to concede special areas to different Western powers (e.g., the United States, Germany, Britain, and France), and in those areas, Western-style architecture dominated, and foreign rules and foreign ways of life prevailed (Whyte and Parish 1984). Because of their foreign trade and as a result of regional resources shifting to the coast, these treaty ports began to grow disproportionately with respect to their former importance in the Chinese urban system. It should be noted here that despite a long history of urban development in China, large industrial cities did not appear in China until the early twentieth century when Russia and later Japan occupied China's northeastern region to extract mineral resources, mainly for their domestic economies. This rise of industrial cities in the Northeast added yet another layer to the urban system in China.

Although the growth of China's coastal cities during the nineteenth century can be attributed to foreign trade, many Chinese believe that the treaty ports had a largely negative influence because of a resultant massive loss of wealth through unfair trade and the influx of opium, which led to an increase in acute urban problems such as drug addiction, inequality, unemployment, and poverty (Hao 1986). These urban problems were further aggravated by the civil wars that followed (1927–1937 and 1946–1949) and the Japanese invasion (1937–1945). As a result, when the Chinese Communist Party (CCP) came to power in 1949, it inherited an uneven urban system with widespread social problems. The CCP was determined during the following decades to build new Chinese cities that would be free of such problems.

Urban Development in the Socialist Era (1949–1977)

Since the establishment of a socialist government in 1949, urban development in China has veered sharply away from its ancient history, and industrialization—instead of administration and commerce—has become its driving force. Despite the country's rapid industrialization and economic growth, its socialist government controlled the growth of the cities, especially large cities, to avoid an "urban explosion" and related urban problems widespread in other developing countries. Having inherited an un-

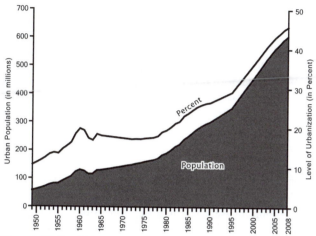

Figure 10.1.   Urban population and the level of urbanization in China, 1950–2008 (National Bureau of Statistics 1995, 2002, 2009a)

even urban system biased toward the coastal port cities, China's socialist government also aimed to create a spatially more balanced urban system. Using the socialist ideology of equality, the notion of "producer cities," and principles of socialist urban planning, socialist China achieved a unique path of urban development.

Underurbanization

Despite rapid urban growth in the 1950s, socialist China still had low levels of urbanization, with less than 20 percent of its population living in cities (see figure 10.1). At the same time, China had achieved rapid industrialization, with industrial output growing by twenty-one times during the period 1952–1982 (Lin 1998). This phenomenon of slow urban development accompanied by rapid industrialization has been called "underurbanization," and is in sharp contrast to the "overurbanization" common in most developing countries, where urbanization is much faster than what can usually be attained given specific levels of industrialization (Castells 1977). The Chinese government's strict control of rural-to-urban migration, its political campaigns with massive deportation of urban residents, and a unique definition of urban population all contributed to underurbanization in China.

In the 1950s, Chinese cities grew rapidly, with China's percentage of urban population increasing from 11.78 percent in 1951 to 19.75 percent in 1960. In addition to a rapid, natural increase in urban population that resulted from postwar stability and an improvement in living conditions, the government was exerting very little control over rural-to-urban migration. China's economic development, by emphasizing industry, resulted in massive rural-to-urban migration in the 1950s (Kirkby 1985). Yet at the same time, one of the most important institutions in China—household registration (*hukou*)—was being set up, and it was fully implemented in the late 1950s to control rural-to-urban migration, among other functions (Chen and Seldon 1994). The *hukou* system divided the population into those with urban (nonagricultural) *hukou* and those with rural (agricultural) *hukou*. The division was based mainly on each individual's birthplace, and there were few ways for people to change their *hukou* status. Further, only people with urban *hukou* were entitled to welfare benefits such as state-supplied grains, free medical care, subsidized housing, guaranteed employment, and a pension. Thus, the system defined an opportunity structure and social hierarchy within China such that it has been called an "internal passport system" (Chan 1994). This system and a related food-rationing system imposed tight control over migration, especially rural-to-urban transfers. In addition to getting certificates of employment, potential migrants had to obtain permission from the local government at both their origin and their destination, often a difficult, lengthy, and bureaucratic process. At the same time, various coupons were required in cities to obtain food, cloth, and other daily essentials, making temporary migration very difficult, if not impossible. During the following two decades, the *hukou* system remained effective, and rural-to-urban migration was tightly controlled.

At the same time, the Chinese socialist government deported million of urbanites to the countryside, resulting in stagnation and even reverses in urban development. The failure of the Great Leap Forward (1958–1960), coupled with bad weather, led to a significant decline in agricultural production and widespread famine. As it became increasingly difficult for China to support a large urban population, urban dwellers were deported to villages during 1961–1963, resulting in a sharp decline in China's level of urbanization, from 19.75 percent in 1960 to 16.84 percent in 1963, a process called "deurbanization" by some scholars. After a brief period of economic recovery, Mao Zedong then launched another radical campaign—the Cultural Revolution (1966–1976), in which urban youth, CCP cadres, and professionals were urged and often forced to move to the countryside to engage in manual labor. While there are no data on the actual number of urban-to-rural migrants, scholars estimate that it could have been as high as 49 million (Kirkby 1985). Despite this massive out-migration, the level of urbanization was maintained at around 17 percent because of a natural increase in urban population and a considerable inflow of peasants to replenish the workforce.[3]

In addition, unique definitions of urban places and urban population contributed to the phenomenon of underurbanization. Because of the *hukou* system and the government's implementation of "city leading counties" (*shi xia xian*),[4] it is a challenge to define urban population in China, as is suggested by the large body of literature on this issue (e.g., Chan 1988, 1994; Ma and Cui 1987; Zhang and Zhao 1998; Zhou and Ma 2003). The Chinese government has used four different definitions of urban population in its four censuses. There are two types of officially designated urban places in China: cities (*shi*) and towns (*zhen*). Both the total population of cities and towns and the total population of cities and towns with urban *hukou* were used to define urban population. Despite the fact that urban residents such as suburban farmers and rural migrants might have rural *hukou*, the latter definition (based on urban *hukou*) was more commonly used (Chan 1994). This definition was less problematic in the socialist era with its relatively small volume of temporary rural-to-urban migrants; yet in the reform era, it significantly underestimates urban population because of massive and ongoing rural-to-urban migration, and a more complex definition derived from actual occupation, residence, and *hukou* status has been recommended (Zhang and Zhao 1998).

Socialist Ideology and a Balanced Urban System

Urban development in socialist China was heavily influenced by the socialist ideology of equality. With the exception of the rapid urban growth that was permitted during the initial postwar recovery period, the focus during most of the socialist era was on achieving controlled and balanced urban development. The official policy for urban development was to "strictly control the development of large cities, moderately develop medium-size cities, and vigorously promote the development of small cities and towns" (Kirkby 1985). To achieve such goals, various measures were used, including the allocation of state investment, the designation of new urban places, the strict control of migration into cities, and even the deportation of urbanites during various political campaigns. As a result, the number of medium cities and small cities with more than 200,000 residents increased more rapidly than that of large cities (see table 10.1). There was also a significant containment of the growth of the very largest cities, such as Beijing, Shanghai, and Tianjin (Pannell 1981).

Because most of China's large cities are located in the coastal and northeastern regions (mainly as a result of foreign influence), a majority of state investment was channeled toward inland cities to achieve a regionally more balanced urban development. During the first Five-Year Plan (1953–1957), two-thirds of the nearly seven hundred large and medium-scale industrial projects established by the government were set up in inland cities (Cannon 1990). The hostile international political environment of the 1960s further encouraged the Chinese government to channel more than half of its state investment into inland mountainous areas, particularly in Sichuan province, a strategy that became known as the Third Front policy.[5] Although mainly used as a military strategy, the Third Front policy effectively promoted urban development in inland China. Thus, a spatially more balanced urban system was created despite the harsh physical environment of the interior and the country's historical bias toward its coastal regions. In 1949, more than 52 percent of cities and 63 percent of all urban

Table 10.1. Number of Cities by Population in China

City Size	1949		1978		2008	
	N	%	N	%	N	%
Large (1 million +)	10	7.58	29	15.03	122	18.63
Medium (500,000–999,999)	6	4.55	35	18.13	118	18.02
Small (200,000–499,999)	32	24.24	80	41.45	151	23.05
Small (<200,000)	84	63.64	49	25.39	264	40.31
Total	132	100.00	193	100.00	655	100.00

Note: Urban population refers to the total population of urbanized areas for each city.
Source: National Statistics Bureau (2009b).

population were located in the eastern region. The ratios between the numbers of cities in the eastern, central, and western regions were 1:1:0.4, and those for the urban population were 1:0.46:0.18. By the end of 1978, the ratios changed to 1:2.2:0.6 for the numbers of cities and 1:0.69:0.33 for the urban population (National Bureau of Statistics 2009b), with a significant increase in both the number of cities and the urban population in the central and western regions.

The socialist ideology of equality had a profound impact on the spatial pattern of urban development in China. Although China's socialist government inherited an uneven urban system that was heavily biased toward the coastal port cities and industrial cities in the Northeast, it made substantial progress in developing a more balanced urban system.

Socialist Urban Planning and Urban Sociospatial Structure

The internal structure of a city often reflects the cultural norms of its society and the ideology embedded in that society's urban-design and planning policies. Over the course of more than two thousand years, Chinese cities developed internal structures that were significantly different from those found in Western cities (Steinhardt 1999; Whyte and Parish 1984). Traditional Chinese cities often had a rectangular, symmetrical layout with an elaborate structure of city walls and gates symbolizing authority and security, providing protection, and representing Chinese cosmology (Chang 1977; Ma 1971; Sit 1995; Steinhardt 1999; Wright 1977; see figure 10.2). Most Chinese cities originally served as political and administrative centers, with government buildings and facilities dominating the city center and commercial activities and ceremonial buildings relegated to the periphery (Skinner 1977). Today, remnants of traditional Chinese city planning are still visible in the inner cities of Beijing, Xi'an, and Nanjing.

Influenced by urban planning in the Soviet Union, the Chinese socialist planning system added another layer to the urban structure of Chinese cities, which, as a result, were characterized by three factors: 1) a focus on the symbolism of the city center, 2) an emphasis on industry, and 3) the desire for a cellular landscape based on work units (*danwei*). First, city centers were symbolically considered centers of cultural and political life. To glorify the socialist state, city centers were devoted to large, public squares for political gatherings, with wide boulevards and monumental public buildings (Sit 1995), in sharp contrast to the intensive development of commercial and business interests at the center of most Western cities. For example, at the center of Beijing, the largest public square in the world—Tiananmen Square—was constructed adjacent to the Forbidden City, the historical center of power. The square was surrounded by monumental architectural structures devoted to political and cultural purposes, such as the Hall of the People, the Museum of History, and the Museum of Revolution, which were matched by the Memorial of the People's Hero at the center of the square (see photo 10.1). While the city center in Beijing is unique because of its capital status, many cities in China mimicked its design, with their public squares and government

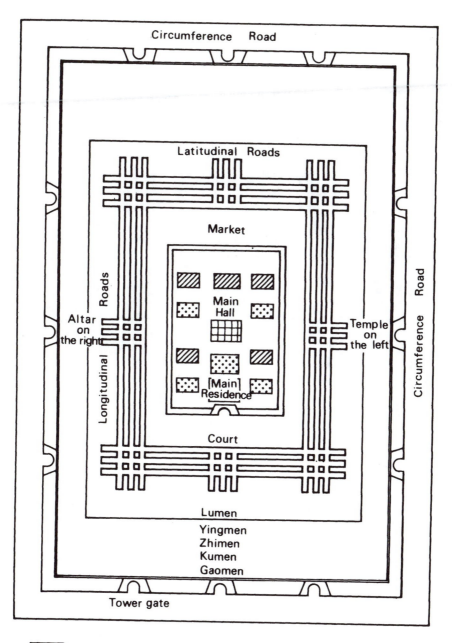

Circumference Road

Circumference Road

Latitudinal Roads

Market

Roads

Main Hall

Altar on the right

Temple on the left

Longitudinal

[Main] Residence

Court

Lumen
Yingmen
Zhimen
Kumen
Gaomen

Tower gate

▨ Minor Hall

▦ Minor Residence

Figure 10.2. A typical design for traditional Chinese cities (Yung Lo Da Din 1425)

Photo 10.1.   Tiananmen Square in the heart of Beijing (Youqin Huang)

buildings located at city centers with wide boulevards radiating outward. Thus, the city centers of Chinese cities during this period often had the lowest intensity of land use, forming a doughnut-shape pattern of land use that is very different from the high-density development in Western urban centers.

Second, an emphasis on industry and the concept of the "producer city" were central to socialist planning (Ma 1976; Sit 1995). Nonindustrial cities like Shanghai were considered by the socialist government to be parasitic and to have been significantly debased by their colonial history. China's socialist government was determined to convert these "consumer cities" into "producer cities" by developing a large industrial sector. Even Beijing, the nation's cultural and political center, was no exception. Most state investment in Beijing went into industry, especially heavy industries such as steel mills, petrochemical plants, and power plants, which imposed tremendous pressure on local resources (especially water and electricity) and created severe environmental problems (Dong 1985). In 1981, Beijing became the second-largest industrial city in China. At the same time, only a very small proportion of state investment (1 to 3 percent of total urban fixed-asset investment, compared to 10 percent recommended by the United Nations) was for civic infrastructure. This emphasis on production resulted in a pattern of land use devoted mainly to industrial facilities, especially at the outer edge of cities, with much less for infrastructure, housing, services, and recreation. Not surprisingly, residential crowding and a lack of amenities and infrastructure were common complaints among residents of socialist Chinese cities (Whyte and Parish 1984).

Third, and related to this massive industrial development, self-contained work-unit compounds were constructed as the basic urban unit. In order to reduce traffic congestion and create "walking-scale" cities, work-unit compounds with public apartment buildings were built adjacent to employment centers, allowing employees from the same work unit to live together (Ma 1981). This was especially important in Chinese cities, where bicycles and buses were the main modes of transportation. In addition, basic social, subsistence, and recreational services, such as cafeterias, public bathhouses, grocery stores, small clinics, kindergartens, and small parks, were provided in these compounds, all within walking distance (Ma 1981). In contrast to the specialized functional zones in traditional Chinese cities and most Western cities, a generalized, functional organization was achieved in socialist Chinese cities through these self-contained work-unit compounds, which provided employment, housing, and a range of other social services. Because of this close link between employment and housing, it was extremely difficult for a worker to move, even within a city. During the 1960s and 1970s, only 1 percent of urban households changed residence in one year, and the average length of residence in the same apartment was eighteen years—an extreme immobility compared to that of China in the past as well as that of other nations (Whyte and Parish 1984).

Corresponding to this cellular physical landscape was a relatively homogeneous society. Because housing was considered part of social welfare, it was provided mostly by the state through work units, which usually built low-rise, standardized apartment buildings in the form of work-unit compounds. Apartments were then allocated among the work-unit employees, who paid only nominal rents. While there were some differences in housing consumption (Logan, Bian, and Bian 1999), people working for the same work unit, including high-ranking cadres and their subordinates in factories or professors and their staff in universities, often lived in the same work-unit compound, if not the same building, and they shared similar housing conditions and amenities. Although most households suffered from poor housing conditions and severe crowding (Huang 2003), social stratification and residential segregation were minimalized in socialist Chinese cities. There were different social areas, but they were based mainly on land use, occupation, and population density instead of on residents' socioeconomic status, as is the case in the West (Sit 2000; Yeh and Wu 1995). A relatively homogeneous society was achieved through the massive provision of public housing in the form of work-unit compounds.

In sum, because of socialist planning that emphasized the symbolism of the central city, the concept of the "producer city," and the combining of residential and work space, urban structure in socialist Chinese cities demonstrated unique features such as low-density land use in the central city, generalized functional organization through work-unit compounds, a dominance of industry, prevalence of uniform public housing, and societal homogeneity—all of which highlighted fundamental differences in urban structure among socialist Chinese cities, traditional Chinese cities, and Western cities.

Socialist ideology and socialist planning were clearly important, but we should recognize that urban development in socialist China was a complex process shaped by political, social, historical, and economic forces (Tang 1997). For example, the constraints of limited resources and a hostile international environment were important to the socialist strategy of urban development. As Lin (1998) points out, "No single factor, whether ideological conviction or rational economic consideration is able to claim sole responsibility for the process of China's urbanization" (109). The socialist Chinese government, which desperately wanted to demonstrate its legitimacy and superiority, had to balance considerations of social and spatial equality against the need for economic efficiency in urban development, and this is never an easy task. The Chinese government often had to choose one or the other, or under different circumstances it sought compromises between the two (Lin 1998). Changes in the urban development strategy from the socialist era to the reform era—although they may appear to be radical—are in fact a continuation of this struggle for balance between equality and efficiency, with the focus lately shifting toward economic efficiency and urban growth.

Urban Development in the Reform Era (1978 to the Present)

After the more pragmatic Deng Xiaoping gained power in 1978, the Chinese government began to shift its focus from class struggle to economic development, from social and spatial equality to economic growth, and China has since experienced unprecedented urban growth and urban transformation. Cities have grown rapidly in both number and size, with 655 cities and an urban population of 607 million in 2008, compared to 195 cities and 172 million only three decades ago (see figure 10.3 and table 10.1). Chinese cities have become the engine of the national economy, with prefectural-level and higher cities alone contributing almost two-thirds of the national gross domestic product (GDP). There are twenty cities (e.g., Shanghai, Beijing, Shenzhen, and Guanzhou) each with more than 200 billion yuan of GDP and another twenty-three cities with more than 100 billion in 2008 (National Bureau of Statistics 2009b). Chinese cities have also demonstrated higher degrees of urbanism with an increasingly cosmopolitan landscape, higher rates of mobility, rising consumerism, and increasing cultural and social diversity. They are dynamic, exciting, and sometimes overwhelming. This new pattern of urban development can only be attributed to changes in the Chinese government's philosophy of urban development and to the effects of economic reforms that have initiated institutional changes on many fronts. These changes include the integration of the urban economy with the world economy through the government's open-door policy, liberation of the nonstate economy, a relaxation in migration control, and reforms in the land and housing system.

Opening Up and Uneven Urban Development

In 1978, the new Chinese leadership initiated an open-door policy to utilize foreign investment and engage in international trade, a policy that has since brought profound changes to urban development in China. Through a strategy of gradualism, the open-

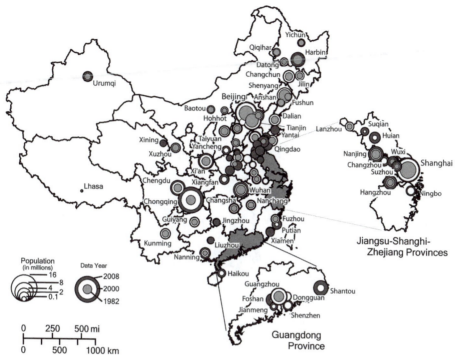

Figure 10.3. Both the number of cities and existing cities grow rapidly over time. (National Bureau of Statistics 1983, 2001, 2009b)

door policy was initially implemented in coastal cities before it was (recently) adopted nationwide. Four special economic zones (SEZs) were established in 1979—Shenzhen, Zhuhai, and Shantou in Guangdong province and Xiamen in Fujian province—all within geographic proximity of Hong Kong and Taiwan, aiming to attract foreign investment from overseas Chinese (see figure 7.2 in chapter 7). With their immediate success, another fourteen coastal cities were opened up in 1984, and in 1985, open economic regions in the Yangtze River Delta, the Pearl River Delta, the Minnan Delta, Shangdong, and the Liaoning Peninsula were designated. Hainan Island and Pudong New District in Shanghai were added as the latest SEZs in 1988 and 1990, respectively. Preferential policies and tax incentives are often given in these designated cities and regions to attract foreign investment. As a result, China has become one of the most favored destinations for foreign direct investment (FDI). During 1979–2008, in total there were about $1.5 trillion of contracted FDI and more than $853 billion of utilized FDI (National Bureau of Statistics 2009a).

Because of the geographically biased open-door policy and the better social and physical infrastructure in China's large cities, the majority of foreign investment has gone to large and medium-size cities in the eastern region, especially the SEZs and coastal open cities. As a result, they have experienced more rapid urban development and higher economic growth than inland cities. Formerly a quiet fishing village, Shenzhen has now been transformed into a modern, bustling, and wealthy city with

Photo 10.2.   The urban landscape of Shenzhen; the left of the river is Hong Kong (Youqin Huang)

a population of more than 8.9 million and per capita GDP of 92,771 yuan ($13,581) in 2009 (Shenzhen Statistical Bureau 2010; see photo 10.2). With three SEZs and two coastal open cities, the Pearl River Delta region in Guangdong province has been one of the most rapidly urbanizing regions in China. Today, 80 percent of its population lives in cities and towns (Guangdong Statistics Bureau 2010), similar to the level in North America. In addition to millions of workers migrating into the region from elsewhere, 60 percent of the delta's local rural population has given up farming and moved into nonagricultural sectors, many in manufacturing industries relocated from Hong Kong and Taiwan, making it the "factory of the world" (Chan 1995; Xu and Li 1990). As a result, gross industrial output in the delta grows rapidly, with almost 3 trillion yuan in 2008, accounting for 10 percent of GDP by all cities in China (National Bureau of Statistics 2009b). With Pudong New District as the latest SEZ, Shanghai has renewed its status as the favorite city for foreign companies, as it was in the colonial period. In 2008, 12 percent of all foreign companies in China chose Shanghai as their host city, and 13 percent of all foreign investment in China went to Shanghai (National Bureau of Statistics 2009a), making Shanghai the third destination after Guangdong and Jiangsu province. Meanwhile, the population in Shanghai reached 19 million in 2009 (Shanghai Statistical Bureau 2010). The phenomenal transformation of Pudong New District from farmland in the 1980s to a financial district with a forest of futuristic skyscrapers (see photo 10.3) is another example of how Chinese cities have been transformed by the open-door policy.

Photo 10.3.   The skyline of the Pudong New District, Shanghai (Youqin Huang)

Compared to the socialist era, the driving forces for and patterns of urban development in these coast cities are very different. "Exourbanization" (Sit and Yang 1997) and "urbanization from outside" (Fan 1995) have been used to refer to this rapid urban development that has been induced mainly by foreign investment. While it is debatable whether the rapid urbanization in China is mainly a result of endogenous or exogenous forces (Friedmann 2005), it is clear that the open-door policy and the consequent integration of China into the global system have changed the course of urban development in China, with the country's focus now shifting from spatial equality to economic efficiency, from inland to coastal cities, from small to large cities, and from self-reliance to globalization. As a result, an uneven urban development, biased toward coastal large cities with much faster urbanization and economic growth, has been achieved. During 1978–2008, 215 new cities were added in the eastern region, compared to 162 cities in the central region and eighty-five in the western region (National Bureau of Statistics 2009b). By 2008, there were 122 cities in China with a population of more than 1 million, and 56 percent of all the urban population lived in the eastern region.

"Urbanization from Below"

Small cities and towns (*xiao cheng zhen*) in China served primarily as marketing and administrative centers throughout the nation's history and remained stagnant until the early 1980s when they became centers of industrial production (Fei et al. 1986; Skinner 1977). The State Council (1984) issued a landmark policy that allowed peasants to move to towns and small cities as long as they could provide themselves with food and shelter. Because of the millions of surplus rural laborers that had resulted from the Household Responsibility System, this policy generated a massive rural-to-urban migration, with more than 5 million new urban residents added during the five years after the policy was adopted (Zhu and Gu 1991). In 2008, about 15 percent of migrants moved to designated towns and another 20 percent to county-level cities, while only 27 percent moved to municipalities and provincial capitals (National Bureau of Statistics 2009b). Still banned from jobs in the state sector, most of these migrants in the early 1980s moved into rapidly expanding nonstate sectors such as services and privately or collectively owned industries, often called township-village enterprises, that manufactured consumer goods ranging from sneakers to electronics. These migrants provided a large pool of cheap labor for the nonstate sectors, and they were cheap because they were denied subsidized housing, medical care, pensions, and the other benefits that were enjoyed by employees in the state sector. Together with local governments' entrepreneurship, the nonstate sectors and especially the township-village enterprises grew rapidly, contributing 72 percent of China's total industrial output in 1998. This influx of migrants and the phenomenal growth of nonstate economies have led to the rapid development of small cities and towns.

To further promote the growth of small cities and towns, the government's criteria for designating towns were relaxed in 1984, leading to a surge in the number of designated towns, which went from 2,781 in 1983 to 6,211 in 1984 and to 19,234 in 2008 (Chan 1994; National Bureau of Statistics 2009b). There were also 462 new cities designated between 1978 and 2008. As a result, the urban population in small cities and towns accounts for 45 percent of the total urban population in 2008, compared to 20 percent in 1978 (National Bureau of Statistics 2009b). The large number of small cities and towns and their important status in urban economy have made urbanization in China rather unique compared to the high primacy in most developing countries where one large city often dominates the national urban system (Clark 1996). This phenomena of indigenous urban development that is town- or small city–based and that relies mainly on local initiatives and resources is often called "urbanization from below" and is in sharp contrast to "urbanization from above," which is state-planned development based on large-scale industries in large cities (Ma and Fan 1994).

Small cities and towns are particularly vibrant in the Chinese coastal regions, but there are large regional variations. For example, because of their proximity to large cities such as Shanghai, Changzhou, Wuxi, and Suzhou, many small cities and towns in the Chang Jiang (Yangtze River) Delta have benefited significantly from the economic, technological, and cultural spillover from these large urban centers (Tan 1993). In Zhejiang province, small cities and towns have grown rapidly because of vibrant private businesses, while in Guangdong province, foreign investment mainly from overseas Chinese has attracted millions of migrants from all over China and has transformed the province's towns and even villages into large cities. Dongguan, once a small town in Guandong province, now is one of the largest cities in China, with 5 million migrants among its 7 million residents. With a large number of vibrant small cities, towns, and industrialized villages, extended metropolitan regions or urbanized regions are taking shape in areas surrounding large cities in the Chang Jiang Delta and Pearl River Delta (Lin 1994; Zhou 2003).

Accompanying this phenomenal growth in small cities and towns are many problems, such as pollution (caused by using old technologies and the lack of environmental regulations) and loss of farmland (Tan 1993). Yet "urbanization from below" has to some extent alleviated problems such as congestion and an overburdened infrastructure in large cities by channeling surplus rural labor into the towns and small cities. The urban development of these small cities and towns, however, emphasizes privatization and economic efficiency, an approach that is very different from the approach of urban development in the socialist era, which focused on equality; yet, ironically, a smaller spatial inequality and phenomenal economic growth have been achieved, especially in the coastal regions.

Massive Rural-to-Urban Migration and a Two-Class Urban Society

In the socialist era, migration, especially rural-to-urban transfers, was tightly controlled through the *hukou* system (see earlier in this chapter and chapter 5 for more on hukou system). Consequently, individual mobility was extremely low, with the annual rate of mobility around 2 to 3 percent in the 1960s and 1970s. The majority of limited migration was state organized through the form of labor recruitment, frontier development, large development projects, and political migration (Chan 2001), and there were few individual migrations. While there were rural-to-urban migrations, many migrations were urban-to-rural and east-to-west.

Since the 1980s, however, there has been an increasingly large volume of migrants, mostly rural-to-urban migrants, who have sought to take advantage of the new opportunities brought about by the reforms. While it is difficult to know the exact volume, there were about 175 million to 225 million migrants looking for jobs in Chinese cities in 2008, compared to 30 million in 1982 (see note 2; National Bureau of Statistics 2009b). Crowded transit centers, such as railway and bus stations and airports, provide the best testimony of the higher mobility in Chinese cities of the reform era. In particular, the annual "spring movement" (*chun yun*) of millions of migrants traveling long distances via trains and buses back to their hometowns before the traditional Chinese Lunar New Year holiday has captured the sensation of the world's media, and it has been called "the world's greatest annual human migration" (Mitchell 2009). Many factors contributed to this massive rural-to-urban migration, including the wide economic gap between the rural and the urban areas; rural reforms that released millions of surplus laborers from the land (see chapter 9); the rapid expansion of the urban economy, especially in labor-intensive industries and private services (see chapter 7); the reform of the *hukou* system and the relaxation of migration controls; and the rapid development of the urban food and labor market due to the phasing out of the rationing (Chan 2001). It was considered the first breakthrough in the *hukou* system when peasants were allowed to move to small cities and towns in 1984 if they could secure their own employment, housing, and food. This generated massive rural-to-urban migration and started a new era of migration in China. The need for cheap labor in cities has made local governments more receptive to migrants over time, although they may periodically demolish migrant settlements and deport migrants for various political and social reasons (Zhang 2001). In addition, the rationing system for food and other daily essentials was phased out in the 1990s, and temporary registration certificates were created to allow migrants to live in cities. As a result, today migrants can go to any city (and the countryside) they desire without changing their official household registration—also called "temporary migrants." It is very common for migrants to remain registered in their hometowns and move back and forth between cities and their home villages and between cities depending on the season and their job situation; thus, not surprisingly, they are often called a "floating population." In contrast, there are "permanent migrants" with their registration changed to destinations who are usually educated and skilled workers and professionals (Chan 1996; Fan 2008), but their volume is much smaller than temporary migrants (Chan 2001).

Because of the open-door policy and consequent uneven urban development biased toward coastal cities, most rural-to-urban migrants come from the central and western regions and move to cities in the eastern region, as the Chinese say, "peacocks flying to the southeast." In 2008, more than two-thirds of migrants settled down in the eastern region. Main destinations include Beijing, Shanghai, Guangdong province, Zhejiang province, and Jiangsu province, which together attracted 75 percent of interprovincial migrants, and Guangdong province alone attracted 44 percent (National Bureau of Statistics 2009b). Noncoastal populous provinces such as Sichuan, Anhui, Henan, Hunan, Hubei, Guangxi, and Chongqing are the main sending areas of migrants, which contributed two-thirds of interprovincial migrants in 2008 (figure 10.4). With geographical proximity and similarity in culture, neighboring provinces of major magnets tend to provide more migrants than provinces afar. For example, Hebei province provides the largest migration flow to Beijing, while Guangxi and Hunan provinces provide the largest flows to Guangdong province. This geography of migration is similar to those in the 1980s and 1990s but with an even stronger eastward trend, yet it is almost the opposite to that in the socialist era (Fan 2008; Chan 2001). While about 20 percent of migrants chose towns and the countryside as their destinations (National Bureau of Statistics 2009b), the majority moved to cities, especially coastal cities. In 2007, there were thirteen cities, all in the eastern region, with more than 1 million migrants (table 10.2). Shenzhen, the first SEZ, is a "city of migrants" with 6.5 million migrants, followed by Dongguan (5.2 million), Shanghai (4.79 million), and Beijing (4.2 million).

Within cities, migrants tend to settle down at the outskirts of cities in formerly suburban villages, forming "migrant enclaves" (Ma and Xiang 1998). Since temporary migrants are not allowed to access subsidized housing in Chinese cities and the formal rental market is very much underdeveloped with large-scale private housing estates for sale only, they have to rent housing from individual households. Suburban farmers often have extra housing, and they are allowed to build housing on collectively owned land,[6] while most urban households still live in relatively crowded housing

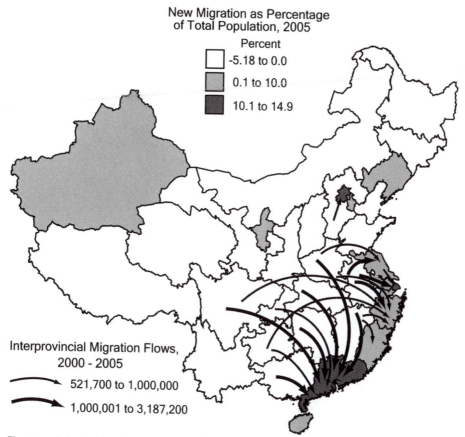

New Migration as Percentage
of Total Population, 2005

Percent

☐ -5.18 to 0.0

▨ 0.1 to 10.0

■ 10.1 to 14.9

Interprovincial Migration Flows,
2000 - 2005

521,700 to 1,000,000

1,000,001 to 3,187,200

Figure 10.4. Major interprovincial migration flows and net interprovincial migration rate (2005 1% Survey)

even though housing consumption has improved significantly and the rate of second-home ownership is rising rapidly (Huang 2003; Huang and Yi 2010). Thus, migrants are accommodated mainly in suburban villages, supplemented with quarter housing, such as factory dorms and individual homes scattered across cities. In addition, social network based on common place of origin—the "Chinese ethnicity"—has also greatly facilitated migration in China (Ma and Xiang 1998; Zhang 2001). Migrants from the same place often settle down in congregation in cities and even engage in similar occupations, forming large migrant settlements based on place of origin and occupation. For example, in Beijing, the largest migrant enclave is often called "Zhejiang village" with mainly migrants from Zhejiang province engaging in the textile industry, while "Henan village" accommodates migrants mainly from Henan province engaging in the recycling business (see photo 10.4).

With economic- and job-related reasons as the main motivations for migration, not surprisingly, migrants are mostly young adults, with more than 55 percent of all migrants aged eighteen to thirty-nine, and there are more males than females with the sex ratio of 160 males per 100 females in 2008 (National Bureau of Statistics 2009b).

Table 10.2.   Number of Migrants in Major Chinese Cities in 2007

City	Migrants (millions)
Beijing	4.20
Tianjin	1.56
Shanghai	4.79
Guandong province	
Guangzhou	2.32
Shenzhen	6.50
Fushan	2.31
Dongguan	5.24
Zhongshan	1.06
Jiangsu province	
Nanjin	1.24
Suzhou	2.58
Wuxi	1.37
Zhejiang province	
Hanzhou	1.14
Ninbo	1.25

Source: National Bureau of Statistics (2009b).

While the conventional migrant profile of the young single male is still true in some way, Chinese women have been active agents seeking to improve their chances through migration, and married couples and families are also very common among migrants (Fan 2008; Fan and Huang 1998; Gaetano and Jacka 2004). This increasingly diversi-fied migrant profile demands cities to provide not only jobs but also better housing, medical care, and other social services, including education for migrant children, a huge challenge facing most cities. As a result, separated families and left-behind chil-dren are quite common among migrants. But a recent study shows that migrants often change their jobs and household organization in order to pursue the best of both the urban and the rural world (Fan 2009).

Rural-to-urban migrants have unquestionably made enormous contributions to rapid urban development by providing their labor, skills, and talents and by filling employment and service gaps in the cities. Yet these migrants are still discriminated against by both existing institutions and local urban residents. The *hukou* system is under reform, and differences in economic opportunities and social benefits based on *hukou* status are shrinking (Wang 1997), but the institutional discrimination against migrants has by no means disappeared. Although "permanent migrants" enjoy em-ployment opportunities and welfare benefits similar to those of urban residents, "tem-porary migrants," who are mostly from the countryside and account for the majority of migrants, are not eligible for state employment and benefits in cities. They have to work in informal sectors and in temporary jobs—the so-called 3D jobs (demanding, dangerous, and dirty; Chan 2001)—and they often cluster in a few undesirable occu-pational niches, such as in construction work (for men) and in sales and food services (for women; Huang 2001). They are not allowed access to subsidized housing, so they often live in much worse housing conditions than local urban residents (Solinger 1995;

Photo 10.4. The largest migrant enclave in Beijing, "Zhejiang Village" (Youqin Huang)

Wu, Weiping 2002). The emergence of dilapidated migrant enclaves in large cities is a result mainly of the limited housing options available to temporary migrants (see photo 10.4). In addition, their children are not allowed to access public education in cities such that many children are left behind in villages with grandparents and other family members. Furthermore, the Chinese urban public generally has a negative attitude toward migrants who have flocked into their cities, and this is further perpetuated by negativity in the media. Migrants are often blamed for rising crime rates, overcrowding, overburdening of the transportation system, and even deterioration of the urban environment such that local municipal governments have periodically attempted to deport them and demolish their settlements (Zhang 2002). Persistent negative public discourse, together with a discriminatory institutional system, offers this floating population little chance to be assimilated into mainstream urban society, and a two-class urban society is emerging in China (Chan 1996; Solinger 1995).

Reforms in Land and Housing System and Urban Sociospatial Restructuring

One of the most visible changes in Chinese cities is the profound and rapid transformation in urban landscape. In sharp contrast to the previously uniform, utilitarian

Photo 10.5.   New private housing in suburban Beijing, with high rises and multi-story apartment buildings in the forefront and villas in the background (Youqin Huang)

buildings that dominated socialist Chinese cities, now commercial buildings and office towers are competing for the world's attention with their daunting heights and bold architecture, and colorful new housing estates with various amenities are emerging like "bamboos shooting after the spring rain" (see photos 10.3 and 10.5). As part of China's overall market transition, land reform and housing reform have resulted in an increasingly mature land market and housing market, in turn contributing to a profound transformation of the urban landscape and the sociospatial structure.

In China, all urban land is legally owned by the state. In the socialist era, urban land was allocated by the government to land users, such as work units, free of charge for an indefinite period, and all land transactions were banned by the constitution (Ding 2003; Ho and Lin 2003). Thus, land was not considered a commodity and had no value, and businesses had no incentive to choose sites and build to maximize profits. Such a land system created enormous land-use deficiencies that were manifested in the presence of huge public squares and bungalows in city centers and the presence of warehouses and factories in central locations. Cities were "low" and "flat." For example, in Changchun, a city in the Northeast with a population of more than 2 million, the average height of buildings was only 1.74 stories (Ding 2003). In the late 1980s, many reforms in the land system were carried out to develop a land market, including the separation of land-use rights from ownership and the commercialization of land-use rights. Thus, since 1988, urban land-use rights have been leased to end users with a fee, and users may transfer their land-use rights to others on the second-ary land market (Ho and Lin 2003). As a result, intensive and more efficient land use has been achieved, and high-rise buildings and skyscrapers have mushroomed, sig-nificantly changing the skylines of many Chinese cities. Beijing, where the city center is a huge basin defined by Tiananmen Square and the one-story Forbidden City, is building a new central business district on the east side of the city center where many high-rise office buildings, large hotels, and convention facilities are already located. Urban planners and decision makers in Beijing clearly envision a Western-style cen-tral business district comparable to those in Hong Kong, Paris, and New York. With virtually no modern high-rise official towers in 1980, Shanghai today has more than twice as many as New York City (Campanella 2008). With more than one hundred foreign banks and financial institutions, Pudong New District in Shanghai has some of the tallest buildings in the world, including the 101-story Shanghai World Financial Center (second tallest in the world) and the 88-story Jinmao Tower (fifth tallest in the world) (see photo 10.5).

The commercial leasing of land-use rights has become a lucrative source of rev-enue to local municipal governments, which then use the revenue to fund mainly infrastructure that further increases land leasing fees. As a result, there has been an explosive development in infrastructure in Chinese cities, with local governments pouring money into roads, subway systems and airports, bridges and tunnels, public squares, museums, opera houses, convention centers, recreation facilities, parks, and waterfront promenades (Campanella 2008; photo 10.6). Local municipal govern-ments also spend millions to beautify their cities. The recently renovated pedestrian street Wang Fu Jing (Beijing's "Fifth Avenue") is decorated with flowers, fountains, and Christmas lights (see photo 10.7). Xidan Culture Square in Beijing is character-ized by postmodern statues dotting well-manicured lawns—in sharp contrast to the bare cement of Tiananmen Square—whereas the skyscrapers in Shanghai's Pudong New District are lit up each night, attracting millions of tourists. Similar makeover projects are taking place in every city. Megaevents such as the 2008 Olympic Games in Beijing and the 2010 World Expo in Shanghai have become trigger events for mu-nicipal governments to further upgrade their cities' infrastructure and environment. For example, to prepare itself for the Olympic Games, Beijing added eight new subway lines, a ninety-six-mile light rail system, two new ring roads, a new airport terminal (Terminal Three), and nineteen purpose-built facilities (Campanella 2008). The re-form in land system and the decentralization of China's fiscal system have given local municipal governments unprecedented freedom in resource mobilization, enabling them to improve their urban infrastructures and environment (Wu 1999). This is in sharp contrast to the ideology of "production first, consumption later" in the socialist era, when the consideration of aesthetics and amenities was never a primary objective in Chinese socialist urban planning.

Related is the privatization of the housing system. As discussed earlier, public rental housing in the form of homogeneous work-unit compounds dominated the housing stock in the socialist era. While the public housing system improved housing conditions for the masses, there were also a number of problems, such as a severe hous-ing shortage and poor housing quality, that were a result mainly of the low-rent policy

Photo 10.6.   Highways and subways in the eastern suburbs of Beijing (Youqin Huang)

(Wang and Murie 1999; Zhang 1998). After pilot experiments in several cities, nation-wide urban housing reform was launched in 1988 to introduce market mechanisms into the housing system, and these reforms continue. First, while existing public hous-ing is being privatized through subsidized sales and rising rents, new private housing built by domestic and foreign developers is being added to the stock rapidly (Huang and Clark 2002; Zhang 1998). China has built more housing in the past twenty-five years than any nation in history, indeed, more than most nations' total housing stock (Campanella 2008). Second, in contrast to the confiscation of private homes in the socialist era, home ownership is now being actively promoted through various meth-ods. For example, public housing is being sold to sitting tenants at a fraction of its market price, and the price for some private housing (e.g., "affordable housing") is government controlled to encourage home ownership among low- to medium-income households (State Council 1998). In addition, a mandatory long-term housing saving system—the Housing Provident Fund—has been set up through which public em-ployees contribute no less than 5 percent of their wages to their own accounts, which is then matched with the same amount from their work units. Recently, commercial housing loans have also become an increasingly important source for financing home ownership (Li and Yi 2007).

Urban home seekers, who in the relatively recent past had few housing choices other than waiting in long lines for public rental housing now can choose between public and private housing and between rental and home ownership (Huang and

Photo 10.7. The Wang Fu Jing Shopping Center in downtown Beijing (Youqin Huang)

Clark 2002; Li 2000). After several decades of suppression during the Maoist era and recent promotion campaigns by the government, home ownership is clearly preferred by most Chinese. About 82 percent of urban households own their homes, making China one of the countries with the highest rate of home ownership in the world. Furthermore, about 15 percent of urban households own second homes (Huang and Yi 2011). This is an impressive achievement considering the fact that the rate of home ownership was less than 20 percent in the 1980s (Huang and Clark 2002). Housing reforms have also significantly improved housing conditions in Chinese cities, with per capita living space increasing from 3.9 m² in 1978 to about 28 m² in 2007 (National Bureau of Statistics 2009b). In addition to better housing condition and diverse housing options, urban households also enjoy more occupational and spatial mobility, as housing is no long strictly tied to their work units. They can move to their preferred neighborhoods and change jobs without worrying about losing their housing as before.

Yet none of these achievements come without trade-offs. First, millions of urban residents have been displaced to give way to new infrastructure and housing development, and many century-old neighborhoods embodying Chinese culture and traditions, such as the *hutong* neighborhoods in inner cities of Beijing, are forever lost. Conflicts between displaced urban residents and developers and sometimes local municipal governments are increasingly common in Chinese cities. How to preserve urban heritage during rapid urbanization and massive urban renewal is one of the major challenges facing Chinese cities. Second, increasing social and spatial inequality is becoming a concern in a previously rather homogeneous urban society (see photos 10.4 and 10.5). Instead of representing similar class backgrounds and living in similar work-unit compounds as was common in the socialist era, urban households are now sorted into different types of housing and neighborhoods. With the emergence of upscale gated communities and dilapidated migrant enclaves, China's relatively homogeneous urban society is rapidly disappearing, and one with social separation and residential segregation is emerging. In addition, the recent real estate boom has made housing increasingly unaffordable in most cities, especially for low- and medium-income households in large cities. The housing price-to-income ratio can be as high as ten to fifteen in some cities, while a ratio of three to five is considered affordable (Man and Ren 2009). In addition to large housing demand and widespread housing speculation, the fact that local municipal governments depend on land leasing for revenue has contributed to increasingly higher land prices and thus higher housing prices. Since 2007, the central government has been aggressively promoting low-income housing with more investment in subsidized rental housing. But local governments have been reluctant, and the result is yet to be seen. Furthermore, millions of migrants remain ineligible for subsidized housing, although experiments are being carried out in cities such as Chongqing with migrants enjoying the same housing rights as urban residents.

New Cities in a New Millennium

Despite the strategy of gradualism and a relatively short period of reforms, profound changes have taken place in Chinese cities. While there are still some vestiges of socialist urban planning, such as large public squares and uniform apartment buildings, Chinese cities are now beginning to display features of modern cities, including high-rise office buildings, suburban housing estates, large shopping centers, and specialized districts for art, entertainment, and high-tech industries. Globalization has further added a touch of internationalism to Chinese cities.

In contrast to a previously cellular urban structure based on work-unit compounds with comprehensive functions, specialization and differentiation are now being pursued in Chinese urban planning, and the results are evident in the spatial restructuring of land use and the transformation of the urban landscapes. First, large housing estates, separate from employment centers, have been developed mostly on the outskirts of cities to accommodate households from different work units. While central cities are still preferred by most urban residents, suburbanization is taking place outside many large cities (Zhou and Ma 2000). People can now live in neighborhoods away from the watchful eyes of their colleagues, yet in general they have to commute farther to work. This increasing separation between residential and work space has made traffic a daily concern in most Chinese cities despite aggressive expansion and upgrading of the road and highway systems. The surge in the use and ownership of private automobiles in Chinese cities has made traffic even worse. China is currently the number one auto market and producer in the world, with thousands of cars added to the street each day. In Beijing, each car is required to be off the road for one day per week as a way to alleviate congestion. In the "kingdom of bicycles," the number of bicyclists dropped 26 percent during 2001–2006 (Campanella 2008).

Second, different specialized districts, such as high-tech zones, foreign enclaves, restoration districts, and art districts, are taking shape in large cities (Gaubatz 1995). To attract foreign investment, promote economic growth, and encourage innovation, municipal governments set up all sorts of high-tech and economic development zones, creating a "zone fever" in the 1990s (Cartier 2001b; Deng and Huang 2004). While many turn out to be not successful, the Zhongguancun High-Tech Zone in northwestern Beijing, with its proximity to many prestigious universities and research institutions and access to a huge talent pool, has become a bustling hub of high-tech business and research-and-development labs, and it is often called the "Silicon Valley" of China (Zhou 2007). While the Zhongguancun High-Tech Zone aims to change the image of "made in China" to "created in China," the Xiantiandi shopping district in Shanghai recasts its colonial history in the era of globalization. With icons such as the Shikumen house of nineteenth-century Shanghai and contemporary brand names such as Starbucks, Xintiandi converted a dilapidated neighborhood into an upscale transhistorical and transnational consumption space in a globalizing Shanghai (Wai 2006). Similar entrainment districts have appeared in other cities as well, such as "1912" in Nanjing and "Xihutiandi" in Hangzhou. In addition, specialized art districts, such as the "798 Art District" in Beijing and the "M50 Moganshan Art District" in Shanghai, which are avant-garde and trendy spaces that feature galleries, art centers, artists' studios, design companies, architectural workshops, restaurants, cafés, and bars, add another layer of diversity to the Chinese urban landscape.

With increasing globalization, Chinese urban residents are also getting a taste of Western consumption and lifestyles. While street-corner convenience stores and traditional farmers' markets are still prevalent, Western-style shopping malls and supermarkets are increasingly common in every Chinese city, and they often dwarf their Western counterparts in size. Seven of the ten largest shopping malls on earth are in China, and the South China Mall in Dongguan (Guangdong province) is the largest on earth with 7 million square feet of floor area, larger than the Pentagon (Campanella 2008). Global chain store giants such as Wal-Mart, Carrefour (France), and IKEA (Sweden) all rush in to take advantage of the increasingly high purchasing power of the emerging middle-class in Chinese cities by setting up branches everywhere. Since opening its first Supercenter in Shenzhen in 1996, Wal-Mart opened 189 stores in 101 cities by August 2010 and is planning for more stores (http://www.wal-martchina.com), Carrefour has 159 hypermarkets (http://carrefour.com), and IKEA has eight group stores in China (http://www.ikea.com) (photo 10.8).

Meanwhile, Chinese urbanites are increasingly brand conscious, and they are keen to purchase Western goods, from imported cars to clothes and from electronics to cosmetics. Shopping malls in Chinese cities are stocked with foreign goods from all over the world. In addition, Chinese urbanites have also embraced the Western (particularly American) fast foods provided by major chains such as KFC, Pizza Hut, McDonald's, and Starbucks. The first KFC outlet opened in Beijing in 1987. It is the largest KFC restaurant in the world, with five hundred seats in a three-story building, and it was an instant hit, setting the record for both single-day and annual sales

Photo 10.8. A Wal-Mart Supercenter on a busy street in Shenzhen (Youqin Huang)

in 1988 among the more than nine thousand KFC outlets worldwide (Yan 2000). In similar fashion, the first McDonald's in Beijing served more than forty thousand customers on its opening day, April 23, 1992. Thousands of similar establishments have since opened in Chinese cities. These fast-food establishments function more as social than as eating places in China because they provide Chinese with an environment in which to experience a Western culture and lifestyle (Yan 2000).

In addition, Chinese cities, especially the large coastal cities, are being transformed by international architectural firms and designers. Major landmark buildings in cities like Beijing, Shanghai, and Guangzhou are often designed by foreign architects—in contrast to the emphasis on self-reliance in the socialist era. For example, the World Financial Center and the Jinmao Tower in Pudong New District (see photo 10.3) were designed by American firms (Kohn Pedersen Fox Associates and Skidmore, Owings and Merrill, respectively), while the pearl-shaped National Grand Theater in Beijing and modern Pudong International Airport in Shanghai were designed by French architect Paul Andreu. Major stadiums built for the 2008 Olympic Games in Beijing, such as the National Stadium (Bird's Nest) and the National Aquatic Center (the Water Cube) (photo 10.9), the first magnetic levitation railway in Shanghai, the Citic Plaza in Guangzhou, and the city hall in Shenzhen, were all designed by Western architects. Even the state-owned China Central Television headquarters in Beijing was designed by Rem Koolhass (Netherlands) and Ole Scheeren (Germany) and is known for its irregular structure with a loop of six horizontal and vertical sections. In addition

Photo 10.9.   The National Stadium (Bird's Nest) and the National Aquatic Center (Water Cube) built for the 2008 Beijing Olympic Games (Youqin Huang)

to individual buildings, megaprojects, such as the new CBD in Beijing, the Pudong Financial District, and the "one-city, nine-town" housing project (eventually housing 500,000 people) in Shanghai, are often designed and planned by foreign firms. China is becoming an exciting playground for the world's renowned architects and designers who in turn are shaping the urban landscape in China in unprecedented and long-lasting ways.

But as Chinese cities such as these have become larger and more cosmopolitan, a whole set of new problems has emerged, including increasing social and spatial inequality, overcrowding, traffic jams, pollutions, homelessness, high crime, and high unemployment rates, despite the fact that the government has constantly attempted to avoid such social problems. Two decades ago, the Chinese government decided to shift its focus from social equality to economic efficiency, and today it is facing the dilemma of balancing these two again in its bustling cities.

Conclusion

As should not be surprising in a nation with the longest urban history in the world, urban development in China has followed a changing path, from a mature urban system before the middle of the nineteenth century to unbalanced urban development biased toward the treaty port cities during colonial times, then to strictly planned and controlled urban development in the socialist era, and now to a more market-driven urban development biased again toward the coastal cities. Despite constant shifts in ideology and changes in urban policies, especially since 1949, it is clear that the Chinese government has striven to achieve both sociospatial equality and economic efficiency in its urban development policies by emphasizing one aspect more than another as time and circumstance demand. While it is debatable whether the Chinese government has been successful in this regard, it is clear that to a certain degree China has avoided "urban explosion" and its associated urban problems, which have been widespread in many developing countries. Characterized by its underurbanization, its changing path, its spatially and hierarchically balanced urban system, and a mixture of traditional, socialist, and modern urban landscapes, urban development in China is certainly unique. The classic theories of urban development based on European experience are inadequate to explain urban development in China, and country-specific perspectives, concepts, and theories need to be developed (Fan 1999; Lin 1994; Ma 2002). While the important roles of history, globalization, and local initiatives in China's urban development should be recognized, socialist institutions and urban policies embodying the government's ideology have been more important in shaping urban development in China, especially since 1949. Ma (2002) argues for a political economy perspective, one that emphasizes the role of the state as the ultimate decision maker, regulator, and participant in urban development in China. Despite a drastic shift in urban policies and different patterns of urban development, the central role of the state has not changed significantly. In the socialist era, the state developed the *hukou* system to control rural-to-urban migration and urban growth, used state investment to achieve a desired urban economy and a balanced urban system, applied its principle of spatial and administrative organization to reduce rural–urban inequality, and initiated political campaigns, such as the Great Leap Forward, the Cultural Revolution, and the Third Front, that had a direct and profound impact on urban development. In the reform era, despite globalization of production, significant privatization, and the infusion of foreign capital, the state continues to play a central role in urban development and urban transformation (Ma 2002). As we have suggested in this chapter, it was the state that designed gradual and spatially biased open-door policies, encouraged the development of nonstate sectors, allowed rural-to-urban migration in a controlled fashion, and blueprinted the privatization of the land and housing system. Despite recent economic reforms, China is still politically a socialist nation with one dominating party, and the state still plays an important role in its urban development. Furthermore, with decentralization of fiscal and political power to local places, local governments also play increasingly important roles shaping Chinese cities, as demonstrated in the rapid development of urban infrastructure, and consequently there are huge regional variations in urban development in China.

Questions for Discussion

1. What are the major characteristics of urban development in China? How is urban development in China different from that in developed countries and other developing countries?
2. What is the spatial pattern of urban development in China, and how has the spatial pattern changed over time from the presocialist era to the socialist era and the reform era?
3. How have government policies and political ideologies shaped the pattern of urban development in China? Please compare and contrast the pre-1949 era, the socialist era, and the reform era.
4. What are the main characteristics of urban structure of Chinese cities in the presocialist era, the socialist era, and the reform era? What are the main forces for the changes between historical eras?

5. Since 1978, when China launched its open-door policy, China has been transformed by globalization in many respects. How have the open-door policy and globalization affected urban development in China? Please refer to both economic and sociospatial dimensions.

6. What is the *hukou* system? How has it affected the pattern of migration and urban development in China? What kind of changes have occurred in the reform era, and how have these changes affected the patterns of migration and urban development? What would happen if China completely abandoned the household registration system and removed all controls over rural-to-urban migration?

7. Compare and contrast "urbanization from below," "urbanization from above," and "urbanization from outside." What are their respective advantages and disadvantages, and how have they affected urbanization in China?

8. What are the main challenges and problems in Chinese cities today?

Notes

1. There are two types of urban places in China: cities (*shi*) and towns (*zhen*). Among cities, there are four municipalities under the jurisdiction of the central government (Beijing, Shanghai, Tianjing, and Chongqing), prefectural-level cities that are under the jurisdiction of provincial governments, and county-level cities.

2. Estimations for the volume of rural-to-urban migrants (defined as individuals who had left hometowns for at least six months) vary. In 2008, the National Bureau of Statistics (2008) estimated that there were 225 million migrants, yet another survey shows that there were 175 million migrants (National Bureau of Statistics 2009b).

3. Because of the different welfare benefit entitlements owed to the urban and rural populations, it was considered a calculated political decision for the government to send part of its urban population to the countryside and at the same time to attract peasants to cities (Zhao 1981).

4. To promote integration between the countryside and cities and to guarantee a supply of food and vegetables to the cities, rural counties surrounding an existing large city were often incorporated into the city proper (Lo 1987). Population in these counties, mainly rural, was therefore sometimes included in the total population of cities and towns and thus became "urban" population, leading to an inflation in apparent urban populations.

5. The First Front was regarded as China's highly vulnerable coastal cities, and the Second Front was a vague intermediate zone between the First and Third fronts.

6. There are two land systems in China: urban land is owned by the state, while rural land is owned by rural collectives. Only rural households are allowed to build housing on collectively owned land for self-consumption, but in suburban villages, households often build for rent seeking.

References Cited

Campanella, Thomas J. 2008. *The Concrete Dragon: China's Urban Revolution and What It Means for the World.* New York: Princeton Architectural Press.

Cannon, Terry. 1990. Region, inequality, and spatial policy. In *The Geography of Contemporary China*, ed. Terry Cannon and Alan Jenkins. London: Routledge, 28–59.

Cartier, Carolyn. 2001a. *Globalizing South China*. Malden, MA: Blackwell.

———. 2001b. Zone fever, the arable land debate, and real estate speculation: China's evolving land use regime and its geographical contradictions. *Journal of Contemporary China* 10, no. 28: 445–69.

Castells, Manuel. 1977. *The Urban Question: A Marxist Approach*. Cambridge, MA: MIT Press.

Chan, Kam Wing. 1988. Rural-urban migration in China, 1950–1982: Estimates and analysis. *Urban Geography* 9, no. 1: 53–84.

———. 1994. *Cities with Invisible Walls: Reinterpreting Urbanization in Post-1949 China*. Hong Kong: Oxford University Press.

———. 1996. Post-Mao China: A two-class urban society in the making. *International Journal of Urban and Regional Research* 20, no. 1: 134–50.

———. 2001. Recent migration in China: Patterns, trends, and policies. *Asian Perspective* 25, no. 4: 127–55.

———. 2007. Misconceptions and complexities in the study of China's cities: Definitions, statistics, and implications. *Eurasian Geography and Economics* 48, no. 4: 383–412.

Chan, Kam Wing, and Ying Hu. 2003. Urbanization in China in the 1990s: New definition, different series, and revised trends. *China Review* 3, no. 2: 49–71.

Chan, Roger C. K. 1995. Urban development strategy in an era of global competition: The case of Hong Kong and South China. In *Globalization and Regional Development in Southeast Asia and Pacific Rim*, ed. Gun Young Lee and Yong Woong Kim. Seoul: Korea Research Institute for Human Settlements, 202–27.

Chang, Sen-dou. 1977. The morphology of walled capital. In *The City in Late Imperial China*, ed. William Skinner. Stanford, CA: Stanford University Press, 75–101.

Chen, Tiejun, and Mark Seldon. 1994. The origins and social consequences of China's *hukou* system. *China Quarterly* 139: 644–68.

Clark, David, 1996. *Urban World/Global City*. London: Routledge.

Deng, F. Frederic, and Youqin Huang, 2004. Uneven land reform and urban sprawl in China: The case of Beijing, *Progress in Planning* 61: 211–36.

Ding, Chengru. 2003. Land policy reform in China: Assessment and prospects. *Land Use Policy* 20: 109–20.

Dong, Liming. 1985. Beijing: The development of a socialist capital. In *Chinese Cities: The Growth of the Metropolis since 1949*, ed. Victor F. S. Sit. Oxford: Oxford University Press.

Fan, C. Cindy. 1995. Developments from above, below, and outside: Spatial impacts of China's economic reforms in Jiangsu and Guangdong provinces. *Chinese Environment and Development* 6, no. 1 and 2: 85–116.

———. 1999. The vertical and horizontal expansions of China's city system. *Urban Geography* 20, no. 6: 493–515.

———. 2008. *China on the Move: Migration, the State, and the Household*. London: Routledge.

———. 2009. Flexible work, flexible household: Labor migration and rural households in China. In *Work and Organizations in China after Thirty Years of Transition*, ed. Lisa A. Keister. Bingley: Emerald Group Publishing, 381–412.

Fan, C. Cindy, and Youqin Huang. 1998. Waves of rural brides: Female marriage migration in China. *Annals of the Association of American Geographers* 88, no. 2: 227–51.

Fei, H.-T., et al. 1986. *Small Towns in China: Functions, Problems, and Prospects*. Beijing: New World Press.

Friedmann, John. 2005. *China's Urban Transition*, Minneapolis: University of Minnesota Press.

Gaetano, Arianne, and Tamara Jacka, 2004. *On the Move: Women in Rural-to-Urban Migration in Contemporary China*. New York: Columbia University Press.

Gaubatz, Piper Rae. 1995. Urban transformation in post-Mao China: Impacts of the reform era on China's urban form. In *Urban Spaces in Contemporary China: The Potential for Autonomy and Community in Post-Mao China*, ed. Deborah S. Davis et al. Washington, DC: Woodrow Wilson Center Press, 28–60.

Guangdong Statistics Bureau. 2010. 2009 nian Guangdong sheng guoming jinji he shuihui Fazhan tongji gongbao [A statistical report on economic and social development in Guangdong province in 2009], http://www.gdstats.gov.cn/tjgb/t20100225_74438.htm (accessed August 20, 2010).

Hao, Yen-Ping. 1986. *The Commercial Revolution in Nineteenth-Century China: The Rise of Sino-Western Mercantile Capitalism*. Berkeley: University of California Press.

Ho, Samuel P. S., and George C. S. Lin, 2003. Emerging land markets in rural and urban China: Policies and practices. *China Quarterly*, no. 175: 681–707.

Huang, Youqin. 2001. Gender, hukou, and the occupational attainment of female migrants in China (1985–1990). *Environment and Planning A* 33, no. 2: 257–79.

———. 2003. A room of one's own: Housing consumption and residential crowding in transitional urban China. *Environment and Planning A* 35, no. 4: 591–614.

Huang, Youqin, and William A. V. Clark. 2002. Housing tenure choice in transitional urban China: A multilevel analysis. *Urban Studies* 39, no. 1: 7–32.

Huang, Youqin, and Chengdong Yi. 2010. Consumption and tenure choice of multiple-home in transitional urban China. *International Journal of Housing Policy* 10, no. 2: 105–32.

Huang, Youqin, and Chengdong Yi. 2011. Patterns of second home in Chinese cities. In *China's Housing Reform and Outcomes*, ed. Joyce Yanyun Man. Cambridge, MA: Lincoln Institute of Land Policy, 91–108.

Kirkby, Richard. 1985. *Urbanization in China: Town and Country in a Developing Economy, 1949–2000 A.D.* London: Croom Helm.

Li, Si-ming. 2000. The housing market and tenure decision in Chinese cities: A multivariate analysis of the case of Guangzhou. *Housing Studies* 15, no. 2: 213–36.

Li, Si-ming, and Zheng Yi. 2007. Financing home purchase in China, with special reference to Guangzhou. *Housing Studies* 22, no. 3: 409–25.

Lin, George C. S. 1994. Changing theoretical perspectives on urbanisation in Asian developing countries. *Third World Planning Review* 16, no. 1: 1–23.

———. 1998. China's industrialization with controlled urbanization: Anti-urbanism or urban-biased? *Issues and Studies* 34, no. 6: 98–116.

Lo, C. P. 1987. Socialist ideology and urban strategies in China. *Urban Geography* 8, no. 5: 440–58.

Logan, John, Yanjie Bian, and Fuqin Bian. 1999. Housing inequality in urban China in the 1990s. *International Journal of Urban and Regional Development* 23, no. 1: 7–25.

Ma, Laurence J. C. 1971. *Commercial Development and Urban Change in Sung China (960–1279)*. Ann Arbor: University of Michigan Press.

———. 1976. Anti-urbanism in China. *Proceedings of the Association of American Geographers* 8: 114–18.

———. 1981. Introduction: The city in modern China. In *Urban Development in Modern China*, ed. Laurence J. C. Ma and Edward W. Hanten. Boulder, CO: Westview.

———. 2002. Urban transformation in China, 1949–2000: A review and research agenda. *Environment and Planning A* 34: 1545–69.

Ma, Laurence J. C., and Biao Xiang. 1998. Native place, migration, and the emergence of peasant enclaves in Beijing. *China Quarterly* 155: 546–81.

Ma, Laurence J. C., and Gonghao Cui. 1987. Administrative changes and urban population in China. *Annals of the Association of American Geographers* 77: 373–95.

Ma, Laurence J. C., and Ming Fan. 1994. Urbanization from below: The growth of towns in Jiangsu, China. *Urban Studies* 31, no. 10: 1625–45.

Man, Joyce, and Rongrong Ren. 2009. Housing affordability in Chinese cities. Unpublished manuscript.

Mitchell, Tom. 2009. Daunting departure. *Financial Times*, January 7. http://www.ft.com/cms/s/0/b3990974-dcf1-11dd-a2a9-000077b07658.html#axzz1EidYtWMN.

Mote, Frederick W. 1977. The transformation of Nanking, 1350–1400. In *The City in Late Imperial China*, ed. William Skinner. Stanford, CA: Stanford University Press, 101–54.

Murphy, Rhodes. 1970. *The Treaty Ports and China's Modernization: What Went Wrong?* Ann Arbor: University of Michigan, Center for Chinese Studies.

National Bureau of Statistics. 1983. *Zhonguo chengshi tongji nianjian 1983* [Urban statistical yearbook of China 1983]. Beijing: China Statistics Press.

———. 1995. *Zhonguo chengshi tongji nianjian 1995* [Urban statistical yearbook of China 1995]. Beijing: China Statistics Press.

———. 2001. *Zhonguo chengshi tongji nianjian 2001* [Urban statistical yearbook of China 2001]. Beijing: China Statistics Press.

———. 2002. *Zhongguo tongji nianjian 2002* [China statistical yearbook 2002]. Beijing: China Statistics Press.

———. 2003. *Zhonguo chengshi tongji nianjian 2003* [Urban statistical yearbook of China 2003]. Beijing: China Statistics Press.

———. 2008. *Nianmo quanguo nongmingong zongliang wei 22542 wan ren* [The national total of migrant workers at the end 2008 was 225.42 million]. March 25, 2009. http://www.stats.gov.cn/tjfx/fxbg/t20090325_402547406.htm.

———. 2009a. *Zhongguo tongji nianjian 2009* [China statistical yearbook 2009]. Beijing: China Statistics Press.

———. 2009b. *Zhongguo chengshi tongji nianjian 2009* [China urban statistical yearbook 2009]. Beijing: China Statistics Press.

———. 2010. *Zhongguo tongji nianjian 2010* [China statistical yearbook 2010]. Beijing: China Statistics Press.

Pannell, Clifton W. 1981. Recent growth and change in China's urban system. In *Urban Development in Modern China*, ed. Laurence J. C. Ma and Edward W. Hanten. Boulder, CO: Westview.

Ricci, Matteo. 1953. *China in the Sixteenth Century: The Journals of Matthew Ricci, 1583–1610*. Translated by Louis J. Gallagher. New York: Random House.

Shanghai Statistics Bureau. 2010. *Shanghai tongji nianjian 2010* [Shanghai statistical yearbook 2010]. Beijing: China Statistics Press.

Shenzhen Statistical Bureau. 2010. Shenzhen shi 2009 nian guoming jinji he shuihui Fazhan tongji gongbao [A statistical report on economic and social development in Shenzhen in 2009], http://www.sztj.com/main/xxgk/tjsj/tjgb/gmjjhshfzgb/201004265740.shtml (accessed August 24, 2010).

Sit, Victor F. S. 1995. *Beijing: The Nature and Planning of a Chinese Capital City*. New York: Wiley.

———. 2000. A window on Beijing: The social geography of urban housing in a period of transition, 1985–1990. *Third World Planning Review* 22, no. 3: 237–59.

Sit, Victor F. S., and C. Yang. 1997. Foreign-investment-induced exo-urbanization in the Pearl River Delta, China. *Urban Studies* 34: 647–77.

Skinner, G. W. 1977. Introduction: Urban development in imperial China. In *The City in Late Imperial China*, ed. William Skinner. Stanford, CA: Stanford University Press, 1–32.

Solinger, Dorothy J. 1995. The floating population in the cities: Chances for assimilation? In *Urban Spaces in Contemporary China: The Potential for Autonomy and Community in Post-*

Mao China, ed. Deborah S. Davis, Richard Kraus, Barry Naughton, and Elizabeth J. Perry. Washington, DC: Woodrow Wilson Center Press, 113–39.

State Council. 1984. Guowuyuan guanyu nongmin jinru jizhen luohu wenti de tongzhi [Circular of the state council concerning the question of peasants entering towns for settlement]. *Guowuyuan gongbao* [Bulletin of the state council] 26: 919–20.

———. 1998. Guowuyuan guanyu jingyibu shenhua chengzhen zhufang zhidu gaige jiakuai zhufang jianshe de tongzhi [A notification from the state council on further deepening the reform of urban housing system and accelerating housing construction]. State Council documentation no. 23.

Steinhardt, Nancy S. 1999. *Chinese Imperial City Planning.* Honolulu: University of Hawaii Press.

Tan, K. C. 1993. China's small town urbanization program: Criticism and adaptation. *GeoJournal* 29, no. 2: 155–62.

Tang, Wing-Shing. 1997. Urbanization in China: A review of its causal mechanisms and spatial relations. *Progress in Planning* 48, no. 1: 1–65.

Wai, Albert Wing Tai. 2006. Place promotion and iconography in Shanghais Xintiandi. *Habitat International* 30: 245–60.

Wang, Feng. 1997. The breakdown of a Great Wall: Recent changes in the household registration system of China. In *Floating Population and Migration in China: The Impact of Economic Reforms*, ed. Thomas Scharping. Hamburg: Institute of Asian Studies, 149–65.

Wang, Ya Ping, and Alan Murie. 1999. *Housing Policy and Practice in China.* New York: St. Martin's.

Whyte, Martin King, and William L. Parish. 1984. *Urban Life in Contemporary China.* Chicago: University of Chicago Press.

Wright, Arthur F. 1977. The cosmology of the Chinese city. In *The City in Late Imperial China*, ed. William Skinner. Stanford, CA: Stanford University Press, 33–74.

Wu, Fulong. 2002. Sociospatial differentiation in urban China: Evidence from Shanghai's real estate market. *Environment and Planning A* 34: 1591–1615.

Wu, Weiping. 1999. Reforming china's institutional environment for urban infrastructure provision. *Urban Studies* 36, no. 13: 2263–82.

———. 2002. Migrant housing in urban China: Choices and constraints. *Urban Affairs Review* 38, no. 1: 90–119.

Xu, Xueqiang, and Si-ming Li. 1990. China's open door policy and urbanization in the Pearl River Delta region. *International Journal of Urban and Regional Research* 14, no. 1: 46–69.

Yan, Yunxiang. 2000. Of hamburger and social space. In *The Consumer Revolution in Urban China*, ed. Deborah S. Davis. Berkeley: University of California Press, 201–25.

Yeh, Anthony Gar-on, and Fulong Wu. 1995. Internal structure of Chinese cities in the midst of economic reform. *Urban Geography* 16, no. 6: 521–54.

Zhang Li. 2001. *Strangers in the City: Reconfigurations of Space, Power, and Social Networks within China's Floating Population.* Stanford, CA: Stanford University Press.

———. 2002. Spatiality and urban citizenship in late Socialist China. *Public Culture* 14, no. 2: 311–34.

Zhang, Li, and Simon X. B. Zhao. 1998. Re-examining China's "urban" concept and the level of urbanization. *China Quarterly* 154: 330–81.

Zhang, Xin Quan. 1998. *Privatisation: A Study of Housing Policy in Urban China.* New York: Nova Science.

Zhao, Lukun. 1981. Zai lun laodong jiuye wenti [A further discussion on the employment question]. *Renkou Yanjiu* [*Population Research*] 4: 18–24.

Zhou, Yixing. 2003. Gaige kaifang tiaojian xia de zhongguo chengshi jingji qu [Urban economic regions in China under reform]. *Dili Xuebao* [*Acta Geographica Sinica*] 58, no. 2: 271–84.

Zhou, Yixing, and Laurence J. C. Ma. 2000. Economic restructuring and suburbanization in China. *Urban Geography* 21, no. 3: 205–36.

———. 2003. China's urbanization level: Reconstructing a baseline from the fifth population census. *China Quarterly* 173: 176–96.

Zhou, Yu, 2007. *The Inside Story of China's High-Tech Industry: Making Silicon Valley in Beijing.* Lanham, MD: Rowman & Littlefield.

Zhu, Baoshu, and Guoqian Gu. 1991. Renkou qianyi fenxizhong de diyu koujing wenti [The spatial specification in the migration analysis]. *Renkou Yanjiu* [*Population Research*] 6: 28–34.

CHAPTER 11

Hong Kong before and after the Return

Christopher J. Smith

A Rosy Future for Hong Kong, or No Future at All?

Immediately after Hong Kong was returned to China in 1997, the new special administrative region's (SAR's) hopes of becoming the region's (or even China's) number one city did not look promising. To focus on just one domain that seemed to offer a way forward for cities aspiring to be world cities—the high-technology field—Hong Kong hardly looked like it would be able to compete with Shanghai or Beijing in the new millennium. The former colony was not overly endowed with technology talent: there was no equivalent to Tsinghua University, for example, and Hong Kong did not seem to possess the dynamism of Shanghai, which Wasserstrom (2008) has recently described as "a city in a hurry." It was looking as though the vast majority of venture capital funds would bypass Hong Kong completely: many onlookers feared that the city's best online technology and marketing talent would begin to look toward the mainland or other destinations around the world.

In a desperate effort to turn the technology tide in its favor, the post-handover government in the new SAR invested heavily to enhance the city's technology infrastructure, first with the development of Cyberport on the Hong Kong side and then with another high-tech cluster adjacent to Chinese University in the New Territories. For years, both campuses stood half empty, but today Cyberport is humming, and some observers are sensing a whiff of self-confidence that had been missing for a decade or more:

> Government support of innovation . . . has grown significantly since the handover. The effort was spearheaded by the establishment of the "Innovation and Technology Fund" (ITF), set up in 1999 with HK$5 billion (approximately US$650 million/€500 million) earmarked to provide funding for projects that contribute to innovation and technology upgrading in both new and established industries. The Innovation and Technology Commission (ITC) was also set up to lead Hong Kong's drive to become a world-class knowledge-based society. The establishment of The Hong Kong Science and Technology Parks Corporation (HKSTPC) is an important

expression of this ambition to increase R&D spending in small firms. The HKSTPC aims to establish a flagship technology infrastructure to provide a comprehensive range of services that cater to the needs of the high technology industry at various stages, ranging from nurturing technology start-ups through the incubation program to providing facilities and services in the Science Park for applied R&D activities. It also provides land in industrial estates for production. (Sharif and Baark 2008)

A brief checklist review of technology and related business trends does indeed point toward a brighter high-tech future for Hong Kong (Denlinger 2010).[1] On the strength of these and other natural advantages, it appears that, for the time being at least, Hong Kong seems to be more producer and consumer friendly in the information technology area than the cities on the mainland, which bodes well for it becoming a technology center for Asia and the rest of China. Since 1997, research-and-development investment and public support for efforts to generate new technologies have come together to help sketch out a future for Hong Kong as an "innovation hub" (Sharif and Baark 2008) with links to and from mainland China and well beyond. Sharif and Baark see Hong Kong becoming a "center of exchange" similar to the earlier development of trade or financial hubs, and they are confident that in the new millennium, Hong Kong's comparative advantage in the realm of innovation may allow it to dominate the region. The SAR's recent emphasis on innovation and technology, they argue, has helped it create strong linkages and knowledge "flows" into neighboring regions. Hong Kong, they argue,

> can serve as an innovation hub through its ability to facilitate innovation-related activities . . . [which] means not only devoting more local resources to innovative activities in a greater commitment to R&D but also, and just as importantly, effectively applying new knowledge produced elsewhere to create additional value in the production chain.

The problem, of course—and this pertains to all hopeful predictions that Hong Kongers might be making about the future of their city—is that all these comparative advantages could so easily come to nothing if the powers that be on the mainland state decide to "ratchet up" control from the center in favor of other potential hub cities. What if, for example, Beijing suddenly were to clamp down on all the freedoms and accessibility advantages that Hong Kong currently holds? China, if it chose to do so, could eliminate the Hong Kong advantage quite easily, for example, by doing away with all content restrictions on the Internet, by treating foreign companies exactly the same as Chinese companies, and/or by completely deregulating its mobile industry. In all (or maybe in any) of these cases, Hong Kong would lose its edge over the mainland overnight.

Economic Development: Looking at Two Sides of the Picture

The post-1949 diversification of Hong Kong's economy in the direction of export-oriented manufacturing was reinforced during the Korean War as a result of the economic embargo on trade with China. The embargo cut deeply into the entrepôt function of Hong Kong: in 1952, for example, the value of exports to China was HK$1.6 billion, but this had fallen to HK$520 million by 1953 and to HK$136 million by 1956 (Ho 1992). As exports from China plummeted, local Hong Kong industries received a boost, especially in such areas as textiles, clothing, light metal goods, footwear, plastics, electronics, and optical instruments, for which it became globally recognized. Beginning in the 1970s, however, Hong Kong's comparative advantage in the production of such goods was seriously challenged, especially by other economies in East Asia, some of which had wage levels that were significantly lower. The result was a gradual but perceptible shift in the direction of more sophisticated and higher-quality value-added production lines but still with a clear emphasis on exports. The restructuring of Hong Kong's manufacturing base at that time was accelerated by the shift of some low-level production activities to factories on the mainland, particularly the production of such items as clothing, shoes, plastic toys, and luggage, which are now made almost exclusively in mainland factories. The net effect of this shift was that during the next two decades, virtually all heavy industry departed from Hong Kong.

The magnitude of this transformation is reflected in the changing structure of exports coming out of Hong Kong. In 1960, clothing represented 29 percent of Hong Kong's export total, textiles 23 percent, electronics 2.5 percent, and precision instruments (watches and clocks) 0.7 percent. By 1988, the proportion of clothing was about the same (31 percent), but the textiles sector had fallen to only 7 percent, while electronic goods had increased to 22.4 percent and precision instruments to 10 percent (Ho 1992, table 4.1, 76). In spite of this shift, Hong Kong's manufacturing sector—unlike Taiwan's—was still dominated in the early 1990s by labor-intensive operations, implying that the colony was not yet making any significant strides toward high-technology production methods. This was still the case by the start of the new millennium, and this situation—which might be interpreted as either oversight or failure—goes a long way toward explaining why some scholars predicted a serious downturn for Hong Kong after the handover in 1997 (see, e.g., Castells 1999).

The growth of Hong Kong's export-oriented manufacturing was accompanied by a long period of rapid economic growth. During the 1960s, for example, gross domestic product (GDP) expanded by more than 10 percent each year, with the contribution of manufacturing to overall GDP increasing from 20 percent to more than 30 percent, and by 1971, 47 percent of Hong Kong's labor force was involved in manufacturing. Nearly a third of all manufacturing workers were in the textile business, popularizing the "Made in Hong Kong" label on clothing that became famous (or infamous) around the world. The driving force of the Hong Kong economy during the 1960s and 1970s was clearly the export sector, which grew at an average of 11.5 percent per year—more than twice the average rate of growth for the world as a whole. For the most part, the expansion of the economy generated a demand for labor that guaranteed nearly full employment and a healthy annual increase in wages. Inflation remained low (around 4 percent per year), and the economy benefited from healthy injections of direct foreign investment.

In addition to its manufacturing base, Hong Kong's economy was rapidly becoming dependent on the development of its service sector, especially in the areas of shipping, banking, insurance, and real estate. Before 1997, Hong Kong was also a major tourist attraction for visitors from all parts of the world, contributing significantly to its local economy. The sharp dropoff in tourism after 1997 has been very worrying to Hong Kong's business elite. It is assumed that the growth in tourism during the late 1980s and early 1990s was based on the desire to visit Hong Kong for "one last time" before it reverted to Chinese rule, which obviously peaked by the time of the actual handover. Tourism statistics circulated by the Hong Kong government indicate that incoming tourists to Hong Kong increased from 2.6 million in 1982 to 8 million in 1992, and although the average length of stay was relatively short (two to three days), per capita spending by tourists increased during this period (especially those coming from Japan and Taiwan). Hong Kong's largest airline, Cathay Pacific, has begun to advertise very heavily, especially on the Internet, to try to turn this tourism downward trend around, offering relatively cheap package deals for tourists who want to use Hong Kong as a base for visiting other countries in Asia.

Hong Kong was able to maintain a healthy balance of payments throughout this period, and its currency remained relatively stable, supporting the consensus view of a "success narrative": Hong Kong had achieved rapid and sustained economic growth with relative price stability while at the same time realizing a substantial rise in real income, maintaining high levels of employment and a reasonable degree of income inequality (Ho 1992).[2] Hong Kong's economic performance from the 1960s through the 1980s had enabled it to catch up with some of the other global economic powers, particularly the United States and the United Kingdom. By 1988, for example, Hong Kong's per capita gross national product was close to 70 percent that of the United Kingdom and 50 percent that of the United States, but by 1995, Hong Kong's per capita gross national product (at US$22,990) was higher than that of the United Kingdom and 85 percent that of the United States (United Nations Development Program 1998, 125, 184).

Structural change in Hong Kong's economy continued throughout the 1970s and 1980s, dominated by three major trends: one was an increasing focus on higher-value value-added goods (such as garments and finished clothing as opposed to unfinished textile goods), and another was a marked shift toward the manufacture of electrical goods and appliances, such as watches and clocks. In the 1980s, another important trend was evident: a massive increase in the financial services sector, which had overtaken manufacturing as the largest employer category in Hong Kong's economy. The new jobs in this sector were mainly in what are generally referred to as producer services, including banking, investment facilities, insurance, real estate, and corporate services, which were providing the essential infrastructure for a growing number of foreign multinational companies located in the territory. By the early 1990s, manufacturing was contributing only 15 percent of Hong Kong's GDP and employed only 33 percent of its workforce (in itself suggesting the low-wage economy prevalent in manufacturing). In the decade from 1982 to 1991, the number of people employed in manufacturing in Hong Kong fell by almost 34 percent (see table 11.1a), and the contribution of manufacturing to overall GDP in Hong Kong during the same period fell by almost 27 percent (see table 11.1b). Not surprisingly, the contribution of the service sectors grew significantly through the 1980s.

Table 11.1a. Hong Kong: Persons Engaged in Selected Industry Sectors (in Thousands)

Occupation	1982	1987	1991	% Growth, 1982–1991
Manufacturing	847.2	867.9	565.1	−33.9
Building and construction	82.1	72.5	59.5	−27.5
Wholesale, retail, and import/ export, restaurants and hotels	517.7	657.4	914.8	+76.7
Financing, insurance, real estate, business services	116.1	212.2	314.8	+230.7

Source: Hong Kong Census and Statistics Department (1993).

Table 11.1b.   Hong Kong: Contribution of Economic Activities to GDP (%)

Occupation	1982	1987	1992	% Growth, 1982–1992
Manufacturing	20.7	21.7	15.2	−26.6
Wholesale, retail, and import/export, restaurants and hotels	19.1	23.2	25.4	+33.0
Financing, insurance, real estate, business services	22.6	18.2	22.7	−0.14
Building and construction	7.3	4.7	5.6	−23.3

Source: Hong Kong Census and Statistics Department (1993, tables 2.3, 2.4, 3.4).

The most significant rate of growth in the last four decades of the 1900s was in Hong Kong's role as a world trader. Exports were valued at US$0.69 billion in 1960, with Hong Kong ranked twenty-seventh in the world, but by 1985, its exports were valued at US$63.17 billion, making it the tenth-largest exporter in the world. The sharp increase in industrial production in the 1950s and 1960s meant that the export of domestic-made goods began to outstrip the reexport of goods made elsewhere (mostly in China), and by 1970, domestic exports outranked reexports by more than four to one. In the 1980s, this situation changed, and with the increasing economic integration of Hong Kong and South China and the shift of many manufacturing plants to the mainland, the reexport business began to grow rapidly, much of it involving repackaging and improved presentation of goods made elsewhere. By the end of the 1980s, reexports had moved ahead of domestic exports, and by the mid-1990s, about one-third of Hong Kong's exports consisted of reexports of goods produced in China, primarily by Hong Kong–based companies (Hong Kong Census and Statistics Department 1993).

Hong Kong has become known as a place where "anything goes" and where it is easy and relatively cheap to start up a business, and this is still true today, as the World Bank's *Doing Business Database* indicates. The database provides objective measures of business regulations and their enforcement in 145 economies around the world. The measures that are employed assess the regulatory costs of doing business and can be used to analyze specific regulations that enhance or constrain investment, productivity, and growth. One of the measures, referred to as "the challenges of launching a business," uses the following criteria: the procedures required to establish a business, the associated time and costs involved, and the minimum capital requirement. In its report, the database indicates that Hong Kong entrepreneurs can expect to go through five steps to launch a business over eleven days on average at a cost equal to 3.4 percent of gross national income (GNI) per capita, compared with a regional average of 100.5 percent of GNI and an average among Organization for Economic Cooperation and Development (OECD) countries of 44.1 percent of GNI. In addition, there is no minimum deposit requirement to obtain a business registration number in Hong Kong. In other words, it is still much quicker, easier, and cheaper to start a business in Hong Kong than in most other places in the region or in Organization for Economic Cooperation and Development countries. Other indicators in the database include the ease of hiring and firing workers, registering property, getting credit, protecting investors, enforcing contracts, and closing a business.[3]

The Economist magazine ranked Hong Kong as first in the world in its Economic Freedom Index for 1997, an index that grades countries on the extent to which government policies restrict or interfere with economic activities (see *World in Figures 1997*). The indicators used to construct the index include trade policy, taxation rates, monetary policy, the banking system, foreign investment rules and regulations, property rights, the proportion of economic output consumed by the government, regulation policy, size of the black market, and the extent of wage and price controls. Each country is assigned a score ranging from 1 (the most free) to 5 (the least free). In 1997, Hong Kong scored 1.25, Singapore 1.30, and Bahrain 1.70, while the United States ranked seventh, with a score of 1.90, and the United Kingdom scored 1.95. At the bottom end of the Freedom Index were Ghana with 3.25 and Algeria with 3.25. Since then, several other measures of global competitiveness have emerged on which Hong Kong consistently performs well in spite of the instability during the first years after the handover in 1997.

Davies (1990) predicted that Hong Kong's business climate for the period 1997–2002 would deteriorate, primarily as a result of the "anticipated deterioration in the operating environment associated with the handover" in addition to "fallout" from the regional (East and Southeast Asian) economic crisis. Davies also predicted that Hong Kong would no longer be perceived as one of the "best" place in the world to do business, and there is some evidence to support such a conclusion, although the Cato Institute reported in 2004 that Hong Kong still had the highest degree of economic freedom in the world.[4] In spite of this, the consensus of opinion is that Hong Kong's relative decline is the result of a combination of factors in addition to the externality effects of the regional economic downturn. The real damage, according to some commentators, has resulted from a perception, globally, that Hong Kong has become much more politicized than it was in the past. One prominent Western banker, for example, was quoted as saying,

> There is a feeling that Hong Kong has lost something. We all know that Singapore has a much more politicized environment than Hong Kong, but it has always been that way. There is a fear that this could be the beginning of an ongoing erosion and that things are going to get worse not better,

whereas in Singapore at least you know things are going to stay the same.
(*South China Morning Post* 1999)

There are also some externality effects that work to tone down somewhat the tri-umphal story of economic success in Hong Kong and its gradual integration with China even before the handover. In the first place, a dangerous precedent was being set by the mass transfer of manufacturing jobs from Hong Kong to the mainland, a transfer involving the export of "dirty" and heavily polluting industries from what amounts to the global "north" to the global "south." Although this may have been a boon for Hong Kong's immediate pollution problems, it was (and still is) extremely damaging to China, and many scholars and local activists have pointed out some of the damaging side effects for and even in Hong Kong. The prevailing winds blow into the former colony from the north, which is thought to be one of the primary factors in the sharp decline in Hong Kong's air quality during the past decade. On many days throughout the year, Hong Kong is shrouded in pollution, making many of its glass skyscrapers barely visible and prompting authorities to warn people to stay inside (Headley 2011) (see photo 11.1).[5]

Although variations in weather patterns, as well as local sources of pollution, are at least partly to blame for these effects, many locals have pointed fingers across the border at the smoke-belching factories in China's neighboring Guangdong province.

Photo 11.1. Hong Kong's new skyline, the International Financial Center shrouded in pollution (Christopher Smith)

In 2002, Hong Kong's air pollution reached a record level on August 28 when the community of Tung Chung, close to the new international airport at Chep Lap Kok, recorded an air pollution index reading of 185 (on a scale where anything over 100 is considered dangerously high). The government issued warnings to people with heart or breathing problems to avoid physical strain and told them not to go to places with heavy traffic. The concentrations of ozone and of factory and vehicle emissions that accumulate on hot, still days throughout the Hong Kong summers have forced the government to begin developing a plan to deal with this problem as part of what it describes as its effort to turn Hong Kong into a world-class city. Environmental officials have been pushing owners of high-emission diesel vehicles in Hong Kong to use cleaner fuels, such as liquefied petroleum gas, and, to date, most of Hong Kong's more than eighteen thousand taxis have made the switch.

In April 2002, Guangdong provincial and Hong Kong environmental officials reached an accord, pledging to reduce emissions of pollutants such as sulfur dioxide and nitrogen oxide from power plants and factories by 2010, although specific actions were not immediately implemented. Guangdong officials, not surprisingly perhaps, have denied that Hong Kong's air-quality problems are caused mainly by pollution blowing in from the north, claiming that there is no hard scientific evidence to support such a claim. Meanwhile, Hong Kong officials are convinced that the most serious offenders in Guangdong are the sixty or more cement factories in the Pearl River Delta region (with almost every community in the delta having several such factories until quite recently during the furious phase of urban and industrial development). Also of particular concern to Hong Kong environmentalists are the production processes (again in mainland China) involved in making printed circuit boards for computers, which use significant amounts of ozone-depleting chlorofluorocarbons as cleaning solvents. There has also been a significant growth in the manufacture of Styrofoam and plastic products, which are used in China to make disposable lunch boxes and now water bottles and which are literally being made by the billion every week for China's hungry and thirsty workers.

On the relatively few days of good visibility, Hong Kong's dazzling skyline reminds visitors and residents alike that the territory has been on the cutting edge of modern (and postmodern) architecture during the past three decades. The modernist impulse has also left its imprint on other parts of Hong Kong, particularly in the New Territories with the development of a series of new towns. To date, eight have been built, located approximately 20 to 30 km from downtown Hong Kong, and a ninth is now fully operating on Lantau Island, at Tung Chung, adjacent to the new airport. The town planners, borrowing many of their ideas from planning strategies implemented in Britain in earlier decades, envisioned Hong Kong's New Towns as self-contained residential, manufacturing, and commercial centers. Similar to those originally built in the south of England, the New Towns were intended to provide decent and inexpensive housing for Hong Kong's crowded population; to alleviate the pressures of high-density living, particularly on Hong Kong Island and in Kowloon; and to provide ample space for industrial development in open-field sites that would allow a switch from the traditional high-rise factories in the crowded residential areas of Kowloon and the central district of Hong Kong Island.

The Hong Kong SAR in the New Millennium

As the handover to China approached, many Hong Kong residents began to watch and become increasingly sensitive to events occurring on the mainland. In spite of the local concern about what sort of a place Hong Kong would evolve into after 1997, the vast majority of evidence suggests that throughout the 1990s, the actual process of economic integration with China was gaining strength and that when the actual handover occurred, it was little more than a formality. As Hong Kong's influence on the mainland continued to grow, it became increasingly evident that China was unlikely to do anything to sabotage or slow down economic growth after 1997. And as more and more private capital entered China from the colony, the local Chinese economies, especially in Guangdong and Fujian provinces, were increasingly becoming disconnected from the planned economy of the mainland and thrust more than ever into the vortex of both regional and global capitalism. With each passing year, Hong Kong and South China became ever-more interlocked and interdependent to such an extent that at least one resident characterized 1997 as little more that a "sideshow."[6]

It was estimated that before the handover, more than US$20 billion had been invested in each direction, and it was widely rumored—although again never substantiated—that Hong Kong entrepreneurs were providing jobs for more than 3 million workers in China and that over one-third of the colony's currency was in common use on the mainland.[7] It was clear that investment penetration was occurring in both directions: Guangdong township-village enterprises, for example, were investing in some of the choicest properties in Hong Kong's central financial district, while Hong Kong businessmen were regular investors in even the remotest rural areas of the mainland: opening restaurants and theme parks, building bridges, and sometimes even restructuring entire cities into small forests of skyscrapers to be leased as office space, hotels, and apartments. As this activity suggested, the handover itself was merely the legal imprimatur on what was rapidly becoming a fait accompli, and in South China the emergence of a huge, integrated economic region—the Pearl River Delta—was an established fact. None of this evidence obscured the fact that there were still some very pressing concerns about Hong Kong post-1997, some of which were cultural, in addition to the well-publicized focus on political and economic issues.[8]

To provide a sampling of these issues, it is useful to focus on three concerns: how do Hong Kongers think about themselves and their role in the world now that they are no longer part of the British Empire? Will Hong Kong be as wealthy in the future as it has been in the past, and, more specifically, how will the economic impact of the post-handover period be felt by the ordinary people of Hong Kong? And can the "one country, two systems" work adequately in the future?

Hong Kong Identities

A discourse on Hong Kong's identity is meaningless because political identity after 1997 can mean only one thing—Chinese citizenship. Identity as citizens of a city is at best subsidiary to the national identity. But in Hong

> Kong itself, a Chinese population has become increasingly attached to their city as their homeland but not necessarily to their new country. (Lee 2010, 250)

At one end of the spectrum of concern in this context might be a nostalgic sense of the past and perhaps even an acute sense of loss as a result of the handover and a belief that the situation is deteriorating rapidly. At the other extreme, we could expect to find opinions like those of Law (2002): people who feel that Hong Kong is doing splendidly as one of China's newest SARs. Representing the sense of mourning for a long-lost past, we might come across the British travel writer Jan Morris, who wrote the following in the version of her Hong Kong book that was reissued for the handover event:

> To have created upon this improbable terrain, among an alien people, so far from home, a society not only stable, educated, prosperous and free, not only self-governed by the imperialists' own high principles, but also standing as a model and an inspiration to its mother China—that might be a last justification for the idea of imperialism itself. And even if that fulfillment were to survive only a generation, to be destroyed by yet another new brutalism, at least it would add a sad majesty to the aesthetic of Empire—a memorial to what might have been, as the shutters close upon the once exuberant colony. (quoted in Y.-L. R. Wong 2002, 141–42)

Far removed from the colonial longings expressed here by Morris but nevertheless still in a sense mourning for a past that seems to be disappearing rapidly, we find the former Harvard scholar Leo Ou-fan Lee, who, although not born in Hong Kong, has returned there to live after spending more that three decades in the United States. In his 2010 book *City between Worlds*, Lee writes what amounts to a personal history of the city, subtitled *My Hong Kong*. In his attempt to get at the core of Hong Kong's culture, Lee focuses attention on the word "play" (*wan* in Chinese), a word that he feels is the most common word in the local cultural vocabulary. As he observes, almost anything—from stocks and shares to postmodern theory—can be played with and is therefore "playable" or even "playful":

> Play is both the companion of and the antidote to work; and like work, it is also a motivating force behind Hong Kong's lifestyle. In Hong Kong's popular culture, from movies to advertisements, hardly anything "serious" has appeared since the "suffering" 1950's and 1960's happened. The former Executive Chief Tung Chee-hwa introduced the official slogan "Tomorrow will be better." But the average Hong Konger seems only to believe in the present, and the thing to do is seize the moment. (Lee 2010, 226–27)

The chief executive paid close attention in 2003 when more than half a million people took to the streets of Hong Kong in response to the introduction of the vastly unpopular Articles 23, which stated the government's intention to impose limits on individual freedom of expression in the interest of national security. Although Lee attributes, in part at least, Tung's resignation in 2004 to the level of dissatisfaction expressed by the people at this time, he also remarks that the success of the demonstration—Article 23 was quickly withdrawn—actually took everyone by surprise, most of all the marchers themselves: "It seems as if 'playing' [in this case playing at or with politics] . . . is the only choice left for people whose political destiny is already decided and whose democratic form of government is now under close watch by the Beijing government. Even the clamor for direct elections seems to get nowhere. Thus, the legislators (about half of whom were not elected) can only play at politics" (Lee 2010, 227).

Few people in Hong Kong today would echo such blatantly imperialist sentiments as Morris or such pessimism as Lee, but it is plausible that many of them might feel conflicted about their identity. Y.-L. R Wong (2002) tried to find out if this was indeed the case for young, middle-class Hong Kongers who (after 1997) had become Chinese citizens but who had grown up in the uniquely privileged position of being able to construct themselves as free subjects, stemming in part from the Western-oriented way of life associated with British Hong Kong. It would be quite normal, Wong felt, if such individuals thought of themselves as being located, in ideological terms, somewhere between the colonized and the colonizing. To test her hypothesis, Wong conducted in-depth interviews with a number of young Hong Kong citizens who were actively engaged in what she refers to as "development" projects in mainland China, programs focusing on such activities as educational campaigns, efforts to improve conditions and possibilities for China's women, and poverty eradication programs. As she had anticipated, most of those interviewed thought of themselves as culturally Chinese, but many of them were strongly opposed to the political system operating on the mainland. As Wong observes,

> China was imagined as their cultural "roots" which they desired to "go back" to—in both time and space. What the imaginary cultural roots, and hence China, represent are culture, history and landscape. Such a romantic imaginary of China, however, could not be sustained when some informants later in their post–high school years began to reach out for actual contemporary China. Being in touch with the realities of contemporary China, many of these individual Hong Kongers began to experience and construct China as the "inside"—a space of oppression and confined vision. The image of Hong Kong, on the other hand, was further affirmed as free. Between their identification with and alienation from China, individual informants experienced ambivalence in their colonial and national identities. (Y.-L. R. Wong 2002, 147)

The differences between Hong Kong and the mainland and the perceived advantages of being born in Hong Kong were obvious to these young people. One of them who was actually born on the mainland told Wong that for her, Hong Kong actually represented freedom. If she had stayed in China, she would probably have been married by then, with children (one or two, depending on where she lived), and that, hopefully, she would have a state job. This was clearly not a thought that appealed to her imagination; in fact, in her words,

> You know roughly what your fate will be inside the mainland. . . . But if you come out to Hong Kong, I think there are many more opportunities. I

> understand why people want to go to the States, or to Hong Kong. You'll
> have a very very different identity. . . . I think I was the only girl who finished
> university in my village. . . . If I was . . . [still] inside the mainland, I would
> not have had these kinds of opportunities. (Y.-L. R. Wong 2002, 150)

Based on her discussions with a number of such young people, Wong concluded
that the opportunity to go to China to be involved in development work created a
"discursive space" for individual Hong Kongers when they traveled to mainland China
(Y.-K. Wong 2002, 152). Most of them, Wong reports, were on what amounted to
a "liberalizing mission" on their side, giving them the opportunity to help with and
speed up the "opening up" of China to the outside world. This is what Wong refers
to as a colonial discourse of "Hong-Kong-as-liberty," in which Hong Kong residents
think of themselves as the liberal modern Chinese whose mission it is to "free up" their
ancestral country.[9]

Hard Times in Hong Kong?

> The former British territory may still have the world's highest ratio of Rolls
> Royces to people, but the boom-town days of the 1980s are long gone. It's
> easy to miss the dark, narrow entrance to 12-year-old Wong Hung-kai's
> home in Hong Kong's bustling Cheung Sha Wan district, once a maze of
> small factories and workshops. The stench of urine and the pitch black of
> the grimy stairway leading to his family's tiny cubicle greet the boy daily
> on his way to and from school. Amid the plenty of one of Asia's richest cit-
> ies, Wong is among a rising number of immigrants from mainland China
> whose dreams have faded into poverty. Wong shares the cramped space
> with his widowed mother and 17-year-old brother. At 3.7m², it is smaller
> than an average Mercedes Benz car. (Asia Child Rights 2003)

Putting aside these essentially positive views of Hong Kong, where the past, at
least in some people's dreams, has certain beneficial effects (Y.-L. R. Wong 2002),
one of the most pressing concerns in recent years has been in the economic realm,
specifically, the fear that Hong Kong's uncertain status after 1997 would amplify the
problems inherent in the region as a result of the Asian economic crisis of the late
1990s. This topic has a relatively short "shelf life," of course, because economic trends
seem to rise and fall in rapid cycles, as we can clearly see at the end of the end of the
first decade of the new millennium with its global economic crisis. Over the course
of the past thirteen years, it has become clear that Hong Kong's economy has been
more up and down than most people—both locals and internationals—would have
preferred, and the general view is that these oscillations were originally a response, at
least in part, to the serious economic challenges Hong Kong faced immediately after
the handover.

A sharp drop in property prices in late 1997 burst the economic bubble and,
together with the emerging economic crisis in Asia, resulted in deflation, increasing
unemployment, and a budget deficit in 1998. Hong Kong's real GDP fell by more

Table 11.2. Hong Kong's Domestic Economy, 1999-2003

	1999	2000	2001	2002	2003
GDP at constant (2000) market prices (HK$million)	1,169,474	1,288,338	1,294,306	1,318,743	1,361,036
Real GDP growth (%)	3.00%	10.2%	0.5%	1.9%	3.2%
GDP per capita at current market prices (US$)	23,177	24,365	24,070	23,800	23,312
Inflation (%)	-4	-3.8	-1.6	-3	-2.6
Unemployment (%)	6.2	4.9	5.1	7.3	7.9
Foreign Exchange Reserves (US$billion)	96.26	107.6	111.2	111.9	118.4
Population growth (%)	2.5	0.9	0.9	0.9	0.2
Hang Seng Index (31.7.1964 = 100)	16,962	15,095	11,397	9,321	12,576

Sources: "Annual Report 2002—Annex and Tables" (Hong Kong Monetary Authority); "Hong Kong Statistics—Frequently Asked Statistics" (HKSAR Census and Statistics Department 2004); "Hong Kong Statistics—Hong Kong in Figures" (HKSAR Census and Statistics Department 2004).

than 5 percent in 1998, though it would bounce back again in 1999 and especially 2000 and then remain static in 2001 (see table 11.2). Nominal investment fell by 13.9 percent in 1998 and 17.4 percent in 1999, and it was not until 1999 that some measure of recovery occurred. In other words, it was difficult to predict Hong Kong's economic trends even within this relatively narrow window, which is in itself a major cause for concern in an economy that had been so strong for so many years.[10]

There is no doubt that Hong Kong experienced steady economic growth before 1997, even though the (then) colony's average growth was slower throughout the 1990s than the average for the other newly industrialized economies, especially Singapore (although all but Taiwan appeared to lose ground in 1998).[11] Growth at these rates implies that people's livelihoods were improving, and this is especially true if we look at the social indicators of development used by the United Nations in the calculation of its *Human Development Report* (e.g., life expectancy, infant mortality, and per capita availability of doctors and nurses).

The slowdowns and Hong Kong's negative turn in economic fortunes after 1997 suggest, however, that for many Hong Kong residents, the immediate post-handover period was not a prosperous time. This observation needs to be viewed in light of a marked trend toward domestic neoliberalism that was occurring at the same time, with the Hong Kong SAR government attempting to reduce the financial burden of social welfare provision. In announcing his economic plans in mid-2000, Hong Kong's chief executive Tung Chee-hwa made it clear that he would be looking for a healthy contribution from the voluntary services sector, in part to take the sting out of the declining role of the state in financing welfare services.

The most visibly evident economic impacts were the rising rates of unemployment and the wage cuts in the years after the handover. In response to the new circumstances, many small businesses closed down or downsized considerably in the years after the handover, and the jobless rate reflected this: with unemployment increasing from 2.2 percent in 1997 to 6.2 percent in 1999, then growing again to 7.3 percent by 2002, and appearing to increase beyond that level in 2003 (see table 11.3). As bad as these statistics looked, the actual situation may have been even worse because although many people were able to get new jobs, in most cases they faced cuts in both salaries and benefits. As Mok and Lau (2002) showed, in fact, the real index of pay per person across all industry sectors was 123.9 percent in 1992, which had fallen to 98.2 percent by 1996 and increased again in 1998 to 105.7 percent; it was 112.5 percent in 2000, 114.6 percent in 2001, and 118.0 percent by 2002. In other words, Hong Kong was showing a steady increase in real wage levels in the fist few years after the handover but had not been able to return to the level of the early 1990s.[12]

One of the most telling outcomes of all this was a steady increase in income inequality in Hong Kong all through the 1990s (Li 2001). Although there are no data available for the period after 1996, it is commonly believed that the trend has continued into the new millennium (Mok and Lau 2002). The *World Development Index for 2002*, a World Bank publication, reports that Hong Kong's Gini coefficient, a traditional measure of income inequality, stood at 52.2, which placed it as the seventeenth most "unequal" country in the world, on a par with such places as El Salvador

Table 11.3. Statistics on Labor Force, Unemployment, and Underemployment

| Period | Labor Force | | Unemployed (1,000s) | Unemployment Rate (not seasonally adjusted) (%) | Underemployed (1,000s) | Underemployment Rate (%) |
	No. (1,000s)	Percentage Change Over the Same Period in Preceding Year (%)				
1997	3,234.8	2.3	71.2	2.2	37.1	1.1
1998	3,276.1	1.3	154.1	4.7	81.8	2.5
1999	3,319.6	1.3	207.5	6.2	96.9	2.9
2000	3,374.2	1.6	166.9	4.9	93.5	2.8
2001	3,427.1	1.6	174.8	5.1	85.5	2.5
2002	3,487.9	1.8	255.5	7.3	105.2	3.0
2003	3,465.8	-0.2	275.2	7.9	121.9	3.5
2004	3,512.8	1.4	239.2	6.8	114.3	3.3
2005	3,534.2	0.6	197.6	5.6	96.3	2.7
2006	3,571.8	1.1	171.1	4.8	85.3	2.4
2007	3,629.6	1.6	145.7	4.0	79.2	2.2
2008	3,648.9	0.5	130.1	3.6	69.0	1.9
2009	3,676.6	0.8	196.7	5.4	86.4	2.3
2010	3,682.5	-0.3	161.8	4.4	71.0	1.9

Source: Hong Kong Department of Statistics; see http://www.info.gov.hk/censtatd/eng/hkstat/fas/labour/ghs/labour1_index.html.

and Papua New Guinea (a Gini coefficient of 0 indicates perfect equality and 100 indicates absolute inequality). The poorest 20 percent of the population in Hong Kong
earned just 4.4 percent of the total national income, compared to the top or richest
20 percent, who earned 57.1 percent of the national income.[13] Some local scholars
have suggested that one reason for this was that people with information technology
skills who are laid off may be able to find new jobs that pay reasonably well, while
those in less skilled categories may not, especially those in the traditional manufacturing sector. This suggests that Hong Kong's fundamental economic restructuring has
put those people who relied on the traditional sectors of the economy at a particular
disadvantage.

In the first years of the new millennium, local newspapers carried many stories
about workers who were laid off from the industrial sector and who have been forced
to take part-time jobs, for example, at fast-food establishments, to try to make ends
meet (*Ming Pao Daily*, January 21, 2000). Mok and Lau (2002) also show that between 1996 and 1999, the only income group in Hong Kong to experience real growth
in income was the top quintile (114), with all other groups showing decreases that
were proportionately greater the farther down the pay scale. By the end of the first
decade of the new millennium, the economic situation in Hong Kong appeared to
have worsened considerably, especially with the publication of two reports, one local
the other global. In 2009, the UN *Human Development Report* announced that Hong
Kong had the highest income inequality among the world's most advanced economies,
with the wealthiest decile (10 percent) of the local population earning over a third of
the city's total income, while the bottom decile earn less than 2 percent.[14] In the same
year a study published by the Hong Kong Council of Social Services reported that
the number of people living below Hong Kong's poverty line had reached a record
number, with about 1.23 million (17.9 percent of Hong Kong's 7 million residents)
officially defined as "poor" (Earth Times 2009). The study found that in the first six
months of 2009, there were twenty thousand more people living in poverty compared
to the same period in 2008.[15]

Three local scholars recently published the findings of their research, attempting to account for such a glaring rise in inequality in a place traditionally associated
with free spending and massive wealth (Lee, Wong, and Law 2007). As they observe,
discussions of poverty and inequality around the world invariably focus on the underdevelopment of industrial capitalism, but what we are now seeing in such places as
Hong Kong (and Singapore according to the UN *Human Development Report*) is a new
version of poverty induced by the development of the latest phase of global capitalism that emphasizes the role of knowledge and information and the global financing
needed for capital accumulation and profit generation. This new phase of capitalism
represents a transformation from country-based economic systems to city-based systems, and the authors indicate that most of the advanced cities around the globe are
experiencing "duality," a situation in which incomes at the top continue to increase
while those at the bottom continue to fall. This appears to fit well with the situation
in Hong Kong since the handover, which has resulted in many constituencies calling
for immediate action, including the establishment of a minimum wage in the SAR.

One Country, Two Systems?

On the issue of whether Hong Kong would be allowed to maintain its political and economic system after the handover, opinions are sharply divided, depending on who is talking, but there was enough concern by 2003 to ring some bells of warning. The Chinese authorities heaped one tribute after another onto Hong Kong after 1997 in an attempt to reward the performance of Tung Chee-hwa as the chief executive. Vice-premier Qian Qichen, for example, the highest official of the People's Republic of China (PRC) responsible for the Hong Kong SAR, said in 2002 that the Tung administration had chalked up "a very great achievement in implementing 'one country, two systems'"; and the official Xinhua news agency noted that during first five years after the handover, the people running Hong Kong "had won the respect of the world" (Lam 2002).

Some local journalists were keen to show the other side of this picture, and one of these was Willy Wo-Lap Lam, formerly the bureau chief responsible for China for the *South China Morning Post*. For newspaper editors and journalists working regularly in China, there is obviously a certain amount of risk attached to being too openly critical of the current regime and its policies, and since the 1997 handover, there have been fears—and much debate—about Beijing's increasing interference with media freedoms in Hong Kong. Around the time of the handover, for example, the *South China Morning Post* was generally considered to be a relatively liberal or left-leaning newspaper in the Western understanding of that term.[16] It appeared to have enough editorial independence to be mildly critical of the regime in Beijing and did not seem to avoid reporting on events and running editorials that might be considered politically sensitive by the state leaders. In reality, it is always extremely difficult to assess the political leanings of a specific newspaper, and in the absence of any reliable data, it is difficult to judge the extent to which the *South China Morning Post* might now be maintaining a significant trend in the other political direction, but judging from what little can be scraped together, there were some worrying trends in the first years after the handover. Many *South China Morning Post* readers in Hong Kong were openly criticizing the paper for excessive "cozying up" to Beijing, especially after the paper was bought by Robert Kuok Hock Nien, a Malaysian businessman who is reported to have close ties to the PRC. Of particular concern was Willy Wo-Lap Lam's resignation in November 2000, which was interpreted in some quarters as a prime example of Beijing's "tightening the screws."[17] On the broader issue of media freedom in Hong Kong, additional concern was voiced after PRC president Jiang Zemin's furious outburst in Beijing in November 2000, which was filmed and replayed in Hong Kong (but not on the mainland). In this outburst, Jiang lambasted the Hong Kong press for being far too eager to criticize the regime, and "too naive" in their questioning of him about the methods to be used to elect (select) a new chief executive for the Hong Kong SAR.[18]

Lam suggests that there are several clear indications that Beijing has tended to favor the "one country" side of the principle rather than the emphasis on the "two systems." Much of the problem, according to Lam, stems from the accountability or ministerial system that Chief Executive Tung Chee-hwa introduced to centralize power in his hands and to further marginalize the *Democrats*, a generic term for all elected politicians who are in favor of a faster pace for establishing democracy. Under the ministerial system, which was a product of close consultation between Tung Chee-hwa's office and Beijing, the chief executive is empowered to appoint fourteen policy secretaries who report to him and serve at his pleasure. Until this development, the secretaries were drawn mostly from the ranks of civil servants and were supposed to be politically neutral. The new system is interpreted (by Lam at least) as evidence of Beijing's distrust of senior civil servants, most of whom worked their way through the ranks under British rule. This was indirectly confirmed by Vice-premier Qian Qichen, who told a Hong Kong television station that because Tung Chee-hwa was obliged in mid-1997 to retain former governor Chris Patten's senior staff, he insisted on running what amounted to a "one-man show" during much of his first term. Qian added that Beijing was confident Tung would be able to do better in his second term because under the accountability system, he would be in a position to appoint his own "people."

In a roundabout way, the vice premier was repeating the familiar charge that certain top civil servants, perhaps out of residual loyalty to the British, were reluctant to work with Tung Chee-hwa (and Beijing). Tung's new team, announced in the summer of 2003, included a coterie of "like-minded businessmen" and the usual collection of "pro-Beijing" individuals. This trend was even more obvious in the full Executive Council, or cabinet, whose membership was expanded from twelve to twenty. Apart from the fourteen secretaries, Tung appointed five cabinet members who enjoyed Beijing's full trust. Prominent among these were two leaders of influential political parties: Tsang Yok-sing, the chairman of the pro-Beijing Democratic Alliance for the Betterment of Hong Kong, and James Tien, who headed the pro-government business-oriented Liberal Party. Since the two parties now controlled at least eighteen votes in the Legislative Council—less than half of whose sixty members are popularly elected—Tung was guaranteed control over the legislature.

The vice chairman of the Hong Kong Democratic Party, Yeung Sum, insisted that the Democrats were not intimidated by these changes: "Our image as the opposition party will become more clear-cut." Willy Lam, however, felt the new development would make it even easier for Tung Chee-hwa and Beijing to sideline pro-democracy politicians. In his interviews with the Hong Kong media, Vice-premier Qian tried to dampen expectations about a faster pace for the establishment of democracy in Hong Kong when he pointed out that "to promote democracy in Hong Kong, one cannot have Hong Kong emulating the systems of other regions," which did not bode well for anyone expecting to see Western-style elections to identify Legislative Council members (or the chief executive) in the near future.

In the middle of the new decade, this impression was reinforced by a self-initiated decision made by the Standing Committee of the National People's Congress in Beijing to preempt local debate on electoral reform and rule out universal suffrage in the 2007 Hong Kong chief executive and 2008 Legislative Council elections. The committee argued that its decision was based on the apparent lack of consensus on this issue in Hong Kong and on Hong Kong's relatively short history of exercising democracy. The Hong Kong SAR government expressed its support for the decision, claiming that Beijing was trying to protect Hong Kong's stability and prosperity for the future.

Most pro-democracy supporters criticized the National People's Congress and the Hong Kong government, suggesting that their response worked to undercut the "one country, two systems" principle, thereby damaging the prospects for or pushing back the date when Hong Kongers would be ruling Hong Kong. The issue featured largely in Legislative Council debates throughout 2004, and a demonstration organized in protest saw an estimated 300,000 people gathering in Victoria Park on July 1, 2004.

A Legislative Council election in September 2004 was generally ruled to have been free and fair, but the pro-democracy forces were unable to record any significant gains in the number of seats they controlled. Tung Chee-hwa resigned as chief executive in March 2005, and the existing chief secretary, Donald Tsang, was selected as his successor. Two governmental reform proposals were introduced but not passed in late 2005 when pro-democracy legislators rejected what they felt was excessive tinkering with the laws governing the election of the chief executive and the size and makeup of the legislature.

Conclusion: Continuing Political Theater in Hong Kong

In 2010, it appeared that a breakthrough in the political situation was imminent, one that might result in a more "democratic" legislature, though, as many locals pointed out at the time, it is wise not to expect too much too quickly after so many false starts. A compromise bill was introduced that would add ten seats to the legislature, five of which would be elected directly by the people, in free elections. Members of the city's District Councils would be able to run for the remaining five seats, which meant that in principle all ten of the new seats would be elected. This appeared to some political commentators as a major break with the past.[19] China's leaders in Beijing appeared to be allowing (or at least ignoring) the compromise, a rare change from their usual stiffness and unwillingness to compromise.[20]

Chief Executive Donald Tsang told the press that he thought the compromise was a major step forward. As he pointed out, "After two decades of protracted argument over constitutional reform, some rapid and encouraging changes have taken place over the last few months." Unfortunately, the pro-democracy movement was split over the compromise, so much so that "scuffles" were reported at a rally held a few days before the bill was introduced. Some among the camp were happy to see progress, however glacial it may be, but others felt that the proposal did not move fast enough or go far enough to introduce real democracy to Hong Kong.

As Leo Ou-fan Lee writes in his homage to the city, political struggle in Hong Kong is different from the struggle virtually everywhere else. People take to the streets, mock their rulers in the press, or take them to court, all of which occurs under the noses, as it were, of the Beijing government. The Communist Party's opposition to radical change in the former territory has given politics there what one reporter calls "the air of a theatrical set-piece . . . [in which] everybody knows his part."[21] As Lee suggests, politics, like almost everything else in Hong Kong's public realm, can be construed as another form of "play." To illustrate this and to show how well everyone "plays" the game or acts in the "plays," Lee described the U.S.-style televised debate in 2006 between Chief Executive Donald Tsang and his main rival from the Civic Party, Alan Leung. The results of a poll of more than eight hundred viewers showed that Leung was the clear winner in the debate; but the majority of the same people polled said that they would, if they could, vote for Tsang, who, as Beijing's choice for the position, was the assured winner. In true theatrical fashion, Lee reports, Tsang went on television afterward and tearfully thanked the Hong Kong people for their support, in response to which a columnist writing for the mass circulating newspaper *Apple Daily* remarked, "I was never allowed to vote for you, why thank me?" (Lee 2010, 228).

In spite of the theatrical nature of the political struggle in Hong Kong, incremental improvements can be measured in the post-handover period. As noted earlier, the mass demonstrations in 2003 caused the government to drop its antisedition law (Article 23). In 2005, the democrats were publicly frustrated about the lack of progress toward universal suffrage, promised by the Basic Law (Hong Kong's constitution set out during the negotiations for the handover), but after Tsang was "reelected" as chief executive in March 2007, Beijing indicated that it might consider allowing the direct election of the chief executive as early as 2017 and for the Legislative Council in 2020.

In 2010, "Long Hair" Leung's League of Social Democrats and the Civic Party, made up mainly of barristers, proposed that a legislative councillor in each of Hong Kong's five geographical districts should resign, generating by-elections in which each would recampaign on the single issue of universal suffrage. They intended the by-elections—assuming that they went in the right direction—to send a "resounding popular message" to Beijing. If that strategy failed, the proposal called for all of the "democratic" legislators to resign en masse. The plan was a huge gamble for the "democrats," and it appears to be splitting the pan-democratic faction, with some of the "young turks" calling their more established partners "traitors." All of this, naturally, plays directly into hands of the leaders in Beijing. As one democrat complained, "For 20 years Beijing has wanted to divide us . . . now we're destroying the democratic movement without Beijing lifting a finger."

Questions for Discussion

1. What are some of the most obvious concerns the people of Hong Kong had in the run-up to the handover in 1997, and to what extent have those concerns come to pass?
2. Why (and how) did Hong Kong, as a small territory with effectively no resource base, become such a beacon of global capitalism by the end of the twentieth century?
3. Why did so much of Hong Kong's industrial base shift to the Chinese mainland in the last two decades of the twentieth century, and what (thus far) have been some of the positive and negative repercussions of this relocation?
4. Many contemporary human geographers focus attention on the way that changes in identity both influence and are influenced by the economic and political changes

that have accompanied the forces of globalization in the second half of the twentieth century. What can you say about the identity of Hong Kong's people, bearing in mind the extraordinary changes they have witnessed during their lifetimes?

5. What are some of the major economic, political, and cultural concerns and challenges Hong Kong people are likely to face in the next few decades of the new millennium?

Notes

1. Some of the advantages Hong Kong has in comparison to China when it comes to innovation in the information technology and high-tech sector include new software, such as apps for iPhone and Android platforms, which have penetrated deeply into the Hong Kong market. High population density makes universal wireless and broadband a reality in Hong Kong, and U.S. broadband seems glacial in comparison. Setting up a business in Hong Kong is far easier and quicker than in mainland China, and bank accounts can be opened in a matter of days, compared to over a month in China. In addition, Facebook is very popular in Hong Kong, but it is still blocked in China, and in general terms the lack of censorship makes everything seem easier. In mainland China, the introduction of new mobile services involves political rather than business decisions, whereas in Hong Kong, mobile operators can introduce new services for consumers as and when they like. With no Internet censorship, the services that sell well in the West are also common in Hong Kong, with few if any restrictions on their content (see http://blogs.forbes.com/china/2010/05/11/why-hong-kong-is-chinas-new-tech-hub).

2. See also Li and Lo (1993) for a more detailed discussion of the development of Hong Kong's economy during this period. Some of the statistics quoted here are taken from Schiffer (1991).

3. For further details, see the database website http://rru.worldbank.org/DoingBusiness (accessed March 1, 2006).

4. In 2004, the Cato Institute published its *Economic Freedom of the World* index, which measures the degree to which the policies and institutions of different countries are supportive of economic freedom. The cornerstones of economic freedom are considered to be measurable by such indicators as personal choice, voluntary exchange, freedom to compete, and security of privately owned property. Thirty-eight components and subcomponents are used to construct a summary index and to measure the degree of economic freedom along five dimensions: 1) the relative size of government, 2) the legal structure and protection of property rights, 3) access to "sound" money, 4) facilities for international exchange, and 5) regulation of the economy (see http://www.cato.org/pubs/efw [accessed March 1, 2006]). Hong Kong has the highest rating for economic freedom on this scale, scoring 8.7 of 10, closely followed by Singapore at 8.6, with New Zealand, Switzerland, the United Kingdom, and the United States tied for third with ratings of 8.2. Among the other top ten nations are Australia, Canada, Ireland, and Luxembourg. Other large economies and their rankings are Germany, 22; Japan and Italy, 36; France, 44; Mexico, 58; India, 68; Brazil, 74; China, 90; and Russia, 114.

5. With no pun intended, one is tempted to say that the effects on air quality in Hong Kong are clearly visible in the sense that on many days very little at all is clearly visible, as indicated in a series of headline articles in the *South China Morning Post* in October 2004 and in a series of photographs showing the decreasing visibility from one side of the much-vaunted Hong Kong harbor to the other. Although a sizable proportion of the blame for this situation is attributed to the polluted air coming in from the north, in the past few years it has become evident that at

least as much of the problem stems from local sources of pollution: emissions from motor ve-hicles, power plants, and industrial and commercial processes within the SAR (see Wong 2002).

6. I attribute this opinion to Ronald Skeldon, formerly professor of geography at Hong Kong University (personal communication, 1995).

7. Although these statistics were often quoted in newspapers and by local politicians (and even some academics), it is impossible to verify them, mainly because China does not appear to keep or publish statistics about the ownership of all businesses.

8. It seems reasonable, in light of the fact that the handover took place fourteen years ago, that the circumstances under which the transfer of Hong Kong to mainland sovereignty came about be omitted from the current edition of this book. The interested reader can go back to the original version for a brief outline of the process.

9. In spite of their individual differences in terms of political orientation and preferred strategies for development, Wong's informants almost universally identified "human-capacity building" as the key to China's development, which involved solving the illiteracy problem of mainland China and also creating civic space for the empowerment of grassroots popula-tion groups. Wong suggests that these two formulations of China's development problem are premised specifically on the Western ideology of liberalism, an idea that they had been able to nurture in the relative freedom of colonial Hong Kong (Y.-L. R. Wong 2002).

10. Some data sets (e.g., see table 11.2) show that Hong Kong's economy grew by more than 10 percent in 2000, but the global economic slowdown of 2001 hit Hong Kong very hard, along with most other countries in the region, and real GDP growth in 2001 was only 0.5 percent. There was then something of a recovery after that, with real GDP growth in 2002 reaching 1.9 percent.

11. The steady economic growth rate in Hong Kong before 1997 helped it earn the label "the world's only industrial colony" (Mok and Lau 2002, 111). More significant, Hong Kong's per capita GDP increased from HK$106,401 in 1991 to HK$120,540 in 1994, to HK$196,565 in 2000, but then fell to HK$194,969 in 2001 and again to HK$189,656 in 2002 (see http://www.info.gov.hk/censtatd/eng/hkstat/fas/nat_account/gnp/gnp1_index.html).

12. Data from the Hong Kong Census Department show that the biggest losses in the actual value of wages over this period were in the manufacturing sector, which fell from 124.5 in 1992 to below 92.4 in 1998, following with a climb again by the early 2000s to 115.8 in 2002. Workers' wages in the wholesale and retail trades followed basically similar patterns dur-ing this period.

13. The United States, for comparative purposes, had a Gini coefficient of 40.8 in 2002, with the poorest 20 percent of its population receiving 5.2 percent of total income and the richest 20 percent receiving 46.4 percent; see http://www.infoplease.com/ipa/A0908770.html (accessed March 1, 2006). The data for 2002 are reported in the World Bank's publication *World Development Indicators 2002*, vol. 1; see http://econ.worldbank.org/external/default/main?pagePK=64165259&theSitePK=469372&piPK=64165421&menuPK=64166093&entityID=000094946_0210120412542.

14. The UN Development Program uses the Gini coefficient to measure income distribution: the higher the coefficient, the greater the income inequality. In the report in question, Hong Kong's Gini score was reported to be 43.4 compared to Singapore, in second place at 42.5, and the United States, in third place at 40.8 (UN Development Program 2010; the data in question are reported in table M, 196–98; see http://hdr.undp.org/en/reports/global/hdr2009/chapters).

15. UN Development Program (2010). The study defined poverty as earning half or less than half of the average monthly income. For a single person, the average monthly income is 3,300 Hong Kong dollars (US$425), 6,750 dollars for a two-person family and 12,650 dollars

for a family of four or more members. The study found that 32 percent of those aged over sixty-five fell below the poverty line. The study also reinforced the inequality reports published in the UN study, suggesting an ever-widening gulf between the rich and the poor in Hong Kong, even though the city is famous for housing some of Asia's wealthiest families. The income of high earners had risen 34.7 percent over the past twenty years, while those of the low-income groups had fallen by 3.3 percent.

16. In Hong Kong, the terms *leftist* and *left-leaning* identify someone with politically liberal views, in this case, someone who would be in support of further democratization and who would call for direct election (rather than appointment by Beijing) of the Hong Kong SAR's chief executive. The most illustrious and outspoken and certainly the best known of these leftists is known locally as "Long Hair." He wears jeans and a Che Guevara T-shirt, and he refuses to cooperate with the chief executive, Mr. Tung Chee-hwa. Against all odds, Long Hair was elected to the Legislative Council in the 2004 elections.

17. The Lam affair was covered widely by the Hong Kong media. According to statements released by the *South China Morning Post*, Lam resigned after a restructuring of the paper's China Division. He claims, however, that he was never consulted about the proposed change, and in a prepared statement said the management's move was "unreasonable and disturbing." An *iMail* story reported that in June 29, 2000, in a letter to the *Hong Kong Post*, the *South China Morning Post*'s owner accused Lam of "exaggeration and fabrication" after Lam had suggested in a column that a group of Hong Kong tycoons, including Mr. Kuok, were told by Beijing to be more supportive of Chief Executive Tung Chee-hwa. Kuok's denial of this charge was supported by the other tycoons who attended the meeting, but Lam stood by his story.

18. Although many of them may be entirely scurrilous, reports posted on a Hong Kong–based website called *NOT the South China Morning Post* (http://www.ntscmp.com) suggested that not all was well at the *South China Morning Post*. Among the stories causing concern was one about the hiring of an editor from the official English language newspaper of the regime in Beijing, the *China Daily*. Another story reported evidence that some *South China Morning Post* staff members had had their columns seriously censored or edited and that others had been fired or forced to resign, presumably for being too critical of Beijing. There were also reports about the elimination of popular features in the paper, purportedly for being too anti-Beijing. This included the locally popular *World of Lily Wong* cartoons, which lampooned many of the top leaders in Beijing (including the so-called Butcher of Beijing, Li Peng). Similar information appeared on another local online forum in 2000, *Hong Kong IceRed* (http://www.icered.com), when some local readers characterized the *South China Morning Post* as being humiliatingly pro-Beijing and far too uncritical of both China and the current chief executive of the SAR, Tung Chee-hwa.

19. To date, half the members of the Legislative Council are elected from geographical constituencies by the public, with the other half representing "professional constituencies," such as bankers, business owners, lawyers, and other categories; most of these, often because they have business interests in the mainland, are pro-Beijing, while the majority of the people voting choose pro-democracy candidates.

20. The *New York Times* reported that a further sign of the mainland's apparent tolerance for political discussion in Hong Kong was evident when it allowed the uncensored broadcast—in Guangdong province—of an election law debate between Chief Executive Donald Tsang and Audrey Eu, the leader of one of the pro-democracy parties, the Civic Party. Interestingly, the Civic Party met on the same night and voted to oppose the compromise because it did not go far enough in introducing democratic changes. See Bradsher (2010).

21. This was reported in a piece written by Banyan (2009).

References Cited

Asia Child Rights. 2003. Hong Kong's "cubicle kids" highlight growing poverty. http://acr
.hrschool.org/mainfile.php/0145/207.

Banyan. 2009. Suffragette City: Little watched, Hong Kong's democratic fever is reaching its
crisis. *The Economist.* November 26. http://www.economist.com/node/14966209?story_
id=14966209.

Bradsher, K. 2010. Hong Kong moves ahead on reforms. *New York Times.* June 21. http://
www.nytimes.com/2010/06/22/world/asia/22hongkong.html?_r=1&ref=asia.

Castells, M. 1999. *End of Millennium.* Vol. 3 of *The Information Age: Economy, Society, Culture.*
Oxford: Blackwell.

Davies, Ken. 1990. *Hong Kong to 1994: A Question of Confidence.* Special report no. 2022.
London: Economist Intelligence Unit.

Denlinger, P. 2010. Why Hong Kong is China's new tech hub. *China Tracker,* May 11.

EarthTimes. 2009. Record number living below Hong Kong's poverty line, study shows. September 28. http://www.earthtimes.org/articles/news/287481,record-number-living-below
-hong-kongs-poverty-line-study-shows.html (accessed November 2, 2010).

Headley, Anthony J. 2011. Panel on Environmental Affairs Subcommittee on Improving Air
Quality Meeting. *HKU Study on Impact of Loss of Visibility on Mortality risks: A Report of
Regional and Global Importance for the Legislative Council,* January 28. http://www.legco.gov
.hk/yr10-11/chinese/panels/ea/ea_iaq/papers/ea_iaq0128cb1-1188-1-c.pdf (accessed March
15, 2011).

HKSAR Census and Statistics Department. 2004. *Hong Kong Statistics—Frequently Asked Questions.* http://www.censtatd.gov.hk/site_map/index.jsp . [accessed April 27, 2011]

HKSAR Census and Statistics Department. 2004. Hong Kong in Figures. www.censtatd.gov
.hk/.../B10100062011AN11E0100.pdf&title=Hong+Kong+in+Figures&issue=2011+Edition
&lang=1 - 2011-02-23 -[accessed April 27, 2011]

Ho, Y. P. 1992. *Trade, Industrial Restructuring, and Development in Hong Kong.* Honolulu:
University of Hawaii Press.

Hong Kong Census and Statistics Department. 1993. *Hong Kong Social and Economic Trends,
1982–1993.* Hong Kong: Hong Kong Census and Statistics Department.

Hong Kong Monetary Authority. 2002. *Annual Report-Annex and Tables.* www.info.gov.hk/
hkma/ar2002/eng/pdf/Annex_Table.pdf [accessed April 27, 2011]

Lam, W. W. L. 2002. Gloomy forecast for Hong Kong democracy. June 26. http://edition.cnn
.com/2002/WORLD/asiapcf/east/06/25/Hongkong.woes/index.html.

Law, J. 2002. Hong Kong's economy is on the up-and-up. http://tdctrade.com/shippers/
vol23_1/vol23_1_ind11.htm (accessed March 1, 2005).

Lee, K., H. Wong, and K. Law. 2007. Social polarisation and poverty in the global city: The
case of Hong Kong. *China Report* 43, no. 1 (January): 130.

Lee, L. O. F. 2010. *City between Worlds: My Hong Kong.* Cambridge, MA: Harvard University
Press.

Li, K. W. 2001. The political economy of pre- and post-1997 Hong Kong. *Asian Affairs* 28,
no. 2: 67–79.

Li, W., and K. W. K. Lo. 1993. Trade and industry. In *The Other Hong Kong Report.* Hong
Kong: Chinese University Press, 109–26.

Mok, K. H., and M. Lau. 2002. Changing government role for socio-economic development in
Hong Kong in the twenty-first century. *Policy Studies* 23, no. 2 (June): 107–24.

Schiffer, J. R. 1991. State policy and economic growth: A note on the Hong Kong model. *International Journal of Urban and Regional Research* 15, no. 2 (June): 180–96.

Sharif, N., and E. Baark. 2008. Hong Kong as an innovation hub for southern China. http://www.hkjournal.org/commentary/index.html.

South China Morning Post. 1999. February 1. http://www.scmp.com/special/hkfuture/index.asp7.html (accessed March 1, 2006).

UN Development Program. 1998. *Human Development Report.* New York: Oxford University Press.

———. 2010. *Human Development Report 2009: Overcoming Barriers: Human Mobility and Development.* New York: Oxford University Press

Wasserstrom, Jeffery. 2008. *Global Shanghai, 1850–2010.* New York: Routledge.

Wong, Y.-L. R. 2002. Going "back" and "staying out": Articulating the postcolonial Hong Kong subjects in the development of China. *Journal of Contemporary China* 11, no. 30: 141–59.

World in Figures. 1997. London: Economist Press.

CHAPTER 12

Macau (Macao)

The origins of the place known for the last 450 years as Macau, or *Ao Men* in Chinese *putonghua*, began in the somewhat isolated, out-of-the-way coastal margins of the Pearl River Delta near the exit into the estuary of one of the major distributary channels of the large Xi Jiang (West River) at what is believed to have been a tiny fishing village (see figure 12.1). The impetus for the growth and change of this place came about through contact and interaction between Europeans, in this case the Portuguese, and the Chinese. The Portuguese were seeking to extend their influence from South and Southeast Asia, where they had established major trading and religious colonies at Goa on India's west coast and Malacca on the west coast of the Malay Peninsula facing the Strait of Malacca. Their motivation was commercial—to profit from trade in various commodities and with various partners—as well as cultural—for they wished to spread the gospel and gain converts to Christianity.

Cheng (1999) has asserted the two goals were complementary for the Portuguese, who sought to capture some of the trade monopoly that Arab traders had established in the Indian Ocean and nearby areas and to counter the spread of Islam by spreading the gospel and converting locals to Christianity. The Portuguese also were competing vigorously with other European powers, such as the Dutch, who were seeking to extend their own commercial interests along the seaward margins of Asia.

By contrast the Chinese, who had in the fourteenth century sent out a major fleet into the Indian Ocean, had redirected their interests into the interior of their empire, as the Ming dynasty sought to reestablish Chinese dominion over its imperial holdings and to revitalize the idea of Chinese hegemony in those interior regions where tributary relations were essential to maintaining Chinese sway over the far-flung and disparate interior regions of the empire. Thus, its leaders sought to minimize contacts with and the effects of the presence of foreign intruders, who in their view had little to offer a celestial empire, and they insisted on maintaining only the most tenuous of links with places far removed and isolated from mainstream China. In this way, the Chinese signaled their disinterest in and disdain for the foreigners and their intrusive ways.

The Portuguese first arrived in this area in the early sixteenth century, perhaps as early as 1503. According to Charles Boxer (1948), the first Portuguese to visit China was a man named Jorge Alvarez, who in 1514 left a stone memorial, or *padrao*, on an

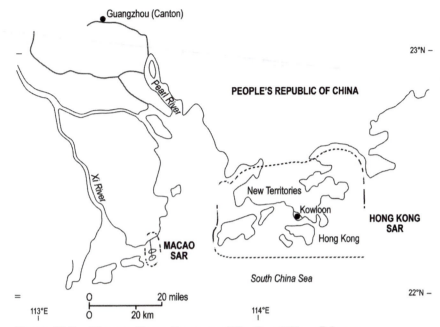

Figure 12.1.   Macau, Hong Kong, and the Pearl River Estuary.

island named Tamao near what is today Macau, and the first Portuguese ships to visit China's great southern port city of Canton arrived at the head of the Zhu Jiang (Pearl River) in 1517. The Portuguese eagerly sought a place to establish a trading post, but the Chinese, despite varying Portuguese efforts at different locations along the coast, would not allow them a permanent foothold. But in 1554, the Chinese finally allowed a temporary trading post on the island of Sanchuan about fifty miles southwest of Macau. This arrangement did not suit the Portuguese, however, and after further squabbling and negotiation, according to most historians and essayists, in 1557 the Chinese allowed the Portuguese to form a settlement on what had formerly been an island but had become attached to the mainland by a narrow neck of land, and this became Macau. Having been settled by the Portuguese in 1557, Macau was the first and oldest outpost of European influence in East Asia, and on December 20, 1999, it became the last such colony in Asia on its return to China as a special administrative region (SAR) of the People's Republic of China (PRC), following the model of the earlier return of Hong Kong to the Chinese motherland in 1997.

A useful way to conceptualize and explain the geography and history of Macau is to use the palimpsest approach, to reconstruct slices of the past geographies of the territory as we seek to present a better understanding of the present (Sauer 1963).

Location and Environmental Setting

Physically, Macau is a tiny peninsula (7–8 km²) jutting off the tip of the southwestern flank of the Pearl River estuary on the southeastern coast of China at the point where

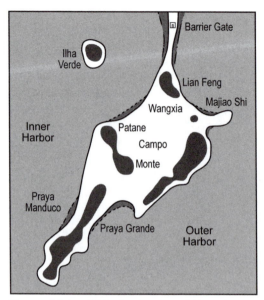

Figure 12.2. Historical map of the Macau Peninsula (adapted from Porter 1996)

several distributaries of the Xi Jiang empty into the South China Sea. The territory also claims two small islands (Coloane and Taipa) just south of the peninsula, to which Macau today is linked by two bridges. Today the total area of the territory is roughly 30 km²; this includes the peninsula, the two islands, and a considerable amount of created fill land and a new island created from fill. Macau may be viewed as one leg or foot on the western, or left, side of a regional triangle of which Canton (Guangzhou) would be the apex at the north and Hong Kong, across the Pearl River estuary, would be the right leg or foot on the eastern side (see figures 12.1 and 12.2).

All three cities lie just within the northern margins of the tropics, although their location on the edge of the huge Eurasian landmass brings cooler weather during the winter months; frost is rare. The climate regime is monsoonal with precipitation mainly during the warm summer months. About 2,000 mm of rain falls per year on average, some of it associated with the annual arrival of typhoons, the great low-pressure tropical storms of Asia's Pacific coast. The growing season is year-round.

The surface terrain and landforms suggested by the descriptive name Pearl River Delta offer a key to the nature of the region's landscape and indeed are the reason for the early settlement and development of this region. This is one of China's major del-taic plains, an area of low-lying, alluvial floodplains characterized in ecological terms by high energy transfers in the natural systems and a considerable degree of biodiver-sity that has rendered the area very attractive for human settlement and exploitation of its natural environment and its rich and constantly renewing resources. The deltaic plain, however, is broken by low but rugged uplands and mountains composed of granite and other igneous and metamorphic rock structures and material. The adjacent coastline is highly indented with numerous islands and coves, a fact that provides safe and convenient harbors and offers great opportunities for fishing, shipping, smuggling, and, formerly, piracy. It is therefore not surprising that South China and the Pearl River estuary became the center for early sailing, shipping, and commerce in China and indeed was one of the oldest centers of China's contact with the outside world, maintaining a history of trade and shipping that dates back more than one thousand years (Edmonds 1989).

Rise of Macau

Macau's origin and importance link it to the age of European exploration and the extension of the Europeans' desire to trade and to spread the gospel, a desire that drove the advance of that exploration to East Asia. It was the farthest permanent extension of Portuguese power and influence, and it was the final point in the Portuguese–Asian Empire that once extended from Goa on the Indian subcontinent to Malacca on the Strait of Malacca and finally to the China coast, though efforts were made to penetrate Japan, efforts that were in fact orchestrated largely from and based in Macau. Macau's story begins in the mid-sixteenth century with the arrival of Portuguese ships seeking to trade with China via Canton and the Pearl River region. These foreigners sought a permanent and commodious post to enhance their commercial efforts, yet the Chinese did not want them as permanent guests. Finally, an adequate if not convenient location was allowed them by the Chinese, as noted above, in an effort to keep the foreigners at arm's length and in a place as far and as isolated as possible from the larger city of Canton and the more civilized parts of the Middle Kingdom. This small, rocky peninsula had only a tiny fishing village. The peninsula was connected to the mainland by a sandbar that eventually formed a narrow isthmus and on which was erected in 1573 a barrier gate to separate the foreigners from the Middle Kingdom.

A visit to Macau should always begin with a stop by the China Barrier Gate, where the visitor will have the opportunity to contemplate Macau's current situation as well as reflect on the last 450 years of European and Chinese relations. At the gate, the visitor can view a series of mosaics that depict the growth and spatial development of Macau, from its earliest encounters with the West to the present. The visitor may also look through the gate and across to the bustling neighboring city of Zhuhai, a special economic zone that has grown explosively in the past twenty years. Perhaps most striking for the visitor will be the steady stream of pedestrians crossing the narrow isthmus as they move between the SAR of Macau and the PRC, just as they have done for 450 years.

In 1582, the Chinese agreed to allow the Portuguese to remain here based on an annual rental payment of five hundred taels of silver, the then valued currency in Ming China. This is an interesting point, as noted by Cheng (1999), who argues persuasively that Macau then was different from virtually all other colonies in Asia because its presence was tolerated and allowed by the Chinese rather than forced on them. Therefore it was not, stricto sensu, a colony. Rather, it was simply a site occupied by the Portuguese and one where legal questions of sovereignty were not raised because the Chinese always assumed that it was their territory and that the Portuguese were in fact staying there at their sufferance. After the rise of Hong Kong in the mid-nineteenth century, Macau was formally ceded to Portugal as a permanent colony with a transfer of sovereignty in 1887 (Cheng 1999, 22–28). This lasted until 1979 when, by mutual agreement, Macau and the PRC described Macau as "Chinese territory under Portuguese administration," a status implying there would be a subsequent return to China, which came to pass in December 1999.

Macau's history and development trajectory as a Portuguese territory in Asia was one of ebbs and flows, linked principally to its role as a center of shipping and commerce for the Europe–China trade and, in addition, to its varying role in spreading the gospel, both Roman Catholic and Protestant. Initially, the settlement grew slowly owing to Chinese neglect and almost disapproval, whereby it remained largely isolated and simply a small garrison that periodically witnessed increased activity when trading was allowed at the neighboring city of Canton upriver. In these early days, the settlement was occupied not only by Portuguese but by others brought by the Portuguese from Malacca, Goa, and even distant Africa. Thus, the cultural makeup of the settlement reflected Portuguese activity and the taking of women from other Portuguese colonies for the Portuguese who had not brought wives. Churches were also constructed. Religion and the effort to spread Roman Catholicism in East Asia and on the Macau Portuguese beachhead were very important to the mission of early Portuguese maritime and colonial activity, as seen in the travels and proselytizing efforts of Fr. Ignatius Loyola and his associates in the Society of Jesus (the Jesuits) and others, such as Fr. Francis Xavier and Matteo Ricci.

The territory prospered and grew in the seventeenth century through its remarkable ability to serve as an intermediary in the China–Japan trade. Because these two East Asian countries were at odds over the rapacious conduct of Japanese pirates, Macau took full advantage of the willingness of the Japanese to permit Portuguese mariners and traders access to Nagasaki, at least until 1640, and in doing so were able to exchange precious products such as Chinese silk for Japanese silver. Profits from this trade provided the wealth that Macau used to improve its infrastructure and construct more and better fortifications and more churches, some of which survive and may be seen today. When the Japanese finally excluded the Portuguese owing to the heavy proselytizing of the Jesuit missionaries, the Macanese continued their silk and precious commodity trade with Manila, wherein the exchanged products were transshipped to the New World. This trade too was eventually declared illegal, but the Macau traders managed through various artifices to maintain some trade, even in the face of hostile or dangerous environments (Boxer 1984).

Eighteenth- and Nineteenth-Century Macau

Continuing with the conceptual framework of the past geographies of Macau, we know that the China trade was the foundation and key to Macau's prosperity and indeed its very survival. This involved not just the Portuguese but also at various times the Japanese, Dutch, French, British, and even the Americans. All these groups competed at times and at other times cooperated out of mutual interest because of a lack of colonial bases out of which these various powers could operate (Berlie 2000; Guillen-Nunez 1984).

The China trade continued erratically into the eighteenth and early nineteenth century, based on the exclusive nature of the Chinese *cohong* system of trade, which was monopolized by and operated out of Canton through an exclusive guild of local traders. Macau had its problems, however, because ships were getting larger and its harbor was increasingly subject to silting from the adjacent Xi Jiang and Zhu Jiang, a problem that ultimately came to be a limiting factor in its ability to compete effectively as a trading port. In 1839, the first of the Opium Wars broke out and was settled several years later by the Treaty of Nanking that ceded Hong Kong to the British. This event was momentous for Macau because the British, with their superior fleet and aggressive merchants, occupied a much larger and deeper harbor (in Hong Kong) that was also well protected from the devastating annual typhoons. Buttressed by their colonial control mechanism, the British largely captured the trade and commerce of the Pearl River Delta, and Hong Kong took over the key function of the European-controlled gateway to China. This new presence led swiftly to the stagnation and decline of Macau's commerce, and the port soon became a moribund backwater.

One way to reconstruct the geography of early and mid-nineteenth-century Macau is through a consideration and study of the visual images of the portrait and landscape painter George Chinnery. Chinnery was a successful English portraitist who painted in British India. He opted to move to the Portuguese territory of Macau and painted there, in what Guillen-Nunez (1997) describes as self-imposed exile, from 1825 until 1852, the year of his death. According to Guillen-Nunez (1997),

> [Chinnery] the artist was captivated by the city's old colonial buildings, squares, and streets, as well as by its lively Chinese population and its culture. The city was by then already divided into the so-called Christian city,

Photo 12.1. Praia Grande Bom Parto Fort (Chinnery watercolor)

Photo 12.2.   View of the Monastery of Saint Francis (Chinnery watercolor)

or the intramural Portuguese quarters, and the Chinese city, partly outside those walls. Equally important, residents of the Christian city included a small but powerful British and American community, providing the painter with a group of potential patrons. (35)

Chinnery painted the local scene—the landscapes, the structures, the boats, the people, and their activities. Through these sketches and paintings, Chinnery the artist revealed the *genre de vie*, the way of life, of Macau's people in the mid-nineteenth century and provided a remarkable visual documentary of the local geography of Macau at that time. An examination of some of his paintings brings key themes into focus. Macau was, on the one hand, a commercial port city of fortresses and churches. These are the important structures seen in photos 12.1 and 12.2 and symbolize the Portuguese colonial influence in Macau despite the absence of true Portuguese legal sovereignty. Chinnery's views of the forts, the front of a Catholic church, and of the Praia Grande (photo 12.1) will all be recognizable to the contemporary resident of or visitor to Macau who knows or studies the city, although the specifics of these scenes have been much transformed in contemporary Macau. These are the tangible expressions and the reality of a foreign presence on Chinese soil in early and mid-nineteenth-century Macau.

Paralleling this reality is the Chinese presence as seen in the still-standing and famous A Ma Temple on the Inner Harbor side of the peninsula, the Chinese shop houses, and the Chinese street scenes with their vendors, sidewalk eateries, workers, and haulers, all depicting the vibrancy of urban street life for the Chinese since time immemorial. In one especially striking image, we see Chinese shop houses and

Photo 12.3.   Macau dwellings with Saint Joseph Church behind (Chinnery watercolor)

street scenes juxtaposed to a Portuguese colonial edifice, a church that reminds us of the challenging duality of nineteenth-century life (especially for the local people) in a bicultural Asian colonial city (see photo 12.3). What emerges as the past geography of mid-nineteenth-century Macau is a city that has a European aspect in its physical appearance but is almost always characterized by the presence of the local Chinese, who are easily and frequently identified in Chinnery's sketches and paintings by their distinctive headgear and hats. The physical, administrative presence in much of Macau was European and Portuguese; the human reality was local Chinese, omnipresent and always visible. This, then, was the geography of mid-nineteenth-century Macau.

Shifting Fortunes, the "Coolie Trade," and Macao's Decline

To counter the decline in trade owing to the growing power of Hong Kong, Macau turned to another, darker aspect of commerce: trade in human beings. Thus began Macau's role in the infamous "coolie trade," the impressing and shipping of contract laborers from the Pearl River Delta to the New World. For Macau, this began around 1846 as trade in human laborers was beginning to be heavily resisted and criticized in other parts of China, such as Amoy. Laborers were recruited from among the poorest Chinese farmers, who were typically duped by agents from nearby cities who then received a commission for each contract laborer, or "coolie," delivered to a way station, or barracoon. These laborers were then sent to waiting ships for transport under unspeakably harsh conditions to places such as Cuba or South America; often they did not survive the journey. This trade was barely better than slavery, and few places in China wanted to be associated with it. Macau, searching for commercial prosperity and already having a reputation for prostitution and other illicit activities, served as a willing participant in this trade, and it continued for more than twenty years to act as a main point for collection and dispatch of coolies headed to the New World. Between 1846 and 1873, Macau accounted for 44 percent of the more than 320,000 contract laborers who were sent out of Chinese ports. There were huge profits to be made in this business, and the Macau traders were indeed guilty of encouraging this nefarious trade in human cargo. Gradually, this human trade became untenable, however, as it was increasingly recognized as grossly exploitative and morally repugnant.

Once this trade declined, Macau's role as a major port diminished rapidly, and it had to turn to other commercial activities to provide an economic base. Its decline continued owing to the growth and dynamism of neighboring Hong Kong, except for a brief hiatus during World War II when it received European and Chinese refugees fleeing Japanese invaders. Macau's economic survival and future prosperity needed a new activity and function, and this gradually evolved into its role as a petty manufacturer and gambling and tourist center, Macau being in all these functions subordinate to and dependent on its association with Hong Kong. French geographer Jules Sion described Macau in 1928 in this way: "*Mais c'est aujourd'hui une fade envasée, une ville morte, remplie de couvents, comme au temps de Camoens. Quel contraste avec Hong Kong!*" "But today it is a harbor silted up, a dead city, full of convents just as in the time of Camoes [the Portuguese poet]. What a contrast with Hong Kong!" (Sion 1928, 137).

People and Society

Macau has always offered a fascinating meeting and mixing of Asian, European, and African peoples—a clear reflection of its colonial heritage. In its early days, there were few women, and the European colonizers who brought women from South and Southeast Asia often took them as wives, quickly creating a Creole or mestizo society. Later, Chinese women entered the cultural mix, and eventually European women did, as well. In the process of this intermarriage and mixing, a Macanese Creole group was created, defined as those of European and Asian or sometimes African ancestry. This group differed from what was seen as a typical resident of Macau. These people created their own Creole or pidgin language, again a mix of Portuguese with various different languages, including local Chinese, Cantonese, and Fukienese. And they remain in Macau today, a small cadre of Macanese somewhat unsure of their own status in a world of changing political allegiances but somehow clinging to a past that has all but disappeared in the rapidly growing and modernizing Chinese city of Macau.

In 1830, Macau was reported to have a population of 12,500, and the Chinese population had already exceeded the number of all other residents. By 1851, when Macau was a leading port for the coolie trade, the city's total population had increased to 26,900. During the nineteenth and early twentieth centuries, Macau's population grew, especially as more Chinese migrants moved in to take advantage of the better job opportunities. Gradually, the social makeup of the city became more Chinese, and Chinese vernacular architecture began to characterize much of the land use and internal structure of the city aside from the fortresses, churches, convents, public buildings, and residential dwellings of the colonial elites.

World War II brought a surge in population as Europeans and Chinese, fearful of the invading Japanese, sought refuge in the neutral city. Many remained here throughout the war, captives in the sense that it was difficult for them to travel but free from fear of incarceration at the hands of the Japanese military. At this time, the city swelled to more than 250,000 but then began to decline after the war with the revival of Hong Kong as a center of British commerce and society.

In the ensuing decades, Macau's social development and population growth in part reflected developments in Hong Kong, but especially in the past twenty-five years, they have proceeded in parallel with events and developments in China associated with the economic reforms of 1978. These reforms were followed by the opening up of China in the 1980s and the establishment of the special economic zones, one of which was the city of Zhuhai. This has galvanized the economy of the territory, and it has also provided the impetus for rapid population growth.

In 1970, the Macau population was 249,000, and this actually declined slightly to 242,000 in 1981. Within a decade, though, it had climbed to 356,000, an increase that mirrored the remarkably rapid development of Zhuhai and an enormous push to expand infrastructure and build high-rise structures as housing units in Macau. The latest census figures (2008) estimated the population of Macau at 552,000, with a birthrate of 0.85 percent, a death rate of 0.32 percent, and a rate of natural increase of 0.53 percent. The total fertility rate had declined to 1.0, however, well below the replacement level, which tells us that much of Macau's recent growth had in fact been driven by migration. Of this total population, it is estimated that 7,000 to 10,000 may be identified as belonging to the Macanese Creole group, although this population is difficult to count with precision.

As the new millennium opens, it is clear that Macau is becoming a Chinese city in culture and society as it is drawn increasingly into the greater Chinese economy and polity. Yet Macau is emerging as a very special kind of Chinese city that is carving out for itself a distinctive economic niche as a recreational and gaming city. As in many cities along China's Pacific coast, there are numerous migrants who live in the city as temporary sojourners. Although it is difficult to know their exact number, estimates place these temporary workers at perhaps 100,000 to 200,000, a quantity that may push the true population of the territory to a figure over 700,000. These migrants are important for they provide labor services in construction, commerce, and low-end service activities (including the sex trade), jobs that locals may in fact avoid or be unwilling to do. Zhuhai next door has a far larger number of these temporary workers, though, as do so many of the cities and towns in the Pearl River Delta. The economy of this region of China, with its very strong links to the global economy, simply could not operate or offer competitive products and services without the labor input of these migrants (Sit et al. 1991).

With its recently completed international airport, Macau has greatly expanded its tourist trade and seeks to become less dependent on the flow of tourists from Hong Kong. The Macau government as early as 1962 had decided to enhance its role as a gambling center and had set about to improve the structure and quality of the monopoly syndicate that had been in control of gambling since the 1930s. A new monopoly license was awarded to *Sociedade de Turismo e Diversoes de Macau* (Macau Tourist and Amusement Company) (STDM). This syndicate was headed by an enterprising local entrepreneur, Stanley Ho, who was to become one of the wealthiest persons in the Pearl River Delta region (Loughlin and Pannell 2010).

Upgraded facilities followed rapidly with new upscale hotels and much higher quality casinos. The new Casino Lisboa was the signature landmark among these improved facilities. Among the more significant changes was the addition of new Western games of chance that provided added diversity to the more traditional Chinese games. In addition, the STDM had agreed to provide funding for infrastructural improvements, such as harbor improvements and high-speed ferry service. New monies were also provided for greater social services to the local population. With increased advertising and more sophisticated marketing to a broader Asian market, Macau's reputation as a major gambling center spread, and business increased. However, in 1966, the Chinese Cultural Revolution broke out, and Macau, immediately adjacent to the mainland, was much affected by the political chaos that followed. The Portuguese colonial authority began to lose its grip on governing Macau. In 1974, the conservative government of Portugal was overthrown, and the Portuguese were on the verge of returning Macau to China, but the Chinese government declined and postponed the return until a more propitious time.

Moreover, gambling and the related "recreational sin" activities of prostitution and so forth do not come without a serious downside because criminals and gangs are attracted to such activities. In the case of Macau, the city has long tolerated the presence of local Triad gangs, a kind of Chinese mafia that grew out of traditional Chinese secret societies. These gangs commit serious crimes, including shootings in the casinos and kidnappings. As Portuguese authority waned and Chinese actors became more powerful, the role of Triads and crime syndicates became more prominent. Triad-related crime in the casinos was common and frequently made front-page stories in the newspaper to the point that Macau's gambling casinos were developing a seedy and questionable reputation. As the handover date for restoration to China approached in 1999, the Chinese government became increasingly involved in cleaning up the criminal activity in Macau and improving social stability and public safety. In December 1999, full political sovereignty of Macau was restored to China, and tighter safety and social stability followed.

The restoration of Macau to China in 1999 was a historic occasion, as this was the last remaining colonial territory in East Asia, and it was returned after more than 450 years of European occupation. The territory officially became an SAR of China with its administrative and legal status modeled on the pattern of Hong Kong, an SAR that had been returned to China two years earlier after 150 years of British control. This allowed for fifty years of governance under a "Basic Law," similar to a constitution, that would permit considerable autonomy for economic and social development.

A New Era for Gaming

The 1999 handover was quickly followed by a government decision to study the gambling industry to see how it might be improved. The old STDM monopoly was seen as outmoded and too powerful. Competition would be good for the industry, and it was decided to allow open bidding from organizations with international gambling experience. Three companies were selected in 2002 and awarded twenty-year contracts to operate casinos. One was a spin-off of the old STDM monopoly, and the other two were American companies from Las Vegas: Wynn Resorts and Galaxy Casinos. A new era was about to begin as the new casinos were to be built as resorts with much-improved hotel facilities and greatly enhanced casinos. "Gaming" was the term now used to describe gambling, and the gamblers were now to be described as "players." In 2004, the first of a number of new casinos opened (Loughlin and Pannell 2010). By 2009, there were thirty-two casinos operating in Macau, and the gross revenues grew from 2003 to 2008 from US$3.6 billion to US$13.7 billion. These gross revenues in 2008 were greater than the combined total revenues of the two largest U.S. gaming markets of Las Vegas and Atlantic City, New Jersey, in the same year. Approximately 48 percent of the gross domestic product of Macau was accounted for in 2008 by the public administrative sector, which included gambling, according to official statistics (National Bureau of Statistics 2009).

The growth in gambling is an extraordinary story, and it has truly transformed the old Macau. New landfill has created a whole new part of the city, Cotai (5.6 km²), between the islands of Coloane and Taipa, and it is here that a number of the new and most glamorous casino resorts have sprung up—Sands, Venetian, Planet Hollywood, and others—while Wynn and the enlarged and much-improved Grand Lisboa continue operations near the Praia Grande. Thus, the urban structure and land use of Macau are being much altered with new highway overpasses and causeways linking the islands with the Macau peninsula to accommodate the great increase in visitors and tourists. From 2003 to 2008, the number of visitors to Macau nearly doubled from 11.9 million to 22.9 million, and more than half come from the nearby PRC, a huge and growing market for Macau's casinos. Between 2000 and 2008, the gross domestic product of Macau nearly quadrupled, most of this due to the impact of gaming and associated recreational activities and tourism (Loughlin and Pannell 2010)

Gambling's extraordinary growth and economic stimulus to Macau and its development would appear to have established this activity as the centerpiece to Macau's future. As noted, Macau's prosperity and fortunes have ebbed and flowed over the years, and the territory has long sought a dependable and steady activity on which to base its survival. Never have its prospects appeared so promising as with the astonishing success of its emergence as perhaps the largest gambling center in the world and with a growing and prosperous China at its doorstep. Even with the appearance of gaming centers in the Philippines, Singapore, and South Korea, Macau has established a strong lead and now well-deserved reputation as Asia's leading center for casino resorts. Its destiny, in large part owing to its proximity to the China mainland, would appear now strongly tied to its role as a gaming and recreation center.

Conclusion

As the Portuguese prepared to conclude their 450 years of occupancy, they attempted to ensure that there would be a lasting remembrance of their stay. This attempt became increasingly controversial because even the vestiges of colonial control were much resented by local Macanese, and those memorials that depicted Portuguese occupiers in perceived acts of violence or subjugation were removed. Despite objections, there are some remaining structures that identify clearly the Portuguese presence, and the colonial imprint in places can still be seen clearly. The remaining fortresses, often on high and prominent locations, as well as the churches and the well-known tourist attraction of the facade of St. Paul's Cathedral, are good examples. The Leal Senado (Loyal Senate) building, the name of which has recently been changed to reflect its new role as the center of municipal government, and an imposing square with its attractive Portuguese colonial-style buildings offer a modern example of a city center that provides human scale and pleasant surroundings for pedestrians. Yet the bulk of the city is rapidly becoming modern, a prosperous and vibrant setting with four-story shop houses in the older center, while many peripheral sites witness a rapid replacement of the older, lower buildings with high-rise apartment houses (owing to the high cost of urban land and the need for high-density living). In this way, Macau is replicating its neighbor Zhuhai. And along the outer harbor of created landfill are many of the large tourist hotels and casinos, which have become an essential part of Macau's present and future.

At the beginning of the twenty-first century, Macau's geography is a fascinating mix of old and new, and, by all accounts, it has been and remains an extraordinary place. It was the first European colonial beachhead and trading post in East Asia and indeed the last, and it offers the historian a somewhat checkered history of trading, proselytizing, gambling, and prostitution, with some fighting, fishing, and manufacturing thrown in, as well as commerce in human beings, as seen in its nineteenth-century coolie trade. The territory has waxed and waned just as the tides ebb and flow on its shores. After thirty years of stagnation following World War II, it is once again prosperous and growing. The look of Macau and its European heritage is quickly being swept aside because its construction boom does not have the time, interest, or financial incentive to maintain the architectural heritage of colonial Portuguese Macau. As we watch the current remarkable and dynamic transformation of Macau, we ask ourselves, what will distinguish this place in the future from any number of other South China coastal cities? Will it be its Portuguese colonial and commercial legacy, preserved in part perhaps for its tourist value or for some other symbolic value, or will it be its distinctive niche as a gambling and recreational city for East and Southeast Asia?

Questions for Discussion

1. In historic terms, how does Macau's role as a European beachhead and eventual colony fit into the larger theme of European colonialism in Asia, and what were the primary goals of the Portuguese in setting up Macau on the doorstep of China?
2. What was the coolie trade, and when and how did it play a role in Macau?
3. What effect did the rise and development of Hong Kong have on Macau, and how has Macau responded in the past half century?
4. What is the future of gaming (gambling) and recreation tourism in Macau, and how is this linked to Macau's proximity to mainland China?

References Cited

Berlie, J. A., ed. 2000. *Macao, 2000*. Oxford: Oxford University Press.

Boxer, Charles R. 1948. *Fidalgos in the Far East, 1550–1770: Fact and Fancy in the History of Macau*. The Hague: M. Nijhoff.

———, ed. and trans. 1984. *Seventeenth Century Macau in Contemporary Documents and Illustrations*. Hong Kong: Heinemann Educational.

Cheng, Christina Miu Bing. 1999. *Macau: A Cultural Janus*. Hong Kong: Hong Kong University Press.

Edmonds, Richard L. 1989. *Macau*. Oxford: Clio.

Guillen-Nunez, Cesar. 1984. *Macau*. Hong Kong: Oxford University Press.

———. 1997. Introduction. In *Georges Chinnery: Images of Nineteenth-Century Macao*. Macao: Macao Territorial Commission for the Commemoration of Portuguese Discoveries, 35–41.

Loughlin, Philip H., and Clifton W. Pannell. 2010. Gambling in Macau: A brief history and glance at today's modern casinos. *Focus on Geography* 53, no. 1: 1–9.

National Bureau of Statistics. 2009. *Zhongguo tongji nianjian 2009* [China statistical yearbook 2009]. Beijing: Chinese Statistics Press.

Porter, Jonathan. 1996. *Macau, the Imaginary City: Culture and Society, 1557–Present*. Boulder, CO: Westview.

Sauer, Carl Ortwin. 1963. *Land and Life: A Selection from the Writings of Carl Ortwin Sauer*, ed. John Leighly. Berkeley: University of California Press.

Sion, Jules. 1928. Asie des moussons. In *Geographie Universelle*, vol. 9, ed. Paul Vidal de la Blache and L. Gallois. Paris: Libraire Armand Colin, 137.

Sit, V. F. S., R. D. Cremer, and S. L. Wong. 1991. *Entrepreneurs and Enterprises in Macau: A Study of Industrial Development*. Hong Kong: Hong Kong University Press.

CHAPTER 13

Taiwan: Enduring East Asian "Economic Miracle"

By any standard, contemporary Taiwan (formally the Republic of China on Taiwan) represents one of the few genuine economic miracles of the modern era. There are many lessons to be learned from Taiwan's experience and from the resilient people who have contributed to this dynamic success story. Success is all the more remarkable for the small, resource-poor island considering its tumultuous past century. The island was colonized by Japan for fifty years, occupied by the Chinese Nationalist Army, and drawn into the Cold War by U.S. containment strategies. Currently, Taiwan still faces a potential if declining threat of invasion by the People's Republic of China (PRC). The relationship between the PRC and Taiwan is extremely complicated—and while the two are technically still at war, the gradual peaceful resolution of their conflict is central to stability throughout the region and world. Paradoxically, Taiwan is currently the largest investor in the PRC, with two-thirds of listed Taiwan companies reporting some level of investment in China. While the numbers are debated, most agree that in 2010, Taiwanese total investment in China ranges between $100 billion and $150 billion. The number of Taiwanese citizens living (more than 180 days per year) in China is estimated at 750,000 to over 1 million (Kastner 2009). Currently, there are 270 regularly scheduled flights between Taiwan and mainland cities each week, and this number may soon increase to over four hundred, as mainland tourists are flocking to Taiwan to explore and visit relatives and economic and commercial ties increase (Chen 2008; Winkler 2007).

Taiwan's position in the international community is unique. In the first decade of the new millennium, Taiwan is as isolated politically as it is integrated economically. The communist government of the PRC will not allow concurrent recognition of the PRC and Taiwan (Republic of China) by other nations and international organizations. As a consequence, only a handful of small nations have formal diplomatic ties with Taiwan, and Taiwan lacks a seat in the United Nations. Taiwan's international status remains unclear, but its economic progress in the past fifty years has been nothing short of spectacular.

Under the cloud of this political uncertainty and in spite of the challenging past, Taiwan continues to prosper, offering many remarkable success stories and much to admire. Full participatory democracy has come of age in the past two decades, with

342

hotly contested multiparty elections held at all levels—from the presidency to leadership positions in cities, townships, and towns. As other nations develop democratic institutions, Taiwan's experience is invaluable. Recent closely contested elections with majority control shifting from long-term Kuomintang (KMT) control to the Democratic Progressive Party (2000–2004) and back to the KMT in the 2008 presidential election reflect a citizenry considerably divided on Taiwan's path for the next few decades. Voting patterns indicate clear splits between those who favor some process of unification with the mainland, those hoping to live with the status quo, and those who prefer independence; unfortunately, the latter is an option unacceptable to the PRC. Of course, elections are about more than cross-straits relations, with economic and environmental issues ranked the highest among voter concerns.

Universal aspects of a modern global economy are matched by the many enduring influences of traditional Chinese culture. Sophisticated international cities such as Taipei and Kaohsiung (Gaoxiong), with some of the best public transportation networks in the world, are some of East Asia's largest and coexist with indigenous settlements seeking greater autonomy and a return to traditional economies. Environmental protection programs have taken off in the past fifteen years. In 2010, 97.8 percent of the population is literate, and the nation's colleges and universities turn out world-class scholars in all disciplines. Further, citizens are all enrolled in the National Health Insurance Program, which provides health care on demand through a mixed system of private and public providers. By any standard, Taiwan is also a vital player in the global economy as the world's seventeenth-largest merchandise exporter and the eighteenth-largest importer in 2009, and Taiwanese-based multinationals are some of the most important high-tech firms in the world (Government Information Office 2009). Taiwan's economy based on gross domestic product (GDP) was the twenty-fourth largest in 2008, largely because of this leading role in advanced information technology (IT) development and manufacturing. Taiwan is also a leader in synthetic textiles, optoelectronics, precision machinery, and biotechnology (for current figures, see Government Information Office 2010). Unlike the economies of much of East and Southeast Asia, excepting the past two years, the Taiwanese economy has continued to grow for much of the past several decades. In 2008, the small island reported a GDP of over US$391.2 billion (per capita GDP of $17,083.00), and in 2009 was the fourth-largest holder of international foreign exchange reserves.

Taiwan's living standards are among the highest in Asia, with a surprisingly equitable distribution of wealth. Certainly, there are very rich and very poor people in Taiwan, but the great majority of its people have shared in Taiwan's economic prosperity and consider themselves middle class, although the income gap appears to have grown under the Ma administration that began in May 2008 (Hsu 2010). Located 160 km off the coast of Southeast China (see figure 13.1) but originally settled by sojourners from Australasia, Taiwan is now densely populated by the descendants of Chinese migrants. Out of a 2009 population of just over 23.069 million, roughly 84 percent claim Taiwanese ancestry, and 14 percent claim more recent mainland roots. Indigenous peoples and foreigners constitute the final 2 percent.

Given the island's relative size, Taiwan might be expected to be cohesive and homogeneous, and in many ways it is a small and accessible place. Yet its accessible

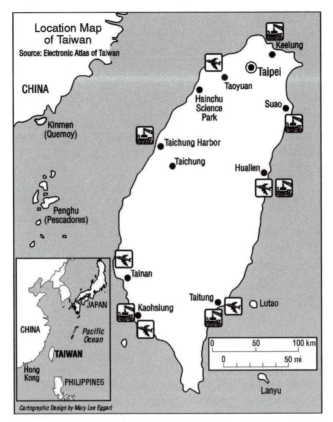

Figure 13.1. Location map of Taiwan (*Electronic Atlas of Taiwan,*** http://sites.inka.de/sites/kajetan/Html/Taiwan.html)**

location and its past century of conflict have made Taiwan an unusual and highly distinctive place within the Chinese cultural realm. The island's role as a refuge for Chinese nationalists (KMT; also called Guomindang) following communist ascendancy on the mainland in 1949 represented only the last of many diverse waves of migration by myriad ethnic groups from Australasia and southern and eastern China. These migrations over the millennia have created a diverse cultural mosaic. At the present time, the island to some extent represents a microcosm of mainland China with major populations of ethnic Taiwanese and Hakka, fourteen self-identified groups of indigenous peoples (Ami, Atayal, Bunun, Kavalan, Paiwan, Puyuma, Rulai, Saisiyat, Sakizaya, Sediq, Thao, Truku, Tsou, and Yami; see figure 13.2), and large numbers of Chinese originally from Fujian and Guangdong provinces. Of course, growing trade and manufacturing ties with the mainland have resulted in many families with members traveling back and forth across the straits on a weekly or monthly basis, just the most recent form of culture exchange to come to Taiwan (China Forum 2010).

Initially, one might be surprised by so much diversity and so many contrasts, for the island is only the size of the U.S. states of Maryland and Delaware combined.

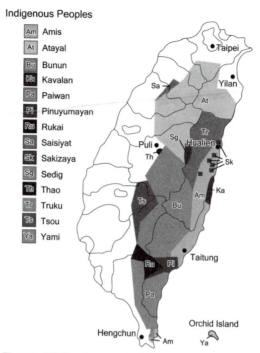

Indigenous Peoples

Am	Amis
At	Atayal
Bu	Bunun
Ka	Kavalan
Pa	Paiwan
Pi	Pinuyumayan
Ru	Rukai
Sa	Saisiyat
Sk	Sakizaya
Sg	Sedig
Th	Thao
Tr	Truku
Ts	Tsou
Ya	Yami

Figure 13.2. Taiwan's indigenous people (Government Information Office 2009, http://www.gio.gov.tw/taiwan-website/5-gp/yearbook/ch02.html#People; Chen 1999)

However, while the island is small, there is also a considerable degree of environmental diversity that is reflected in population densities ranging from below 100 persons/km² in the mountainous eastern county of Hualien to over 9,000 persons/km² in Taipei (see figure 13.3).

Taiwan and its future are linked to China by culture, history, politics, and, most important, economics. Contemporary political and economic relations between China and Taiwan indicate that Taiwan's destiny will remain closely tied to that of Greater China. Additionally, the long-standing security relationship between Taiwan and the United States also ensures that the ultimate resolution of the relationship between Taiwan and the mainland will be one of the decisive foreign policy issues in the near future of the United States.

Taiwan's Physical Environment

Like most of the islands that rim the eastern flank of the Eurasian landmass, Taiwan is mountainous. Lying 160 km east of the Chinese province of Fujian across the shallow Taiwan Strait, the egg-shaped island resembles a giant block tilted up on its

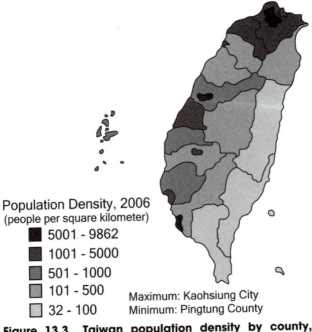

Population Density, 2006
(people per square kilometer)

■ 5001 - 9862
■ 1001 - 5000
■ 501 - 1000
□ 101 - 500
□ 32 - 100

Maximum: Kaohsiung City
Minimum: Pingtung County

Figure 13.3. Taiwan population density by county, 2000 (http://www.gio.gov.tw/Taiwan-website/5-gp/yearbook/chpt02-1.html)

eastward margin to form a sharp dip into the Pacific. A series of north–south-trending ridges form the interior of the island (the Central Mountains) and slope more gently westward, blending into a series of flanking basins and plains (see figure 13.4). These plains are widest in the Southwest and taper off toward the North. About two-thirds of the island is composed of rugged uplands, and numerous peaks crest above 3,000 m (Hsieh 1964). The cross-island highway is popular among hikers from all over the world, and the scenery in the mountains is some of East Asia's best.

Climatic Patterns and Implications

Taiwan's location straddling the Tropic of Cancer and the high elevations of its main mountain range result in a distinctive climate that played an important role in its history and allows the production of a remarkable diversity of agricultural products. Taiwan lies on the northeastern margin of the tropics and is heavily influenced by East Asian monsoon patterns. In general, weather is influenced by northeasterly winds during the winter and southwesterly winds during the summer months. In both cases, these winds are marine influenced and bring considerable precipitation. Many locations report 2,560 mm (~101 inches) or more of rainfall annually. The northern tip of the island receives much of its rain in the winter. In contrast, the western and southern parts of the island, lying in the winter orographic shadow of the major mountain chain that runs the length of the island, have a comparatively dry and pleasant autumn and

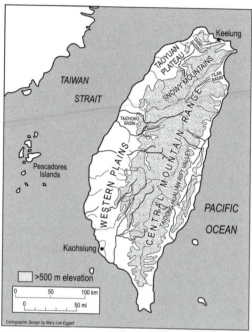

Figure 13.4. Topographic summary of Taiwan. (Pannell and Ma 1983, 295)

winter and experience significant precipitation in the spring and summer when it is needed for intensive rice cultivation. Frost and snow occur only at the higher elevations in the center of the island.

Characteristic late-summer and autumn typhoons annually sweep across Taiwan. These great tropical storms, originating east of the Philippines, bring enough rainfall to generate a secondary seasonal peak of precipitation. Unfortunately, the intensity of the accompanying winds and rain still wreak considerable damage. Development has intensified both the economic losses and the physical effects of these storms (flooding and landslides) as deforestation, urban sprawl, and uncontrolled construction of legal and illegal luxury hillside housing increase the financial losses associated with these storms.

Most of the island's rivers are short and swift flowing. None is more than 105 km in length, and all are useless for navigation except by shallow-draft vessels. Still, many of these streams possess considerable hydropower potential. Hundreds of dams, beginning with Chinese efforts during the Qing dynasty, have been constructed to harness this energy and water. Most projects are multipurpose and are generally tied into water storage, irrigation systems, and flood control. The river and channel management systems are vital to the agricultural plains and basins of western Taiwan. Further, the resulting reservoirs are now critical to the supply of drinking water to Taiwan's major cities, again some of the largest in East Asia.

Taiwan is poorly endowed with strategic minerals—again making the economy's remarkable transformation to a major industrial and trading nation all the more amazing. Almost all industrial inputs must be imported. Aside from limited deposits of coal, copper, sulfur, and gold, all of which have been mined heavily since Japanese colonial times and are largely exhausted, only very modest quantities of petroleum (oil and gas), salt, and high-quality building stone exist for commercial exploitation. The extent of offshore petroleum resources remains uncertain, but preliminary surveys suggest promise for the Taiwan Strait and East China Sea (see chapter 9). Of course, along with many other nations, Taiwan and the PRC jointly claim both of these areas, and exploitation in the coming decade—even joint exploration—does not seem realistic.

Historical Background

Archaeologists have found evidence of human settlement in Taiwan that dates back twelve thousand to fifteen thousand years. Distinctive artifacts indicate that a variety of Australasian cultures settled in the southern and eastern portions of the island. Other settlers came from the Asian mainland and apparently settled at slightly later dates along the northern and western coasts (Stainton 1999). Chinese from the southern provinces of Fujian and Guangdong have migrated to Taiwan and the adjacent Pescadores Islands (a handful of small islands within the Taiwan Strait separating Taiwan from the mainland) for many centuries, although only in the past three and a half centuries have their numbers been significant enough to justify the incorporation of Taiwan as a part of the Chinese polity. Historical records indicate that Chinese migrated to the Pescadores as early as the ninth century.

In those ancient times, the resources on these islands were shared and sometimes contested by three groups: Chinese immigrants, Japanese pirates, and Taiwanese natives from a variety of indigenous culture groups. Probably the first settlers of Chinese origin to establish permanent habitation on Taiwan arrived in the thirteenth century seeking refuge from invading Mongols. During that century, the first imperial administrative office was established on the windswept Pescadores. Despite these early administrative claims, rampant piracy forced the closing of the local administrative office in 1388 and resulted in waning Chinese influence for two centuries. It was not until 1558 that the Chinese reestablished an administrative presence in the Pescadores and assumed control over Taiwan as well.

Substantial permanent Chinese settlements emerged on the island of Taiwan in the late sixteenth or early seventeenth century. From 1620 on, Chinese entrepreneurs visited Taiwan regularly and promoted settlement on the southwestern coast. Trade in camphor, deer antlers, and tea linked indigenous peoples in the interior to these growing Chinese commercial settlements.

Conflicts between the Chinese and the indigenous people were common as agricultural expansions—supported by Chinese troops—drove a number of these groups of indigenous peoples into the interior mountains. Complicating this settlement history during the seventeenth century were periods of coastal occupation by two European colonial powers. Taiwan has been occupied and colonized by three foreign powers in its long history: the Dutch (1624–1662), the Spanish (1626–1642), and the Japanese (1895–1945). The Dutch controlled significant portions of Taiwan from 1624 to 1662, but settlement was focused on a few major coastal settlements along the southwestern periphery. Later the Dutch occupied the port city of Keelung (Jilong) after expelling the Spanish from the northeastern coast. In many ways, the Europeans treated Taiwan as a stopping point on the way to other places, taking on meat, water, and fruit. Few inroads were made beyond the coasts. Though significant today in terms of contemporary politics, the interior uplands of the island were then inhabited almost exclusively by largely autonomous indigenous groups. Even under Japanese administration that began in December 1895, the mountain dwellers were recognized as semiautonomous. Some of Taiwan's indigenous peoples who lived along the southern coasts were relocated to interior reservations once the Japanese instituted "modern" pacification strategies unfortunately copied from the United States in the early 1900s.

Internal politics and the overthrow of the Ming dynasty by the Manchus in 1644 resulted in far-reaching changes in Taiwan. The island became a refuge for Ming loyalists under Cheng Cheng-Kung (Koxinga), who was able to cast out the Dutch colonials (Copper 2009). Cheng also promoted extensive immigration of Chinese exiles to Taiwan. Cheng's rule established a Chinese majority on Taiwan that has never been threatened. Records indicate that toward the end of Cheng Cheng-Kung's rule, nearly 100,000 Chinese had settled on the island with the major concentration in the Southwest.

The Beginnings of Modernization

The Ming loyalists held out for two decades, but in 1683 the island was incorporated into the Manchu empire (Qing dynasty). For two centuries thereafter, Taiwan remained a prefecture of Fujian province and a frontier of Chinese settlement. The island occupied a strategic Pacific location for potential defense of the mainland, but China's inward orientation of that period (chapter 3) gave the island little commercial importance, and, in truth, it received scant attention. By the late nineteenth century, however, an increasing threat of maritime invasion by Europeans and the Japanese illustrated Taiwan's special importance and its military vulnerability to the Manchu imperial court. Consequently, in 1888, the Manchu emperor established Taiwan as a separate Chinese province and upgraded its administrative position to keep pace with the island's growing commercial and strategic significance.

The Sino-Japanese War (1894–1895) again radically altered Taiwan's history. The Chinese quickly lost the war, and this surprising outcome exposed China's many strategic weaknesses—pitting a modern Japanese army and navy against antiquated Qing ships and coastal defenses. The terms of the Treaty of Shimonoseki (1895), which settled the conflict, relegated all Chinese living in Taiwan to the status of Japanese colonial subjects. A short-lived uprising by the island residents was met with brutal force by the new Japanese government. Taiwan loyalists were quickly and violently subdued. By the end of the year, resistance had been crushed, and Taiwan would remain a Japanese colony for the next fifty years. Under colonial status and control, the potential benefits of indigenous political and economic reform were lost. Japan's efforts to more efficiently administrate the island and its promotion of export agriculture have implications to the present.

The Japanese Period: Economic Growth and Political Stagnation

The Japanese colonial occupation (1895–1945) was a harsh time for Taiwan's people. The Taiwanese, like colonial peoples everywhere, were disenfranchised politically and given second-class status, largely excluding them from the educational and administrative opportunities available to the city-dwelling Japanese. Residents were forced to adopt Japanese names and to learn and use the Japanese language and were conscripted into the Japanese military, all in the name of cultural assimilation. Over time, many of the island's young men were shipped to mines and factories throughout the far-flung Japanese Empire as the demands of World War II escalated and labor shortages increased.

Although it is controversial, some scholars still maintain that the colonial period was somewhat beneficial for Taiwan. There is considerable evidence that Japanese efforts to promote agricultural modernization and commercial development were successful in setting the stage for later gains achieved in the post–World War II period. Examples of this are most dramatic in two sectors: capital improvements and agriculture. Capital improvements in Taiwanese infrastructure completed during the Japanese colonial period included the construction of a north–south railway and highways that constituted a modern and integrated transportation system by the mid-1920s. With the exception of a few coastal areas, mainland China would wait more than sixty years for similar infrastructure. Closely linked with these transportation projects was the construction of modern deep-water ports at the northern (Keelung) and southern (Kaohsiung) ends of the island and the construction of related cargo-handling facilities (see figure 13.1).

Fundamental alterations to the agricultural system were equally important. After the Japanese introduction of scientific agriculture, Taiwan's farmers produced ever-more rice and sugar for export to Japan because of improved land management and irrigation systems, new crop varieties, and land reform. Greater efficiency in agricultural production also freed labor for agroprocessing factories located in the Taipei-Keelung region, leading to Taiwan's first urbanization "boom." Advances in agronomy, marketing, and the distribution of commodities led to rapid increases in production and the commercialization of the agricultural system. Colonial Japanese administrators also introduced new techniques in medicine and public health that resulted in a rapid decline in the island's death rate. Without a concurrent decline in birthrates, the net effect was to produce a rapidly growing population. The underpinnings, then, for a modern economy may have been laid down during the fifty years of Japanese colonial rule but at considerable psychological and economic cost to the population.

Restoration to China: 1945

Following the Japanese defeat in World War II, Taiwan was restored to Chinese control as part of the Potsdam Agreement, the blueprint for the new postwar order among the Allies. Oddly enough, relations between the Taiwanese and the mainland KMT Chinese administrators and troops that came to liberate them from the Japanese were tense from the outset. Corruption was rampant, and the native Taiwanese remained second-class citizens just as before. On February 28, 1947, a spontaneous revolt broke out, initiated by the cruel murder by customs police of a Taiwanese woman selling a handful of contraband cigarettes. For ten days, the Taiwanese rioted, seizing control of many government buildings in many cities and towns in protest of the heavy-handed rule of the KMT. Secretly, KMT governor Chen Yi cabled for troops from the mainland to subdue the rioters. Two divisions were sent, arriving on March 9, to occupy Taipei. In the end, from ten thousand to fifty thousand Taiwanese were killed (depending on conflicting estimates). Each year to the present, these martyrs are commemorated on what is now known simply as "2-28," for the date of the first murder. The fact that this event is still memorialized fifty years later exemplifies the powerful animosities and mistrust that remain between some Taiwanese and mainlanders even to the present.

After 1949 and the victory of the communists on the mainland, the KMT government and more than 1.2 million refugees and soldiers retreated to Taiwan and established an exile government in Taipei as the Republic of China on Taiwan. The costs of integrating this great influx of people and soldiers into an inflation-riddled economy were significant. Expectations of Taiwanese elites for self-autonomy, after the defeat of Japan, had been high, and after 2-28 the Taiwanese had both hated and feared the KMT army. The arrival of the often unruly and undisciplined KMT troops led to many armed conflicts and frequent rioting. Many Taiwanese leaders were killed or jailed.

Taipei and its surrounding regions absorbed most of the mainland refugees despite efforts to relocate the soldiers to agricultural areas in the south of the island. The concentration of power and investment in the city at this time created its dominant status. To the present, Taipei and its suburbs represent Taiwan's largest concentration of people, with 2.6 million persons in 2010; roughly 12.8 percent of Taiwan's population is located within the city limits. Perhaps more significant, preferential industrial and infrastructural investments within the Taipei-Keelung Metropolitan Region, which began in the 1950s, soon made northern Taiwan the wealthiest and fastest-growing region in Taiwan, with a population of more than 6.31 million residents (2010). This dense concentration of linked cities and suburbs currently accounts for 42.9 percent of Taiwan's urban population (30.6 percent of the island's total population) and is one of the most densely populated areas in the world.

International politics have favored the PRC over Taiwan in recent years. In 1971, Taiwan was expelled from the United Nations as the legitimate government of China. In 1979, the United States officially recognized the PRC and no longer recognized Taiwan as a sovereign state, although a close, unofficial relationship was established through the creation of the Taiwan Relations Act, also enacted in 1979. This close relationship, it is safe to say, continues to the present. The pragmatic decision to recognize the PRC, based largely on trade opportunities and the realization that the PRC is an emerging global power, officially left Taiwan largely isolated, but Taiwan's vital role in the global manufacturing and research/development systems ensures that, one way or another, all major nations have developed ways of working with Taiwan without unduly irritating the PRC. Over the past several decades, with the exception of the early years of the Chen Shui-bian presidency (2004–2008), Taiwan's leaders have typically treaded lightly on issues related to unification or independence—recognizing the delicacy of the situation and the mixed views of the population. As noted previously, there are considerable differences in opinion among voters about what should come next. Based on a survey conducted in 2004, there is evidence that approximately 10 percent of voters in Taiwan would prefer to quickly force the issue of independence (*The Economist* 2005, 5). But most prefer a more cautious road that recognizes the close and complex relationship between Taiwan and the PRC and the realization that the problem may well sort itself out in time as economic, political, and cultural ties grow.

Environmental Issues

Environmental issues in Taiwan run the gamut, from industrial pollution to air pollution from cars in cities to concerns about the protection of agricultural land and nature preserves. They are receiving increasing attention, in part because of the island's size (35,960 km²) and population. The island is incredibly crowded—more so even than Japan, with an average population density of 636.6 persons/km², second only to Bangladesh for states with a population over 10 million persons in 2009 (National Statistics 2010).

Cities, home to 69 percent of the population, are extremely crowed, and as a consequence urban pollution problems associated with air, water, and noise are major issues to voters (see figure 13.2). The development of modern rail and light rail systems is in part a response to air-quality and congestion issues. Economic growth, especially from 1950 to 1980, often came at the cost of localized environmental pollution of air and water. Further, the impact of crowding has also been great. Much arable land has been lost in Taiwan's drive to modernity—paved over by urban sprawl and industrial/transportation projects. Foul air from factories and automobile exhaust remains a great problem in all of Taiwan's large cities. Water quality in Taipei and other major cities is also still below acceptable standards. Some streams and farmland are heavily polluted with industrial effluents. Regular flooding, including spillover from inadequate drainage systems in many low-lying areas of the largest cities, sends polluted water into residential areas. The loss of wetlands mirrors that of most nations, as encroachment along the coasts, where reclaimed land accounts for a significant portion of new construction, is most extreme. Further, past deforestation by large state-connected mining and forestry concerns and the construction of elite residential housing on slopes have led to landslides and the premature siltation of many small and medium-size reservoirs. Regulations related to environmental protection are now very stringent, but enforcement, while improving, still remains too irregular and often has political overtones. In response to these problems, environmental groups brought these issues to the general public with great success in the 1990s, and Taiwan is now aggressively addressing a wide range of environmental concerns. For example, in 1974, only 0.2 percent of Taiwan's land was incorporated in nature preserves, but by 2010, this had increased to 20 percent, including a multitiered system of protected regions comprising seven national parks, nineteen nature preserves, six forest preserves, seventeen wildlife refuges, and thirty-two major wildlife habitats.

Modern Life

Urban Taiwan, in particular, is very cosmopolitan. There are few major international retail or commercial firms from around the world not represented in Taipei, Kaohsiung, Taichung, or Tainan. In turn, Taiwan's major corporations are some of the world's largest—with far-flung holdings in North America and Europe as well as an increasingly important presence in China and Southeast Asia (Callick 2003). Recently, Taiwan has developed vibrant film, music, and fashion industries, and popular music and movie "stars" play to audiences throughout the Pacific Rim and are idolized as they are in the West. Mass consumption is increasingly the norm, although the savings rate remains higher than in North America or Europe. Social issues such as the generation gap, widening disparities between rich and poor, urban–rural conflicts, and the splintering of the political system by special interests are as great as in any other postindustrial state. To some extent, Taiwan's global culture vies with its local culture, and the island's people live in two worlds. Life in Taiwan is fast paced; the pressure to succeed is great. Competition begins even in elementary schools as students are faced with weeklong islandwide examinations that will determine their academic future.

Although most citizens maintain that Taiwan's society and culture, above all, are Chinese in origin and character, unique features still make Taiwan a separate, identifiable segment of China's sphere of influence. A significant portion of younger Taiwanese, for example, prefer to see themselves as Taiwanese rather than Chinese. To an outsider, this may seem like splitting hairs, but for many, this reflects different worldviews about Taiwan's place in the region and world.

Among the many identifiable ethnic and social groups on Taiwan, two groups are probably most significant, the Taiwanese and the mainlanders, with place of family origin (*laojia*) rather than place of birth the controlling distinction. Mainlanders, although coming from every province of China, are often treated by the Taiwanese as a single political group. While not homogeneous by any stretch of the imagination, once on Taiwan, they are all often seen as mainlanders because of their origins. This group, composing roughly 14 percent of the population, continues to be more or less distinct, although intermarriage has softened the distinctions. During the long period of martial law, mainlanders came to dominate Taiwan's military, security, and police apparatuses and until recently also controlled the national government assembly and bureaucracy.

To assume that Taiwan's remaining population is a cohesive, homogeneous group in opposition to the initial dominance of politics by the KMT party that has been in power for all but eight years of Taiwan's post–World War II history is also incorrect. While two political parties dominate, as of July 2009, there were 148 political parties registered with the government. In addition to the so-called native Taiwanese, as of 2009, there were officially 499,500 self-identified indigenous people living in Taiwan. As noted earlier in the chapter, many of these indigenous people were originally living along the coasts but relocated to poor and isolated mountain areas and to reservations established by the Japanese. Others lived in the mountains for centuries before the Chinese arrived. Increasingly, these aboriginal Taiwanese today seek work and advancement in Taiwan's major cities. An emerging indigenous rights movement, following similar movements in many nations in the 1980s and 1990s, has bloomed following the lifting of martial law and subsequent political reforms. The three largest of these groups are the Ami, the Paiwan, and the Atayal. The politics of the indigenous people's rights movement are complex and tied closely to green issues, student issues, and the political reform movement (Chen 1999).

Modernity and Mobility

The importance of landforms and topography to economic activity is clear and underscored by the population distribution throughout the island. As might be expected, Taiwan's population is concentrated predominantly in the alluvial plains and basins of the northern and western parts of the island. The most densely populated sections are the western and northern coastal areas that were settled by the Chinese over five hundred years ago. As a consequence of this concentration of wealth and later political power, major transportation networks emerged to service these important urban/industrial and agricultural areas. As the economy changed, Taiwan's major cities expanded time and time again, claiming ever more of this originally agricultural hinterland. Recent infrastructural improvements have opened up more of the interior, but it still remains relatively isolated, and recent stringent environmental regulations work to ensure that these areas will be as pristine as possible for coming generations (see figure13.5).

Within the densely populated western portions of the island, major cities are dotted at intervals from Keelung, Taoyuan, and Taipei in the north to Taichung, Tainan, and Kaohsiung in the south, all linked by modern high-speed rail and road networks. The largest concentrations of people are still found in the Taipei Basin in northern Taiwan and in the large plains in the Southwest that contain Tainan and Kaohsiung. Recent migration has been primarily rural to urban in nature, with people moving to the largest cities of the Taipei-Keelung Metropolitan Region and the important industrial port city of Kaohsiung.

It is a paradox that despite extremely high population density, Taiwan is facing genuine labor shortages that are currently being met with a variety of guest-worker policies that admit workers from Southeast Asia and the Philippines and with a wholesale exodus of manufacturing operations of many types to overseas factories (outsourcing). Of course, temporary status is unpopular with the guest workers, who wish for more security in their lives. Regardless of the outcome, the arrival of low-income workers

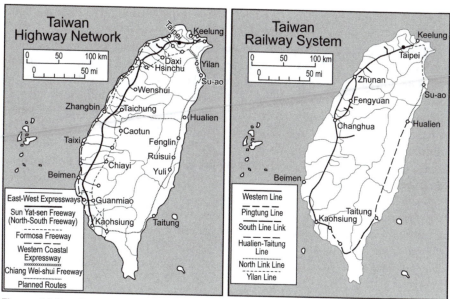

Figure 13.5. Contemporary rail and highway transportation network of Taiwan (Government Information Office 2009, http://www.gio.gov.tw/taiwan-website/5-gp/yearbook/ch13.html#Transportation)

from other countries and "illegals" from mainland China poses new challenges to the social fabric of Taiwan. The lives of these workers are often dreary; there is not yet a secure place for these people in Taiwanese society.

Changing Social Conditions

The primary unit of social organization in Taiwan, as in China, traditionally has been the extended family—where three or even four generations live together. In the face of Taiwan's economic development, it is only natural that there have been major changes in traditional social relations, especially in the past few decades. Housing in the cities is expensive, and space is limited. In addition, long school days, evening "cram schools" to aid in college entrance examinations, and increasingly popular preschools reduce the need for grandparents to help with children. These changes are more obvious in urban areas, which increasingly dominate Taiwan society. In contrast to the past, most families in urban areas now exist as nuclear units composed of parents and children. Social services targeting the increasing elderly population have increased as the population ages and residence patters change. Concerns similar to those in Japan and South Korea related to the "graying" population are in play here as well. Taiwan has one of the lowest total fertility rates in the world (1.05 children per woman in childbearing years), and as a consequence, Taiwan's population will peak at 23.84 million in 2026 and then fall to approximately 20.29 million in 2056 (Government Information Office 2009). New policies promoting and supporting child rearing and development have been instituted in recent years. The decline in population obviously has important implications for society and the economy.

Rural society is also changing with equal or greater speed. Modern mass media extends everywhere in Taiwan, and the images of modernity are ubiquitous. Cable television provides access to stations from Japan, Hong Kong, and the West. The draw of modern city life is strong, pulling rural workers to seasonally migrate to factories and construction work in the cities. The potential for greater economic and educational opportunities in the neighboring cities and towns has drained off many rural workers, especially the young. Many farm families send children to better-equipped urban schools to increase their chances for advancement in the highly competitive educational system. Once you have lived in the city, however, it is difficult to go back to the often hard and uncertain life on the farm. Efforts are under way to allow young farmers access to more land and resources to encourage some to stay and maintain the agricultural economy. During planting and harvest seasons, labor shortages are now common in many rural areas. Foreign labor (Thai and Filipino and even mainland Chinese), once confined to the island's factories, is now common in farm areas as well.

Agriculture

Less important at the present time, agricultural production in Taiwan was long the foundation of its stable and prosperous economic system. Still, until the past several decades, steady gains in agriculture dating back to the Japanese occupation provided the means to feed a rapidly growing population and to finance ongoing development projects. Indeed, Taiwanese agriculture still offers many lessons to other nations and is a model of productivity under conditions of extreme crowding and limited resources. Under the Japanese, more than 90 percent of the workforce was engaged in agriculture. In the years from 1953 to 1960, the agricultural sector accounted for an average of 28.5 percent of gross national product. By 1995, this share had fallen to only 3.6 percent of gross national product and has declined to just 1.68 percent by 2009 (Government Information Office 2009; Howe 1998, 129). Times have changed, and farmers face significant challenges as they are forced to respond to changing domestic and international market conditions that developed after entry into the World Trade Organization. While rice is still the most valuable single crop, grain production has declined, but the value of output of the agricultural sector has never been higher because of significant increases in meat production for the domestic market as well as vegetables, fruit, cut flowers and plants, and betel nut for both domestic and international markets. For example, Taiwan is a major exporter of orchids. As with other postindustrial economies, the agricultural sector is more productive than ever before because its specialty crops—required by an increasingly affluent society or for export— earn much more cash per unit area than grain. Still, Taiwan's very real security issues, including the threat of a naval embargo, still require management of foodstuffs. This means that subsidies to encourage grain production must be provided in relation to opportunity costs for labor and in recognition of the continued importance of the rural vote. Under government sponsorship and the direction of the Taiwan Agricultural Research Institute Taiwan's agricultural research program has emerged at the millennium as one of the most creative and sophisticated in the world. Techniques of rice and fruit cultivation, vegetable propagation, biotechnology research, aquaculture, and animal husbandry developed by this institute have been studied and diffused throughout the world.

The Commercial and Industrial Sectors

Paralleling the radical changes in the agricultural economy over the past half century have been a series of much greater and more important changes in the nature of commerce and industry in Taiwan. Although set in motion six decades ago, the most spectacular changes in industrialization and commerce are very recent; most have occurred in the past two decades with Taiwan becoming a dominant player in IT components. Prior to the 1980s, Taiwan's history of industrialization closely mirrored that of other successful nations. Success in IT, however, makes contemporary Taiwan distinct even when compared to the nations of Europe. This success in IT has earned Taiwan the name Silicon Island, but increasingly, IT investments are made by Taiwanese firms operating in mainland China. China is now the world's largest exporter of IT hardware, but approximately 60 percent of these exports are manufactured in China by Taiwanese companies. Acer, Taiwan's largest IT firm, is the world's fifth-largest producer of personal computers. In 2008, Taiwan ranked first or second in the production of twelve different IT product categories, including notebook PCs, monitors, motherboards, scanners, keyboards, and wireless computer mice.

Copper (2009) argues that there are seven factors that should be given credit for Taiwan's industrial success: 1) the expansion of industrial employment; 2) increases in labor productivity; 3) U.S. economic assistance; 4) privatization; 5) a high rate of local savings and foreign investment; 6) a solid economic infrastructure, including transportation and port facilities; and 7) excellent planning by both the government and the business community (133). Early investments in science parks (specialized industrial parks) might be added to the list. Taiwan currently has more than seventy "industrial clusters" where specific industries are targeted for investment and development, but the science parks represent the crown jewels in the system. There are three megaparks: Hsinchu Science Park, Central Taiwan Science Park, and Southern Taiwan Science Park. Strategically located to promote regional development in the three regions, the combined revenues from firms operating under the parks' auspice amounted to $58.43 billion in 2009 (Government Information Office 2009; Tselichtchev and Debroux 2009). The role of the government in directing industrial location is a hallmark of Taiwan's commercial strategy, but the firms are private with the same range of ownership types as firms in the European Union, Japan, or North America.

The foundation of Taiwan's recent "economic miracle" actually developed over the past fifty years and again offers many interesting policy examples for other nations. Progress in industrialization under the Japanese was limited and predictably associated with extractive industries (e.g., smelting) and the processing of agricultural commodities and forest products, but infrastructural improvements were critical to later success.

These infrastructural investments continue to the present with the ongoing development of new Internet technologies (i.e., WiMAX broadband) supported by both government and venture capital. Taiwan's revenues from WiMAX equipment production could top $3.17 billion in 2010 (Government Information Office 2009). The role of state-directed investment is an important characteristic of Taiwan's industrial strategy. During the 1970s when industrial production exploded, aggressive state-initiated investment policies were made within a select number of targeted industries. Initially, in the 1950s and 1960s, the production of labor-intensive goods was dominant under a strategy of import substitution. Over time, however, Taiwan's success can be credited to a balanced policy of import substitution and export-oriented development despite the fact that these policies are often treated as mutually exclusive in college textbooks. Economic planners understood that a small island peopled by consumers with limited purchasing power could hardly sustain a major industrial development policy. Foreign trade is essential for Taiwan. State-directed manufacturing efforts were combined with high levels of capital reinvestment targeting the production of exports that could quickly return the investment (Clouse 2009).

Massive state investments were also made in education in anticipation of the need for a highly skilled technical workforce. High levels of reinvestment and a balanced distribution of large and small firms that can react quickly to changes in the marketplace remain important characteristics of Taiwan's overall development strategy. As labor and production costs rise, many Taiwanese firms have shifted production to China and other nations in Southeast Asia. Obviously, there are significant political overtones and environmental implications in the relocation of sunset industries from Taiwan to other countries.

Somewhat unique to Taiwan among the world's major industrial nations, however, is the reliance of even its major firms on outsourcing production to smaller subcontractors. The relationship between the large firms that invest in research and development and the medium and small firms that actually produce the product is interesting. In 2008, 85 percent of Taiwan's IT output was actually manufactured by small and medium-size firms that were linked to larger firms by short-term contracts or that had been spun off from these firms. The importance of several hundred thousand small and medium-size firms to Taiwan's economy continues to be a distinct feature of its manufacturing history. Still, the industrial giants that are known as *chaebol* in Korea and *keiretsu* in Japan are not without parallel in Taiwan. Evergreen is currently the second-largest container-shipping firm in the world, and Acer controls a major share of the European Union's computer market. Taiwan Semiconductor Manufacturing and United Microelectronics are hardly household names, but these two Taiwanese companies account for around 70 percent of the world's foundry production of silicon chips, a foundry being a semiconductor manufacturer that produces chips for other manufacturers (*The Economist* 2005, 10). Other large Taiwanese firms with strong government connections (e.g., Koo's Group, Formosa Plastics Group, and Shinkong Group) are multinational, with plants, commercial offices, and distribution systems throughout Europe and North America.

The changes inherent in the post–World Trade Organization global economy are easily observable in Taiwan. At the beginning of the new millennium, the share of the economy and of the workforce controlled by industry and manufacturing in Taiwan is declining, just as agriculture declined in the 1970s. Taiwan's experience is important, possibly foretelling coastal China's evolution but also reflecting the future of most developed industrial nations as well. Further, Taiwan's transition to a post-Fordist (nonmanufacturing) economy reflects the powerful impact of globalization on even a strong domestic economy. Capital-intensive manufacturing and recent growth in the service sector have replaced labor-intensive industries. In 1986, industrial production still accounted for more than 50 percent of Taiwan's GDP. By 2009, it had declined to 24.78 percent of GDP. As the share of employment devoted to agriculture and manufacturing declines, services including insurance, banking, education, tourism, and government employment will increase in importance. As was the case for agriculture in the past, these shifts in employment and economy will again have far-reaching effects on the fabric of Taiwanese society in the future.

Taiwan's commercial and financial institutions grew in equal measure with the industries that initially supported the island's economic modernization through export trade and manufacturing. Now, even as domestic manufacturing declines in importance, the banks remain. In 2010, Taiwan's banking sector plays a major role in international investment linked less to fluctuations in Taiwan's economy than to those of the global economy.

The Challenges of the Future

Taiwan's political history has always been tumultuous. This is a simple fact but one that demands that we again quickly review the role of absolute location in political histories. Taiwan has always been contested space. It was first settled by tribal people of proto-Malay origin and then eventually colonized by Chinese farmers with superior force—seeking alluvial land and escape from crowded conditions in Fujian and Guangdong. The indigenous peoples were pushed to the interior, resulting in conflicts among these peoples as well as a growing Chinese hegemony along the coasts. A half century of European colonial control—again projecting superior force—followed in the early seventeenth century. This was succeeded by the restoration of the island to Chinese control when Koxinga defeated the Dutch. In 1888 during the Qing dynasty, the imperial court became convinced of Taiwan's strategic value and upgraded the status of the island, only to lose it to Japan's growing empire. Fifty years of Japanese colonial rule began in 1895. In 1945, Taiwan was returned to the Chinese. Regulations promulgated by the KMT but with clear Japanese roots created rigidly hierarchical internal political units and subunits that are only now being readjusted. This high level of centralized control must be credited for the efficient use of development capital when Taiwan's economy expanded rapidly in the 1970s and 1980s.

As Taiwan's economic growth has increased, its people have demanded greater political freedom and local representation. Recently, independent investigations associated with these growing demands for local autonomy, a more politically involved public, and an increasingly vigilant press have uncovered serious abuses in the past and at the present time that have challenged voters' faith in the central government. Still, the political reforms beginning in the mid-1990s would probably not have been realized without the prosperity fostered by the state corporatism characteristic of the KMT era. As Taiwan's multiparty political system matures, it is difficult for us to anticipate how domestic politics will influence the country's international diplomacy, but to be sure this is a factor that did not exist until the late 1990s.

In 1972, President Nixon journeyed to China, and representatives of the U.S. government signed the now famous Shanghai Communiqué. Both U.S. and Chinese communist positions toward Taiwan are enumerated in this document, and these positions suggest something about the future status of the island. In brief, the PRC affirmed that Taiwan is Chinese territory and will be returned to the fold in the future. It is a matter of domestic politics, according to the Chinese, and the United States has no legitimate basis for involvement. The PRC thus allowed no possibility of the interpretation of a "two-China" policy or an independent Taiwan. The United States for its part acknowledged that Taiwan is Chinese, reaffirmed its interest in a peaceful settlement of the Taiwan question, and agreed to phase out its military support through the sale of defensive weaponry and information-gathering technology as conditions permitted. Since the signing of that communiqué, the United States has officially recognized the PRC and dropped its recognition of Taiwan. But the United States continues to have close relations with Taiwan through a formula for nongovernmental relations and a law (the Taiwan Relations Act) that established two semiofficial organizations: the American Institute in Taiwan and Taiwan's Coordination Council for North American Affairs in the United States. These two organizations, staffed by diplomatic personnel either retired or temporarily on leave from government agencies, have worked effectively during the thirty years since 1979 to promote continued economic, academic, scientific, and cultural exchanges in a pragmatic but unofficial manner. At the same time, the United States has terminated its bilateral defense treaty with Taiwan and withdrawn official diplomatic recognition of Taiwan in favor of recognition of the PRC (U.S. Department of State 1980). There is little that Taiwan can do to affect the politics of other countries. Further, it is unlikely that either the government, as presently constituted, or the collective will of the people would support political reintegration with mainland China at this time without very significant guarantees for long-term autonomy. Finally, Taiwan is strong and rich enough to defend itself against any kind of attack except full-scale military assault. Moreover, even if the communist government could mount the amphibious force necessary to invade the island, such a solution would be prohibitively destructive for the invaders, for Taiwan, and for the coastal provinces. With seventy thousand Taiwan-funded firms operating in the mainland and probably more than a million Taiwanese businesspeople and their families living there, war would have devastating effects for all (Kastner 2009). Thus, a military solution to the question, at least in the short term, would appear to come at too great a cost.

The alternatives are complex and politically charged on the domestic front as well as in the international arena. Given the mainland's explicit rejection of an independent state, a formally separate Taiwanese state is also unlikely in the short run. Domestically, many Taiwanese would oppose efforts to declare the unilateral independence that would result in the creation of an independent nation. Still, Taiwan is de facto an independent nation. From a U.S. point of view, one can affirm the "Chineseness" of Taiwan, as did the U.S. State Department in the Shanghai Communiqué, while maintaining that the problem be resolved peacefully. The U.S. position appears to be a wait-and-watch posture in which the United States plays a passive role by allowing Chinese of different political outlooks to decide and resolve their own political destinies (U.S. Department of State 1980).

Unilateral unification also seems problematic, even if force could be used. Domestic problems in the PRC must also be considered. Tensions manifest as core or periphery conflicts challenging the national government suggest that the current political system in the PRC still has problems of control and governance. A fractious Taiwan would be difficult to control. Democracy has taken fifty years to emerge in Taiwan, and few would be willing to return to the fold without sound and credible guarantees for the long-term maintenance of this system.

Conclusion

Taiwan remains a place of great promise and horizons as well as a place of pressing uncertainties. The people of Taiwan in many ways have made their good fortune and have used the island's meager resources to their advantage. Lessons from Taiwan are important to many nations for different reasons. Still, location is also important. Situated along the southeastern flank of the Chinese mainland, Taiwan is highly accessible and lies across the main shipping and air lanes connecting East Asia with Europe, South and Southeast Asia, and North America. The ambitions of Taiwan's people and the role Taiwan played in the Cold War, coupled with this location, have propelled Taiwan into the global economy. Such a location and an economic situation have led to the rapid economic development of the island in recent decades. Yet these rosy economic conditions cannot shield it from the troubled political situation, a difficulty that may yet destroy the progress of the recent past if reason and patience do not prevail.

Questions for Discussion

1. What are some of the important economic and political changes that have occurred in Taiwan over the past twenty years?
2. What is meant by a "postindustrial state"? What does this mean in terms of Taiwan's economy?
3. What is the current status of the "cross-strait" relationship between China and Taiwan? What other nations are involved in these issues and why?
4. How does Taiwan's physical geography relate to the population distribution, transportation systems, and economic activities?
5. What are some of the "useful lessons from Taiwan" for other nations?

References Cited

Callick, Rowan. 2003. Taiwanese exodus to the motherland. *Australian Financial Review*, January 17, new section, 1.

Chen Chien-Hsun. 2008. *The Economics of Taiwan's Three Direct Links with Mainland China.* East Asian Institute brief no. 415. Singapore: National University of Singapore November 27.

Chen, Yi-fong. 1999. Economic, social, and political geography of the indigenous people's movement in Taiwan. PhD diss., Louisiana State University, Baton Rouge.

China Forum. 2010. How people of mainland China and Taiwan increasingly mingle. *China Daily Online.* http:/bbs.chinadaily.com.cn/viewthread.php?gid=2&tid=674900.

Clouse, T. 2009. Country Report: Taiwan. *Global Finance.* January. http://www.gfmag.com/archives/21-january-2009/34-features.html#axzz0xA5EcBbb.

Copper, John F. 2009. *Taiwan: Nation-State or Province?* Boulder, CO: Westview.

Government Information Office, Republic of China. 2009. *Republic of China Annual Yearbook 2009.* http://www.gio.gov.tw/taiwan-website/5-gp/yearbook (accessed August 21, 2010).

Government Information Office, Republic of China. 2010. *Republic of China at a Glance.* http://www.gio.gov.tw/taiwan-website/5-gp/glance/ch7.htm (accessed August 19, 2010).

Howe, Christopher. 1998. The Taiwan economy: The transition to maturity and the political economy of its changing international status. In *Contemporary Taiwan*, ed. David Shambaugh. Oxford: Clarendon, 127–51.

Hsieh, Chiao-min. 1964. *Taiwan-Ila Formosa: A Geography in Perspective.* Washington, DC: Butterworths.

Hsu, Crystal. 2010. DPP uses DGBAS data to attack Ma. *Taipei Times.* August 21. http://www.taipeitimes.com/News?pubdate-2010-08-21 (accessed August 24, 2010).

Kastner, S. L. 2009. *Political Conflict and Economic Interdependence across the Taiwan Strait and Beyond.* Studies in Asian Security series. Stanford, CA: Stanford University Press.

National Statistics. 2010. *National Statistics Republic of China 2010.* http://eng.stat.gov.tw/lp .asp?CtNode=2815&CtUnit=1072&BaseDSD=36&xq_xCat=02 (accessed August 24, 2010).

Pannell, Clifton W., and Laurence J. C. Ma. 1983. *China: The Geography of Development and Modernization.* New York: Halsted.

Stainton, Michael. 1999. The politics of Taiwan's aboriginal origins. Chap. 2 of *Taiwan: A New History*, ed. Murry A. Rubinstein. Armonk, NY: M. E. Sharpe, 27–44.

The Economist. 2005. Dancing with the enemy: A survey of Taiwan. January 15, special nation report.

Tselichtchev, Ivan, and Philippe Debroux.2009. *Asia's Turning Point: An Introduction to Asia's Dynamic Economies at the Dawn of the New Century.* Hoboken, NJ: Wiley.

U.S. Department of State. 1980. *Review of Relations with Taiwan.* Current policy no. 190 (June 11, 1980). Washington, DC: Bureau of Public Affairs.

Winkler Partners. 2007. Review: Taiwanese investment in China. http://winklerpartners/com/a/comment/taiwanese-investment-in-china (accessed August 24, 2010).

Index

Page locators in italics indicate figures, photographs, and tables.

About the Authors

Youqin Huang is associate professor in the Department of Geography and Planning and research associate at the Center for Social and Demographic Analysis at the State University of New York, Albany. She teaches courses related to population, urban development, China, and statistical and spatial methods. Her research has focused on two areas: one is housing behavior, residential mobility, neighborhood change, and urban structure; the other is migration and urbanization, in both China and the United States.

Clifton Pannell is professor of geography and Associate Dean of Arts and Sciences Emeritus at the University of Georgia. He served as visiting professor of geography at the University of Hong Kong (2009 and 2011) and Carroll Visiting Professor of Urban Studies at the University of Oregon (2010). His China research focuses on economic, political, population, and urban geography, and he currently serves as an editor of *Eurasian Geography and Economics.*

Christopher J. Smith is jointly appointed as professor in the Department of Geography and Planning and the Department of East Asian Studies at the State University of New York, Albany. He is an urban geographer whose research has been focused on East Asia, especially China. His recent work has been concerned with the social and cultural consequences of China's transition away from socialism.

Gregory Veeck is professor of geography at Western Michigan University, teaching courses related to China and East Asia, research methods, environmental studies, and economic geography. His research interests include agriculture, rural economic development, and environmental issues in China, East Asia, and the United States.